Political Descent

Political Descent

Malthus, Mutualism, and the Politics of Evolution in Victorian England

PIERS J. HALE

The University of Chicago Press
Chicago and London

Piers J. Hale is assistant professor in the Department of the History of Science at the University of Oklahoma.

The University of Chicago Press, Chicago 60637
The University of Chicago Press, Ltd., London

Printed in the United States of America

23 22 21 20 19 18 17 16 15 14 1 2 3 4 5

ISBN-13: 978-0-226-10849-0 (cloth)
ISBN-13: 978-0-226-10852-0 (e-book)

DOI: 10.7208/chicago/9780226108520.001.0001

Library of Congress Cataloging-in-Publication Data

Hale, Piers J., author.
Political descent : Malthus, mutualism, and the politics of evolution in Victorian England / Piers J. Hale.
pages ; cm
Includes bibliographical references and index.
ISBN 978-0-226-10849-0 (cloth : alkaline paper)—ISBN 978-0-226-10852-0 (e-book) 1. Social evolution—Political aspects. 2. Evolution (Biology) and the social sciences—England—History—19th century. 3. Political science—England—History—19th century. 4. Social Darwinism—Political aspects. 5. Evolution (Biology)—Political aspects. 6. Malthusianism—Political aspects. 7. Mutualism—England—History—19th century. 8. Malthus, T. R. (Thomas Robert), 1766–1834—Influence. 9. Lamarck, Jean Baptiste Pieree Antoine de Monet de, 1744–1829—Influence. I. Title.
JC336.H254 2014
320.01—dc23

2013044863

♾ This paper meets the requirements of ANSI/NISO Z39.48-1992 (Permanence of Paper).

For Sativa and Raven

You ask whether I shall discuss "man"; —I think I shall avoid the whole subject, as so surrounded with prejudices, though I fully admit that it is the highest and most interesting problem for the naturalist.

CHARLES DARWIN to ALFRED RUSSEL WALLACE,
22 December 1857

Contents

INTRODUCTION

The Politics of Evolution

Human nature is a political problem as much as it is a philosophical one. However, since the publication of Charles Darwin's *Origin of Species* in 1859, it has also become very much a question of biology. This remains controversial. Currently, sociobiology seeks to explain the evolution of social behavior, in man as well as in other social organisms, while evolutionary psychology offers insights into human cognitive choices. Both disciplines have met with objections from social scientists, who argue that we are a species that cannot be understood through our biology alone and that our culture and society are more than mere biological phenomena. Indeed, concerns abound that evolutionary accounts of human social behaviors are reductive and serve to falsely naturalize often disputed social relationships as well as to undermine human agency. The attempt to explain not only physiological but also psychological sexual differences in terms of Darwinian sexual selection is just one of the more contentious issues in this "biological turn"; critics maintain that sex and race, like class, are socially constructed categories that require social, cultural, and historical analysis beyond any insight that biology might provide.

While there have always been those who have doubted the explanatory reach of evolutionary biology, the fact that today we find social scientists among them bucks a historical trend. From its birth in the middle of the nineteenth century, sociology was wedded to evolutionary explanation and practitioners of the other social sciences were no less enamored by the insights that evolution seemingly offered. The fact of our evolution was also taken to be of the utmost significance in nineteenth-century psychology, history, and political economy. In short, all aspects of the human experience were colored by developments in evolutionary biology. It was in this period that questions about the kind of beings we had become, about how we live and how we

FIGURE I.1. Thomas Robert Malthus, 1766–1834. (From *Popular Science Monthly* 74 [April 1909], 412; courtesy History of Science Collections, University of Oklahoma Libraries)

might live, became questions that were believed to have biological answers. These answers were disputed, of course, and not less hotly than they are today. At issue from the start was the question of which human behaviors were thought to have been "fit" in our evolutionary past and which were not, and thus which might remain so in the present. Of course, the tacit presumption in these debates, both in the past and in the present, is that what is "fit" in evolutionary terms is also "right" in moral terms; the study of human evolution has been and remains as much a prescriptive enterprise as a descriptive one. In this debate it has been a commonplace to derive an "ought" from an "is," despite the objections to doing so that have been raised by philosophers from David Hume to G. E. Moore and that have been repeated by modern-day critics.[1] This has occurred for compelling reasons. Competition or cooperation, self-interest or altruism—across the history of our species, and certainly across the history of our study of our own evolution, these have become key issues in how we make sense of ourselves, of how we might live, and ultimately, of how we think about what it means to be human.

Even among those who believe that biology does have a lot to contribute to how we understand our culture and society, there remains significant disagreement as to exactly what our evolution might mean for us. These debates revolve around which evolutionary processes should be invoked to explain certain behaviors, and while there is agreement that the roles that natural and sexual selection have played in our evolution are important, there remains disagreement about exactly what it is that is being selected in these processes. Further, in recent years, what has become known as the "levels-of-selection" debate in biology has become central to this question. This is a debate over whether natural selection acts upon, or selects, genes, individuals, or groups, and it has been taken to speak directly to the kinds of evolutionary behaviors that have become natural to us and—most significantly—to whether humans have evolved the capacity for genuinely altruistic behaviors or whether we are essentially self-interested beings, open at best to what the theoretical biologist Robert Trivers has termed "reciprocal altruism."[2]

Ever since the publication of *Origin of Species*, evolutionary explanations of altruistic behaviors have proven particularly problematic because selflessness seems to go against the very essence of natural selection. If natural selection works in such a way as to favor the individual that is the best equipped to prevail in a struggle for existence, then it seems only logical that any organism that is altruistic—that gives up some of its own resources to benefit another—will be at a disadvantage. All things being equal, one would expect that altruistic organisms would very quickly be driven to extinction. The suggestion that selection worked upon individuals made it logical to conclude that only self-interested behaviors of one sort or another could be fit behaviors. However, such conclusions have not gone unchallenged, and ever since Darwin, defenders of the idea that genuinely altruistic behaviors can indeed evolve have appealed to various theories of "group selection" in which organisms are presumed to have acted not for their own good but for the good of the group—or of the species—as a means by which altruistic behaviors might have become established as evolutionary stable strategies. Most histories of group selection highlight the work of the English ornithologist Vero Copner Wynne-Edwards. As the historian Mark Borrello has pointed out in *Evolutionary Restraints*, his own study of the contentious history of group selection, Wynne-Edwards recognized that many of the birds in the populations he studied did not reproduce in a given year even though they were sexually mature. Wynne-Edwards explained this phenomenon in his 1962 book, *Animal Dispersion in Relation to Social Behaviour*, as being what he called an "epideictic" behavior that resulted from group selection.[3] He suggested that during flocking individual birds were able to assess the size of their nesting group in relation to the availability of resources and limit their reproduction accordingly. This behavior, Wynne-Edwards argued, had developed because it worked for the overall benefit of the species. Wynne-Edwards's views were initially well received; the emphasis upon cooperation and collectivism resonated with prevailing political sentiments and also appeared to explain observed phenomena. However, only shortly after the publication of *Animal Dispersion*, a number of critics attacked the basic presumptions Wynne-Edwards had made about the processes of evolution that informed his work. In 1963 and 1964, the theoretical biologist William D. Hamilton published two very significant papers under the title "The Genetical Evolution of Social Behaviour" that quickly set mainstream biology at odds with theories of group selection. Describing a gene-based theory of selection, Hamilton's papers set a new norm in theoretical biology that prevails to the present. Describing what has subsequently been termed a "gene's-eye view" of evolution, he argued that what appeared to be genuinely altruistic social behaviors were in fact the product of natural selection targeting the genes that coded for the behaviors that would ensure that those genes would make it into the next generation. From this perspective, Wynne-Edwards's views were simply naive. Hamilton's conclusions coincided with those of a number of other theorists—the American mathematician George Price, the English biologist John Maynard Smith, and the American evolutionary ecologist George C. Williams, in particular. In 1966 Williams had published *Adaptation and Natural Selection: A Critique of Some Current Evolutionary Thought.* The subtitle was telling, for Williams's intention was to undermine exactly the kind of group selection that Wynne-Edwards appealed to in his attempt to explain the adaptive value of altruistic behaviors. Williams was keen to point out that while the kind of group-selectionist explanations made popular by Wynne-Edwards relied upon natural selection working on different groups of individuals, "the natural selection of alternative alleles within populations will be opposed to this development."[4] This fact meant that in all but the most exceptional circumstances, altruistic behaviors would be undermined by self-interest. Behaviors that appeared to be altruistic might certainly evolve between organisms that shared a close enough genetic relationship—Hamilton had established as much; it was this that John Maynard Smith termed "kin selection"—but, Williams insisted, any occurrence of altruism between non-related organisms had to be the result of cases of mistaken identity. More often than not, he wrote, such actions were indicative of "imperfections in the mechanisms that normally regulate the timing and execution of parental behaviour," conceding only that "benefits to groups often arise as incidental statistical consequences of individual activities just as harmful effects may accumulate in the same way."[5] It was the views of Hamilton, Williams, and Maynard Smith that Richard Dawkins popularized in his best-selling book *The Selfish Gene.* From this "gene's-eye view" of selection, behaviors that had earlier been thought of as altruistic were subsequently described as only apparently so. Apparent altruism between kin could be explained as being in the self-interest of the genes these organisms had in common, while actions that appeared to benefit non-related organisms were thought to only evolve as an "evolutionary stable strategy" in the context of a reciprocitous community.[6] Cases of "genuine altruism," in which an organism gave up its own resources for the benefit of a non-related organism, were, to restate Williams's position, merely rare cases of natural selection getting it wrong, evidence of the trial-and-error processes at work in natural selection. Over time we might expect such instances to diminish as the evolutionary process eradicated its own inefficiencies.

Whereas Dawkins popularized the various approaches to gene selection in his *Selfish Gene,* Matt Ridley has more recently returned to the subject, specifically in relation to the evolution of human behavior, in his *Origins of Virtue.* Ridley's book is particularly interesting. The majority of those biologists who have attacked advocates of group selection have made it a point to suggest that group selection is not just wrong but that it is the result of bad science, the product of men who have allowed themselves to be persuaded by what they want to be true rather than to accept the cold hard truth of scientific fact. It would be nice if altruistic behaviors were genuine, they say, but they are not. Group selectionists are romantics who want to see cooperation throughout nature, just as they hope to see it prevail throughout society.

Not only was this charge leveled at Wynne-Edwards, but so too present-day advocates of group selection, such as the philosopher Elliot Sober and the biologists David Sloan Wilson and E. O. Wilson, have been similarly vilified.[7] Although the opponents of group selection continue to charge its champions with being misled by their own desire to see their personal preference for a politics of cooperation rather than conflict legitimated by finding the same motivations in nature, those who have made this claim have made little or no acknowledgment that their own conception of evolution, which highlights competitive self-interest as the only legitimate explanation of social behaviors, fits very nicely with the individualist and competitive ethos of liberal capitalism.[8] While there is much that is implicit in their work that suggests that the presumptions they make about how biological processes work are colored by their own political preferences just as much as Wynne-Edwards's views might have been, Ridley makes the politics of the gene's-eye view of natural selection explicit: "We are not so nasty that we need to be tamed by intrusive government, nor so nice that too much government does not bring out the worst in us . . . : government is the problem, not the solution. The collapse of community spirit in the last few decades and the erosion of civic virtue, is caused in this analysis not by the spread and encouragement of greed but by the dead hand of Leviathan. The state makes no bargain with the citizen to take joint responsibility for civic order, engenders in him no obligation, duty or pride, and imposes obedience instead. Little wonder that, treated like a naughty child, he behaves like one."[9]

With so much at stake in these evolutionary accounts of social behaviors, especially when it comes to our own species, it is gratifying to see that scholars have started to contextualize and historicize these debates. Mark Borrello's *Evolutionary Restraints* and Oren Harman's biography of George Price, *The Price of Altruism*, have been important recent contributions to this field, as have Thomas Dixon's *Invention of Altruism* and Lee Dugatkin's *Altruism Equation*, as well as his more recent *Prince of Evolution*, a short study of the Russian geographer and biologist Peter Kropotkin. *Political Descent* is my own contribution to this growing literature.

In *Political Descent* I do not engage with this recent episode in the politics of evolution directly, however. Rather, like Dixon, I focus on an earlier era. I do so in order to show that the history of evolutionary theory has been deeply political from its inception and that the arguments about the evolution of cooperation and competition—on both sides—have been especially so. I contend that from 1859 there existed two rival traditions of evolutionary politics in Victorian England: the one, deeply Malthusian, which focused upon the adaptation of the individual through struggle as a means to a progressive social evolution; the other, radical and predominant Lamarckian and anti-Malthusian, that tended to emphasize the role of social cohesion as a means to the social evolution of a society in which individual interests tended to be subordinated to the welfare of the group.

By viewing more-recent arguments in a historical perspective, we can see both that group selection has enjoyed a much richer and longer history than many in the present seem willing to acknowledge and that, while much of the rhetoric of these arguments has appealed to what is and is not good science, the arguments are just as much about competing political visions of human nature and society.

People have appealed to biological evolution to advance and justify a range of politics since the publication of the evolutionary ideas of the French naturalists Étienne Geoffroy Saint-Hilaire and Jean-Baptiste Lamarck in the early nineteenth century. To date, though, some of the most prominent work by historians and political scientists has sought to highlight the perceived connections between either Darwinian biology and capitalism or between evolution and the Fascist politics of Nazi Germany. Richard Hofstadter's *Social Darwinism in American Thought* has been the most influential of the former trend, while Richard Weikhart, a Discovery Institute Fellow, continues to push the latter view.[10] However, neither of these perspectives gain much traction in the context of a study of what evolution meant in nineteenth-century England since even the connections between evolution and capitalism need to be qualified with references to the contingencies of time and place. Rather, and as I show here, the predominant debate about the political meaning of evolution took place within English radicalism, the broad-based movement for political reform.

Throughout the eighteenth century, the landed aristocracy had dominated English politics. Two parties, the Whigs and the Tories, had vied for government positions, and elections were often decided by bribery and intimidation that frequently saw the local landholder returned to Parliament. Many electoral boroughs contained only a few electors, and yet, some, such as Penryn or East Retford, returned two members to the House. Critics called them the "rotten boroughs," signifying the character of "Old Corruption," as the institutions of English government were popularly known. In contrast, other regions, including the populous cities of the North—Manchester, Liverpool, and the like—had little or no representation.[11]

Of the two parties, it was the Whigs who were more amenable to reform, and indeed, many of the moral concerns that had occupied Whig politicians resonated with the moral values of many of the new industrialists. It was the prominent Whig statesman Charles James Fox, a member of Parliament for

Midhurst, West Sussex, who had done the most to ensure that this was the case. Fox was ardently opposed to what he perceived to be the tyranny of absolutist monarchy, and he had been outspoken in his support for both the American and later the French Revolutions, declaring the constitution of the French Republic a "most stupendous and glorious edifice of liberty."[12] He initially saw the revolution in France as a victory for constitutional monarchy over absolutism, and defended it on these grounds. He also championed Catholic emancipation, the tolerance of religious dissent, and was an ardent advocate of free trade. In 1791, he stood alongside William Wilberforce in support of a measure that would abolish the slave trade. The motion was defeated, but in taking a stand on the matter Fox brought abolition into the fold of Whig politics.

Even after the French Republic beheaded Louis XVI in January 1793 and in the following month declared war on England, Fox still argued that the Republic was preferable to the tyranny of an absolute monarch. His adherence to this position caused many of his erstwhile supporters and friends to cross the House to sit with the Tories, but even in such politically volatile times Fox continued to speak out for the rights of English Catholics and dissenters. Without payment for members of Parliament, politics had long been the preserve of the landed aristocracy, but one consequence of the vast profits to be made from industry was that finance was no longer a barrier to those of dissenting opinion. In 1810, the young lawyer and editor of the *Edinburgh Review*, Henry Brougham, took up a seat in the House of Commons as a Whig. While Brougham came from a well-established family, he was very much a product of the changing economic and social forces that were transforming England into the workshop of the world. The *Edinburgh Review* was just one publication among many new journals, reviews, and weeklies that gave voice to the opinions of a developing middle-class reading public. With the reduction of the stamp duty—the "tax on knowledge," as it was disparagingly called—from 4 d. to 1 d. on newspapers, and the removal of taxation from pamphlets altogether in 1836, the number and quality of newspapers and journals increased exponentially.[13] The first decades of the nineteenth century witnessed the birth of a liberal "marketplace of ideas"; coffee shops and salons erupted throughout the cities of the nation, and clubs and societies were founded to advance any number of progressive ideas and ideals in both the sciences and the arts.[14]

The Whig ideals of liberty, tolerance, and free trade that Fox had espoused and which Brougham promoted as editor of the *Edinburgh Review* quickly served to make the Whigs the party favored by the middle class of industrialists and self-made men. The rise of a financially wealthy and educated middle class led others to follow Brougham into politics. These new Whigs, many of whom were fired with an evangelical zeal, became ardent and effective campaigners for political reform and, as Boyd Hilton has argued, the establishment of a new moral order.[15] They sought the extension of the franchise and a redistricting of electoral boundaries to reflect the equal representation of the people by Parliament, as well as Catholic emancipation and toleration of dissent. It was a campaign that united reforming middle-class liberal industrialists with workingmen and dissenters with Catholics, and it culminated in the Great Reform Act of 1832.

The passage of the 1832 act had been accepted by much of Parliament as an unwilling compromise, and although it had somewhat modestly doubled the electorate from a meager 366,000 to 642,000, as Richard Reeves argues, "the real 'greatness' of the Great Reform Act lay in the abolition of the fifty-six 'rotten boroughs' and the halving of the representation of another thirty, with their parliamentary seats being redistributed to growing industrial towns such as Manchester, Birmingham and Sheffield."[16] If reform had eradicated the worst of the rotten boroughs that had given "Old Corruption" its name, there were many issues that it had left unaddressed. As Chartists took their dissatisfaction with the settlement to the streets, John Stuart Mill could not help but point out that more often than not, when men demanded a leveling down of society in pursuit of justice, they almost invariably only sought a leveling down as far as themselves, and he was aware that many had been sold short.[17] E. P. Thompson has argued that this compromise was a signal moment in the making of the English working class. "To step over the threshold, from 1832 to 1833 is to step into a world in which the working-class presence can be felt in every county of England," he wrote.[18] Indeed, the 1832 settlement marks the culmination of Thompson's magnum opus, *The Making of the English Working Class.*

Debate continues over whether it really was the case that from this time forward those who worked with their hands came to see their interests as lying with others like themselves and opposed to the interests of their employers. But regardless of where one stands on this particular issue, it is certain that 1832 represented the coming of age of the English middle class.[19] Industry, commerce, and manufactures usurped land, rent, and title as the real basis of power, and even if it did not happen quickly or cleanly—this was no violent coup d'état—there was no going back. The world's first industrial revolution was no French affair, despite the Francophile sympathies of many English radicals. Many of those who rose in society on the back of industry sought to buy their way into the lifestyle of those they had for so long declaimed—the stately home becoming a significant marker of "cultural capital."[20] Others, financiers like Robert Darwin, Charles's father, made their own fortunes by bankrolling the lifestyles of local landed elites who were foundering in this sea of change. Instead of beheading the sons of privilege and title, the new English middle class married their daughters. New money wedded old title, and however much this might have been a marriage of convenience, it was a union that would last. There would be no guillotine erected in Trafalgar Square.[21]

The pace of change in science and in society at large was matched only by that of the newly reformed Parliament. Within a year of the passage of the 1832 Reform Act, the Whigs were in power. The year 1833 saw the passage of the first of a series of Factory Acts intent on restricting child labor and the worst excesses of owners who exploited their workers through truck-shops and factory stores, and with 1834 came the passage of the neo-Malthusian Poor Law Amendment Act, which established the workhouse system that Brougham and the journalist and political economist Harriet Martineau had done so much to promote.[22] Passage of the Prison Act and the Municipal Reform Act followed in 1835. Change was the watchword of the day and progress the presumption. This was not to suggest that change came either willingly or with ease. Thompson's grueling history of social dislocation, injustice, and class prejudice, of allegiance and betrayal, is testament to this much, but even the long-established parties of tradition, stability, and privilege found themselves caught up in the tide: both Whig and Tory governments ushered in reforms in this period, and in doing so they brought about the conditions that would force an identity crises for each of their parties.[23]

It was amid this clamor that serious questions were asked about the nature of man and his place in the world. In addition to debate over the proper relation between land and capital, in the newfound forum of public debate questions were asked too about the proper relation between the individual and the state—and ultimately about the proper relation between one man and another. In the last decades of the eighteenth century, the opposition to slavery became a pressing issue in Parliament and the anti-slavery cause was championed by the outspoken evangelical member of parliament for Yorkshire, William Wilberforce, with dissenting industrialists rallied behind him. Among those moved to action was Charles Darwin's maternal grandfather, Josiah Wedgwood. Wedgwood became a driving force in the Society for Effecting the Abolition of the Slave Trade, of which he was a founding member, and he directed his pottery to produce a medallion by the thousands for distribution to supporters of the cause. It portrayed a slave kneeling below the plea "Am I not a Man and a Brother?" and quickly became an internationally recognized symbol of the movement.[24] Beyond slavery, other questions were pressing too. Even though it was hardly a priority of the newly elected MPs, questions were also raised about the proper place of women in society and in the polity. Indeed, following Jeremy Bentham's lead, a minority voice even demanded that humanity, to be worthy of the name, had a moral obligation to their non-human brethren in pain and suffering: sentient animal life.[25] As Bentham had phrased it, when it came to defining the moral community, "the question is not, Can they reason? nor, Can they talk? but, Can they suffer?"[26] Little could he have known that the nature and significance of the animal-human boundary was shortly to become one of the century's most pressing questions.

Where radicals embraced Lamarckian ideas about the heritable effects of environmental factors and of a progressive development from below to ground their politics in nature, Charles Darwin's great achievement lay in making evolution respectable in the eyes of the new generation of liberal Whigs. Robert Chambers's *Vestiges of the Natural History of Creation*, which was published anonymously in 1844, may well have prepared the soil, but it was Darwin's emphasis upon Malthusian competition that put his theory well within the intellectual framework of those who championed free trade; and as Adrian Desmond and James Moore have noted, Darwin's views on the common ancestry of all men fit well with Whig antislavery politics, too. Indeed, Darwin was not only inspired by Malthus, he was keen to make this fact known. Careful to avoid the charge of wild speculation that still dogged the author of *Vestiges*, Darwin sought to tie his ideas about evolution to the political views of the new middle class. That he did so successfully was arguably his greatest achievement. In doing so, however, he redefined the identity of evolutionary politics—of the perceived relationship between humanity and nature and of the relationship between one man and another. Radicals had embraced evolution as grounds for collectivism and social change; Darwin was taken to be advocating for competitive individualism. The politics of evolution were thus hotly contested.

Specifically, in *Political Descent* I argue that the debates about the evolution of cooperation and competition took place among those dissenting liberals and radicals who were both receptive to evolution and who saw it as the key to defining a broader picture, not only of how the world works but of how they believed it should work. Among them, I identify two distinct and mutually hostile traditions that were split over the importance and veracity of the ideas of the political economist Thomas Robert Malthus—and the significance of his 1798 *Essay on the Principle of Population*, in particular.

Malthus's essay was controversial. He had written it as a rejection of the kind of conclusions that had been advanced both by England's most famous radical, William Godwin, in his *Enquiry Concerning Political Justice* (1793), and by the French Enlightenment philosopher Nicolas de Condorcet in his posthumously published *Sketch of a Historical Picture of the Progress of the Human Mind* (1794). Malthus argued that because population increased exponentially—or "geometrically," to use the term of his own day—while resources could at best increase in only a linear, or "arithmetic," ratio, then scarcity, starvation, and struggle would always be a part and parcel of the human condition.[27] Historians of science have certainly recognized the importance of Malthus in the history of evolutionary thought. Darwin's acknowledgment of Malthus's influence on his own thinking ensured as much: *Origin* was "the doctrine of Malthus applied to the whole animal and vegetable kingdoms."[28] However, we have had comparatively little to say on the anti-Malthusian evolutionary tradition. As Adrian Desmond has made clear in *The Politics of Evolution*, his own study of English radicalism, throughout the eighteenth century English radicalism was rife with evolutionary ideas. Significantly, Desmond shows evolution in the early years of the nineteenth century for what it was: a deeply contentious and darkly political set of ideas. Materialists, atheists, radicals, and revolutionaries cited evidence of progressive change spontaneously erupting from below in the natural world in support of their aspirations to see a similar transformation in society. Evolution and revolution were particularly prominent ideas among nonconformist and free-thinking medical students who, having trained in Edinburgh in the 1820s, later practiced in London and the growing provincial cities in the 1830 and 1840s. It was this radical and deeply political association that made transmutation so controversial, and not simply its theological implications. Certainly, conservative Anglicans railed against the theological heterodoxy of transmutation and its suggestion of man's animal origins, but it was the political challenge they feared most. In looking to understand the context of the later evolution debates, Desmond is quite correct when he says "We ignore these political aspects at our peril," for in doing so we risk missing the always present but always contingent connections between the ways in which we construe our biology and the ways in which we construct our politics.[29] As I shall show in *Political Descent*, this Lamarckian and anti-Malthusian influence persisted into subsequent generations even as the political landscape of the country was changing.[30] Indeed, while the Malthusian thinking that was central to Darwin's theory of natural selection appealed to many of those who, from the 1850s, identified themselves as liberals, or liberal radicals, others continued to emphasize Lamarckian evolutionary processes in a way that was quite at odds with the competitive presumptions of Malthus. It is telling that they could do this while still thinking of themselves as "Darwinian," for the inheritance of acquired characters, much associated with Lamarck, was by no means absent from Darwin's work. Indeed, as Paul Elliot has pointed out, in radical circles the Darwin name had been associated with Lamarckian transmutationist ideas long before Charles Darwin published *Origin of Species*. Erasmus Darwin, Charles Darwin's paternal grandfather, had anticipated Lamarck in a number of ways in his *Zoonomia* (1794), and had tackled the origin and development of human society in *The Temple of Nature; or, The Origin of Society* (1803).[31] Anti-Malthusian dissenting radicals thus felt able to identify themselves as Darwinians without contradiction, choosing to ignore the Malthusian elements of the younger Darwin's theory and the liberal Whig politics with which they were associated.

In the early post-*Origin* days of the 1860s, many of these differences were glossed over as dissenting evolutionists of all stripes found common cause in their defense against the attacks of the established church. However, once the initial battle for evolution had been won and the Anglican Church had been forced to make its peace with the new biology, the common cause that had served to paper over the cracks in English radicalism finally fractured. As contemporaries struggled to make sense of the changes that swept aside the bulwarks of the established social order, they looked to the natural order to find a surer foothold. In this context, understandings of the nature of evolution, of Darwin, and of Malthus, became fundamental aspects of British politics, and they remained so well into the twentieth century.

There are a number of reasons why historians of science have given comparatively little attention to the anti-Malthusian radical-cum-socialist tradition. First and foremost, perhaps, is the fact that they have simply been interested in other things. This period in the history of evolutionary thought has long been termed "the Darwinian revolution," a natural consequence of which has been a focus on Darwin and his discovery. The wealth of materials now available to Darwin scholars has facilitated this preoccupation; with Darwin's notebooks and letters now accessible both in print and online, it is possible for historians to reconstruct the steps that Darwin took to the discovery of what might not inaccurately be described as the most significant development in the history of Western thought. A second facet of the present that has drawn the attention of historians to other things has been our continuing obsession with the relationship between evolutionary thought and religious belief. This has been so particularly in the United States, where science continues to be attacked by anti-modernist religious interests. Historians have quite rightly taken it upon themselves to bring a historical perspective to these ongoing debates. A third reason why these evolutionary political radicals have all but escaped our notice is the fact that, as Adrian Desmond has pointed out, historians of science have tended to focus upon those people who are more readily identifiable as bona fide scientists—those men who dedicated their lives to the study of natural history, paleontology, geology, and biology. For this reason, we know a lot about the anti-evolutionary views of the Cambridge dons who dominated mid-nineteenth century science, but little about the political radicals who studied transmutationist works in quite different venues. As Desmond has pointed out, in looking for evolutionists before Darwin, "historians have consistently looked in the wrong place."[32]

There are further factors which have helped to obscure the story I tell here, but which have more to do with the historical actors than with the predilections of historians. First is the simple fact—as I have noted above—that after the publication of *Origin* evolutionists of all stripes tended to refer to themselves as Darwinians, even those who rejected the Malthusian aspects of natural selection. While Peter Bowler long ago pointed out that there was a general lack of enthusiasm for natural selection among even those members of the scientific community who were convinced of evolution by Darwin's work, and while Greta Jones has provided a broad survey of British political appropriations of evolutionary ideas in her *Social Darwinism in English Thought*, it has only been in light of the work of Bernard Lightman and other scholars interested in the popularization of science that we have started to take seriously the full extent to which there were many competing notions of what it meant to be a Darwinian. Indeed, as Lightman has pointed out, Darwin's own view of the matter was far from being either the most widely held or the most influential.[33] A second point is that Darwin was concerned to distance his own ideas from those of both Lamarckian radicals and the author of the *Vestiges of the Natural History of Creation*, and he was very successful in doing so.[34] He repeated his assertions of the difference between his views and those of Lamarck and the author of *Vestiges* in his autobiography, stating there too that he had not gleaned anything of significance from his grandfather's transmutationist works either. Further, in crafting *Origin* Darwin sought to present himself as a good scientist, which in the mid-nineteenth century still meant to be a good Baconian. He did not want to suggest that he had gone on the *Beagle* voyage with his head full of preconceptions taken from his grandfather or anyone else; rather, he sought to make it clear that he had patiently accumulated facts about the natural world, and only later, after many years of careful reflection and study, allowed himself to speculate on their meaning. Of course, while speculative works like *Vestiges* were considered beyond the pale of good scientific endeavor, this was not so much because the work was perceived to be political, but rather because it was perceived to be associated with the wrong kind of politics. Anglicans like William Whewell, the Master of Trinity College, Cambridge, who, while in favor of moderate political reform, were opposed to the radical political program, policed what passed as good science and what did not, and Darwin, keen to have his views accepted as legitimate science, played down his own political and philosophical commitments, going so far as to excise all but the merest mention of mankind from the early drafts of *Origin* for fear he might compromise the reception of natural selection.[35]

A reconsideration of Darwin's own politics and the place they had in the formation of his theory was the subject of Desmond and Moore's *Darwin's Sacred Cause.* They have argued that Darwin was motivated in his pursuit of the "mystery of mysteries," as the astronomer Sir John Hershel described the question of the origin of new species, by the desire to prove the common ancestry of all of the races of mankind in support of the antislavery campaign that generations of Darwins and Wedgwoods had made their own sacred cause. Desmond and Moore's thesis has been criticized by a number of historians. Most are simply not convinced that this was Darwin's motivation for pursuing the species question, however much he might have sympathized with his family's antislavery views or recognized that evolution might have supported an antislavery argument from common human ancestry.[36] Whatever the truth of the matter, though, it is clear that Darwin was a lot more interested in political issues than he later admitted.

Darwin was well aware of the politics of evolution. He had seen its materialist associations castigated in Edinburgh as well as in Cambridge in his youth, and he knew that political radicals found his grandfather's work attractive. It was the radical associations of evolutionary ideas that spurred the establishment's reaction to the anonymously published *Vestiges of the Natural History of Creation*, which, as James Secord has noted, made such a Victorian sensation when it first appeared in 1844.[37] The book was published even as Darwin was writing up his own ideas, and despite the author's intention of making evolutionary ideas acceptable to a broader Whig public, the majority of the men of science who deigned to review the work rejected it as bad science, bad theology, and the harbinger of a world turned upside down. While *Vestiges* would do a lot to draw the sting from evolutionary ideas in some regions and among some readers, it failed to do so in others, and, significantly, the book was considered a scandal among those whose opinions Darwin most valued.

Whewell was the most prominent and outspoken of the Cambridge men of science. He had been appointed Master of Trinity in 1841 and thereafter dominated the College, the University, and Cambridge, both intellectually and politically, until his death in 1866. In addition to his position at the university, Whewell held offices in the most prominent and influential scientific societies of the day: he had been a founding member of the British Association for the Advancement of Science and enjoyed periods as its president, as well as, variously, president and vice president of the Geological Society of London, the most active and influential of England's scientific societies.[38] Whewell was chief among those who sought to establish a place for the natural and moral sciences in the university curriculum, where hitherto mathematics and classics had predominated, but he was careful to legislate the science and morals that were acceptable.

To Whewell, central to policing the boundaries of what constituted good science was what he referred to as "the recognition of final causes in physiology," and he made this the main theme of his popular 1845 work *Indications of the Creator*, although he had articulated the same in his earlier, three-volume *History of the Inductive Sciences* (1837). Despite its title, Whewell wrote *Indications* for the same popular audience as had read *Vestiges*. He even ensured that its binding was similar and that it was available at a comparable price. In it, he decried the speculative nature of *Vestiges*, but it was clear that he was concerned about its political associations.[39] That the natural and moral sciences were seen to be inextricably linked throughout this period is telling, as is the observation that they spoke to the kind of political economy, and thus the kind of society, that should be implemented.

Given the political connotations of evolution throughout the nineteenth century, I begin *Political Descent* with a consideration of Darwin's politics and how they shaped his worldview. To this end, I begin the first chapter with Darwin's *Beagle* voyage and with his encounter with the natives of Tierra del Fuego. This is a well-known episode in Darwin's circumnavigation of the globe that brought him to reflect upon the differences that separated "savage from civilised man," as well as the similarities that united them. Darwin had no doubt that here were creatures akin to his own ancestors, but what is particularly telling is that when he speculated upon why the Fuegians had not progressed beyond the dreadful existence they endured, Darwin invoked ideas from political economy in explanation. It was their lack of private property that had prevented them from developing the kind of social hierarchy that would nurture individualism and provide the incentive for innovation, he thought. Echoing radical and Lamarckian sentiments, Darwin confessed his belief that habits were formed in response to circumstance, and thus modified, were hereditary. "Nature, by making habit omnipotent, and its effects hereditary, has fitted the Fuegian to the climate and productions of his miserable country," Darwin concluded.[40] In the course of recounting this episode, I will fill in necessary background on the Darwin and Wedgwood families and their political beliefs.

As Donald Winch and Margaret Schabas have each pointed out, conceptions of nature's economy, of political economy, and of moral economy, themselves share a common ancestry.[41] When Darwin wrote *Origin* and made his own contribution to the "species question," he was well aware that evolution already had political associations with radicalism and revolution. Given that he, like other naturalists, was already primed to make connections between natural, political, and moral economies, it is unsurprising that when he read Malthus's essay in 1838 he instantly recognized its political economy as significant for his own theory of nature's economy. He had already noted that the scarce resources on Tierra del Fuego led to a condition of almost constant warfare between the tribes who inhabited the islands.

Historians have had a lot to say about the significance of Malthus for both the development and the reception of Darwin's work. I argue that Darwin was not only aware of the political implications of embracing Malthus, but that he emphasized his debt to England's first political economist, both to hitch his own evolutionary ideas to the Whig politics then associated with the neo-Malthusian "philosophical radicals" Harriet Martineau and John Stuart Mill and to distance himself and his ideas from the revolutionary elements in English radicalism. His mentor, the geologist Charles Lyell, was sympathetic but could not be convinced to follow Darwin so far as to see man as subject to only the same laws of development as the rest of the natural world, and thus, at the last minute, Darwin excised his comments on mankind from what was to have been his big species book. At *Origin*'s publication only one sentence remained: "Light will be thrown on the origin of man and his history."[42] I contend that it was this decision to remove man from his conclusions that led many of Darwin's contemporaries to misread the implications that Darwin believed evolution to have for mankind. As Darwin would only later explain, in *The Descent of Man*, mankind was a social species and thus, whereas in *Origin* Darwin had described natural selection as largely a competition between one individual and another, he did not believe that this was the case among men, or at least not the dominant factor. Rather, like other social species, it was more significant that humans had cooperated throughout their evolutionary history. Evolution thus gave no endorsement to any "devil-take-the-hindmost" politics or philosophy.

If Darwin had thought that he had done enough to avoid controversy, he was very much mistaken. At the heart of Lyell's doubts was a concern that resonated with a great many of those who read and reviewed *Origin.* What did the idea of human evolution say to the sort of beings we are? The suggestion of common ancestry between man and apes had theological implications, of course, but the main point of contention very quickly became whether, despite this fact, there might be more-limited grounds for human exceptionalism. Could the processes of natural selection account for the evolution of those most-human characters—mind, compassion, and conscience—or were they a sign of man's divinity? Darwin was shocked to hear that his former ally, Alfred Russel Wallace, had reached this conclusion.

Even before Darwin, the question of both the nature and the origin of the moral sentiments taxed some of the greatest minds of the Enlightenment—the associationist philosophers Adam Smith, David Hume, and Erasmus Darwin, as well as Nicholas de Condorcet and the English utilitarians William Godwin and Thomas Robert Malthus, to name but a few. These men were followed both in this inquiry and in this tradition by Jeremy Bentham, James Mill, and Mill's son John Stuart Mill, founders of what became known in the mid-nineteenth century as "philosophical radicalism." Investigation of the moral nature of man was no idle curiosity; rather, it had become a fundamental question regarding what constituted a right and just polity. Darwin had read each of these authors; until he published *Descent* in 1871, though, he confined his speculations to his notebooks.

Given that evolutionary ideas were rife throughout English radicalism, in chapter 2 I turn to consider Herbert Spencer. Spencer was brought up in the radical dissenting tradition and eventually became one of the most influential philosophers of the age. Spencer's radical journalism was steeped in Lamarckian ideas about a natural progressive development that drove all of nature onward and upward. Government interference by "Old Corruption" only served to prevent the natural outcomes that might otherwise be possible. In this chapter, I echo Robert J. Richards's point that Spencer was not the Social Darwinist that, following Hofstadter, many have presented him to be. More-recent studies have echoed Richards's call for a more sympathetic and contextualized understanding of Spencer and his work, but here what I want to emphasize is the extent to which Spencer was influenced by the Godwinian radical ideas that he was introduced to in his youth. Even though Spencer has become closely associated with individualism, competition, and laissez-faire economics—an association only strengthened by his description of Darwin's theory of natural selection as "the survival of the fittest"—attention to the context of his words provides us with a very different picture. Spencer's first forays into journalism are telling, as is his first book, *Social Statics.* Here, Spencer mapped out his hope of seeing the birth and development of a utopian-socialist future as a result of the evolution of human social instincts unfettered by corrupt government. This is a far cry from the Spencer whom some would portray as the grandfather of Social Darwinism. It is true that Spencer reined in his socialistic conclusions as socialism became a force in the country in the 1880s; he ultimately made a stand in defense of capitalism, as socialists increasingly looked to government to implement a socialist society rather than to individual citizens to develop their own socialistic qualities. The importance of independent action was central to Spencer's views about moral, individual, and social development—views that were firmly rooted in Lamarckian biological beliefs about the inheritance of acquired characters.

By the 1870s, there had been a Darwinian revolution of sorts. Certainly, in the realm of politics, there were few people who did not invoke evolutionary ideas in support of their own views or to undermine those of their opponents. Darwin had noticed to Lyell that the *Manchester Guardian*, a paper that had a large readership among the industrialists of the region, had hailed *Origin* as a naturalization of out-and-out competition. This was too much. Darwin had excised man from *Origin*, but he returned to the subject in 1868 in what was to become *The Descent of Man, and Selection in Relation to Sex*. As I show in chapter 3, in *Descent* Darwin carefully explained the material origins of morality as well as morphology, excusing himself for venturing into such lofty metaphysics on the grounds that of all the great minds that had thus far tackled the subject, "as far as I know, no one has approached it exclusively from the side of natural history."[43] Darwin's intention was to show that natural and sexual selection could in fact account for the development of even these ennobling faculties. Thus, far from endorsing an ethic of rampant individualism—as some of his early and most enthusiastic supporters had argued—Darwin argued that man had evolved to hold exactly the sort of progressive Whig politics he held dear and that he shared with those most-eminent liberals of his day, William Ewart Gladstone and John Stuart Mill. In contrast to the competition and struggle that dominated nature in *Origin*, when it came to accounting for mankind, a real Darwinian ethic was thus far-removed from what Darwin termed the "low motive" of self-interest.

Darwin was not allowed the final word, however, and by the time *Descent* ran off the presses, there were plenty of other opinions about exactly what evolution might have to say about human nature, our origins, and our destiny. Not only had the radical journalist Herbert Spencer argued that the mechanisms of evolution demanded a laissez-faire economic policy, but from the other end of the political spectrum, Karl Marx was writing to his friend and sponsor, Friedrich Engels, that *Origin* was "the book which in the field of natural history contains the basis for our view."[44] Neither Marx's nor Spencer's views went uncontested, of course, and in chapter 4 I turn to consider how liberals and socialists of different stripes articulated their own various appropriations of evolution. They borrowed from Spencer as well as from Darwin, but it is significant that for the majority of those involved, this was not merely a means to take from biology what supported their preexisting politics, but rather was an earnest attempt to ensure that their politics would guarantee the progressive evolution of the nation and its citizens. It is telling here that this period ushered in what Michael Freeden has called a "new liberalism," as the Industrial Revolution revealed weaknesses in laissez-faire that many liberals came to consider untenable. As a result, liberals moved away from the classical liberal political economy that the likes of Spencer had championed. This period also witnessed what historians have called the "socialist revival." In the decade and a half from the mid-1880s to the end of the century, a plethora of organized socialist organizations were established. Many had their origins in the London radical clubs as old Chartists joined up with a younger generation of discontents in order to press the possibility of building a new heaven on earth. While in the early days of the movement members were clearly excited by the diversity of views that passed for socialism in England, they quickly ossified into partisan differences over strategy and leadership. Echoing the tensions that had wracked liberalism, socialists too wrestled with the question of whether socialism was something that needed to be lived as a means to its accomplishment or whether it could be legislated into being. Socialists also turned to evolutionary ideas in order to press their agendas, and here, just as Malthus had been controversial for an earlier generation of radicals, he was equally so for many socialists. As the historian Daniel Todes has noted regarding the work of the Russian anarchist, naturalist, and geographer Peter Kropotkin, many strove to theorize a conception of Darwin without Malthus. Others, notably from within the more-reformist socialist Fabian Society, embraced Malthus as a factual statement about the natural laws of life and thus the starting point for their social theory.

It is in the context of this national debate about the politics of evolution that in chapter 5 I revisit the differences between Thomas Huxley and Peter Kropotkin. Having welcomed *Origin* as a "veritable Whitworth gun in the armoury of liberalism," and having utilized evolutionary arguments to undermine Herbert Spencer's laissez-faire politics in his essay "Administrative Nihilism," Huxley went on to write "The Struggle for Existence," which appeared in the journal *The Nineteenth Century*. In it he argued that nature was very much red in tooth and claw, and that in consequence the vast majority of our own evolution had depended upon individualism and competition.

While Huxley did admit that in the relatively recent past humans had formed society in opposition to these long-extant competitive instincts, he concluded that because human sociality had arisen so recently, it was much weaker than the deeply ingrained individualism that marked our animal origins. This tension was a theme he would go on to explore in his famous Romanes lecture, "Evolution and Ethics," in 1893.

Kropotkin thought Huxley's article an atrocious misrepresentation of Darwinism. In response, he wrote a series of articles that also appeared in *The Nineteenth Century.* He argued that mutual aid and cooperation were not the recent phenomena in the evolution of mankind that Huxley had claimed, but that, rather, they had been a fundamental aspect of evolution from microorganisms on up. From this point onward, Kropotkin made it his life's work to document the evolution of mutualism across the animal kingdom—and across the history of human development in particular. In contrast to Huxley's "Evolution and Ethics" (1894), Kropotkin's *Ethics, Origin, and Development* (1924), although it remained unfinished at his death in 1921, was intended as the summation of his view that humans had evolved to be cooperative and altruistic beings and that this was the proper lesson that Darwinism had for humanity.

In speaking to the nature and morals of mankind, evolution both set the boundaries and raised new possibilities for how we might live. As George Bernard Shaw noted, Darwin "had the luck to please everybody who had an axe to grind," and in this regard Darwin's name and works became fundamental political reference points.[45] Socialists like Shaw and H. G. Wells vied with each other for the public ear over what Darwin meant for mankind and what evolution meant for socialism. Each had been one-time admirers of William Morris, the socialist artist, craftsman, and author of the socialist-utopia *News from Nowhere*, but from the late 1890s and in light of Darwin, they argued over whether such a future was possible anymore. As I show in chapter 6, this was not only a debate about the veracity of Malthusian political economy; the neo-Darwinian ideas of the German biologist Friedrich Leopold August Weismann seemed to undermine the Lamarckian ideas that were central to the anti-Malthusians' evolutionary and political strategies for change. Theories of evolution were evolving.

Weismann's views were as significant as they were controversial. If true, he had ruled out the inheritance of acquired characters, with dire consequences for all Lamarckian political schemes, as these had been as important for socialism as they had been for radicalism. Neo-Darwinism reigned in the timescale available for any hopes that mankind might evolve into a more socialistic animal, and further, Weismann's theory of "panmixia"—the degeneration of an organism that he believed would follow the suspension of natural selection—was deemed equally problematic for socialism. In the final chapter of *Political Descent*, I show that while this was hotly debated among scientists in the pages of the journal *Nature*, Benjamin Kidd's popular 1894 book *Social Evolution* took the debate to the public. Kidd argued that the threat of panmixia ruled out all hope of socialism. It was in this context that the promising young mathematician, Karl Pearson, entered the fray, arguing that statistical analysis gave the lie to panmixia, and to Kidd's conclusions as well. Pearson had been impressed with Morris in his younger days, as well as with Marx. He combined his socialist insights with what he took from Darwin and advocated his own particular form of evolutionary socialism as a result. Kidd was wrong about panmixia and so was Weismann, Pearson argued. He went on to claim that in fact socialism was the inevitable outcome of social evolution, and he set out to prove it statistically. However, in the process, research that came out of his own lab revealed to him the extent of what came to be known as the differential birthrate. It was the working class and the poor—those who were increasingly being thought of as the "unfit," the "residuum," or the "social problem group"—who had fathered the most children across the previous half-century. This fact was enough to change his mind about panmixia. Panmixia might not engender the continuous degeneration that many of his contemporaries feared, but it would certainly undermine the quality of the English race in the context of growing international competition. This was particularly significant as Europe headed toward war, and Pearson increasingly favored eugenic measures to counter panmixia and increase national efficiency. Although many individuals on both sides of the conflict went into the war wielding Darwinian arguments to bolster their own position, in the aftermath of years of trench warfare it became increasingly difficult to appeal to Darwinian struggle as a means to social progress. The sheer number of the dead and wounded was a tragedy that finally defeated the flagging optimism and faith in progress that had become a hallmark of the long nineteenth century.

Before proceeding, it will be useful to explain a little about my approach to writing this history. *Political Descent* is an intellectual rather than a social history. As such, I intend this work to be an example of what Stefan Collini has described as a study of "the context of refutation." Collini was talking about his own methodological approach to writing *Liberalism and Sociology: L. T. Hobhouse and Political Argument in England, 1880-1914*. Collini's work was not a biography, but rather an attempt to uncover what he referred to as "a certain level of discourse, the medium, as it were, in which his [Hobhouse's] thought moved and had its being." He continued: "This involves supplying what might be called 'the context of refutation'—that is, an account of the theories he was attacking, the arguments he was rebutting, the assessments he was challenging. But it also involves trying to identify the forensic resources at his disposal—the overriding force of certain arguments, the emotional resonances of key terms, the exploitable tensions within accepted beliefs. As my title suggests, it is to the political argument of the three decades before 1914 that I think we must, in the first instance, look for this context."[46]

It is my hope that the narrative I tell in *Political Descent* has enough biography and social context to enable the reader to get a feel for the characters and the debates, although, echoing Collini, I do not intend this work to serve as an exhaustive biography of any of the characters that appear in these pages. Rather, I am concerned to give an account of the debates that occurred in a period in which it was a common cultural presumption that biology and politics were mutually informative subjects. Malthus, Smith, Lamarck, Weismann, and, of course, Darwin, were just some of the key intellectual resources that were central to how people made sense of what it meant to be an evolutionist, a liberal, a radical, or a socialist. These were questions about nature, human nature, and political and moral economy, as well as ultimately being questions about what it means to be human and, in consequence, of how we might live.

Some readers might find it frustrating that I do not say more than I do about causation. Causation is, after all, a good part of the historian's fare. Is it the case that nineteenth-century conceptions of biology molded contemporary politics, or vice-versa? As the reader will detect, I believe this is a two-way street and I have my sympathies with both sides of this question. That said, I think that this either/or way of framing the relationship between science and society is problematic. Rather, it seems more productive to see science as just one aspect of broader social and cultural assumptions. From this perspective, the explanatory emphasis is not upon what the causative effect of science is upon society, or, conversely, of society upon science, but rather, as Collini explains, historical explanation comes to focus upon uncovering the intellectual resources that historical actors had available to them and to which they appealed as they attempted to make sense of and order their world. This is not to avoid addressing the question of causation in history, however. Rather, it is a strong assertion that the relationship between biology and politics is culturally and intellectually contingent. This is what makes it a fit subject for the historian, and it is why scientists and those interested in the ways in which science shapes our sense of our own place in the cosmos should also be interested in this history.

1

Every Cheating Tradesman: The Political Economy of Natural Selection

I have received in a Manchester Newspaper rather a good squib, showing that I have proved "might is right," & therefore that Napoleon is right & every cheating Tradesman is also right.

CHARLES DARWIN TO CHARLES LYELL, *4 May 1860*

Mockingbirds and finches were all well and good, but when it came to thinking seriously about the possibility of transmutation Darwin had had man in mind from the first. His notebooks and diary from his circumnavigation of the globe aboard H.M.S. *Beagle* and after show clearly that he had been deeply affected by what he had seen of the various native peoples he had encountered during the voyage—and by none more so than the natives of Tierra del Fuego.[1] Within a year of returning to England, Darwin was speculating upon exactly where the logic of human evolution might lead. "Origin of man now proved," he had written in Notebook M. "Metaphysics must flourish.— He who understand baboon would do more toward metaphysics than Locke." His mind racing, he had no time for full sentences, correct spelling, or punctuation. "The mind is a function of the body." "Oh you materialist!" he condemned himself.[2] What the historian of science Jonathan Hodge has called Darwin's "notebook programme" was to prove the basis of Darwin's work for the rest of his life.[3]

Darwin had joined *Beagle* as naturalist in December 1831 after the ship's captain, Robert FitzRoy, had lamented not having someone aboard with this type of talent and experience on his previous voyage. Darwin was more than adequately qualified; he had not only studied geology, marine biology, botany, chemistry, and entomology during his time at Edinburgh and Cambridge, but medicine as well. As the Darwin scholar John van Wyhe has pointed out, it is a myth, albeit a well-established one, that Darwin joined *Beagle* merely to be FitzRoy's gentleman companion—as historians have long maintained.[4] Certainly, conventions of class and naval discipline meant that months at sea could be a lonely experience for a ship's captain, despite the cramped conditions. On this voyage the ship's cramped quarters were occupied by more than seventy men, and so it was clearly important to FitzRoy that Darwin was a gentleman.[5] However, as Van Wyhe contends, FitzRoy's stipulation that the naturalist who accompanied him should be a gentleman was indicative not only of his own preference for company at dinner, but it also spoke more deliberately to the status of the position he was eager to fill. As historians of science are aware, in the mid-nineteenth century "naturalists" might be either gentlemen practitioners or collectors of an altogether different breed.[6] As Van Wyhe concludes, "Continuing to portray Darwin as 'companion' rather than 'naturalist' obliterates the most conspicuous example of the long, gradual transformation towards scientific professionalization in the life sciences."[7] Thus, while Darwin was certainly taken on board as naturalist, as FitzRoy made clear to John Stevens Henslow and George Peacock, through whom the opportunity was relayed to Darwin, the position was for a "gentleman naturalist" who was also to dine at the captain's table. FitzRoy was both a talented seaman and a firm disciplinarian, and he and Darwin got on tolerably well, with only one recorded exception: whereas Darwin came from a liberal Whig family that had long been involved in the movement to abolish slavery, FitzRoy was a Tory who thought slavery a benevolent paternalism, and so despite the fact that to question a captain's opinion would have been a mutinous act for any of the crew, Darwin could not let FitzRoy's views on the matter go unchallenged and on one occasion they argued vehemently enough about the subject for Darwin to fear he must leave the ship.[8]

In this chapter, I recapitulate those aspects of Darwin's 1832 encounter with the natives of Tierra del Fuego that illustrate the ways in which his beliefs about political economy informed the way he made sense of the relationship "between savage and civilised man." Historians have noted the connections that have often been made between political economy and natural history, and that Darwin made these sorts of connections was not unusual.[9] Indeed, Darwin scholars have addressed this issue directly, noting Darwin's debt to Malthus in particular. Here, though, I suggest that, at least before his conversion to Malthusian thinking, Darwin was very much indebted to his grandfather, Erasmus Darwin, for his political-economic as well as his natural-historic worldview. Further, while historians have long recognized the importance of Malthusian political economy for the development of Darwin's evolutionary ideas, there remains debate as to exactly what Darwin took from Malthus. Darwin read Malthus's *Essay on the Principle of Population* in 1838, not long after his return from the voyage, and doing so certainly led him to recognize the importance of individual variations within a population. However, here I argue that Darwin cited Malthus as vehemently as he did for explicitly political reasons as well. Not only was he determined to distance his own evolutionary views from those of the politically radical Lamarckians who were being decried as heretics in theology and mere speculators in science, but he also intended to ground his own liberal politics in nature as well. As Adrian Desmond and James Moore have pointed out in their biography of Darwin, "Darwin's biological initiative matched advanced Whig social thinking" and natural selection was "a mechanism that was compatible with the competitive free-trading ideals of the ultra-Whigs."[10] But I do not believe that this is an adequate account; there is more that needs to be said on the politics of Darwin's evolutionary views. Although Darwin had initially intended to include an account of human evolution in *Origin*, he made the last-minute decision to leave man out for fear that he might otherwise prejudice his readers against the validity of his work. This had significant implications for the reception of his theory, for in excluding man he also excluded any reference to the point that Malthus had made in the later editions of his essay, that humans might avoid the dire consequences of struggle by a reasoned evaluation of their circumstances, by working hard, and by exercising moral restraint from reproduction. As scholars have long noted, the Malthus that Darwin invoked in *Origin* was the Malthus of the 1798 first edition, not the 1826 sixth edition of the work, which was the one Darwin actually read. Thus, while Darwin clearly viewed man as a moral as well as a rational animal—a species that not only secured its own progress by each man evaluating his actions in light of their likely consequences—mankind was also a species that had evolved to have deeply "other-regarding" ethical sentiments, a point to which I shall return in chapter 3. This was not something that was borne out in *Origin.* Thus, when his contemporaries applied the insights of *Origin* to humanity, they did so in quite a different way than Darwin had had in mind. Initially relieved that he had succeeded in distancing his views from the revolutionary politics of the radicals, Darwin was later dismayed to see his work then held up as a vindication of the most heartless political individualism and an endorsement of "every cheating Tradesman." In the ensuing furor about apes and angels, Darwin looked to the geologist Charles Lyell, and then to the anatomist Thomas Huxley, to publish an account of human natural history that reflected what he really thought about the nature of man and man's place in nature. However, as the evolution debates moved on from comparative anatomy to focus upon the origin of mind and morals, it was Alfred Russel Wallace who eventually came up with the goods, in a paper he presented to the Anthropological Society of London in 1864: "The Origin of Human Races and the Antiquity of Man Deduced from the Theory of 'Natural Selection.'" Darwin was ecstatic. Having championed Wallace's article as "the best paper that ever appeared in the Anth[ropological] Review!," Darwin was thus doubly disappointed when Wallace later rejected the adequacy of his own argument.

Refitted as a survey vessel, *Beagle* left Devonport dockyard on 27 December 1831. She did not drop anchor in English waters again until she sailed into Falmouth Harbor on 2 October 1836.[11] By all accounts, when Darwin left England he was as conventional as the next man when it came to his views on religion, creation, and transmutation. At least this is what he would later tell people. And this might well have been so if the "next man" had a political radical and avowed transmutationist for a grandfather, a skeptical Unitarian for a father, and had studied primitive life forms under the auspices of an internationally renowned Lamarckian naturalist at a medical school in the most radical city in Britain. It was Doctor Robert Grant who had taught Darwin during his time in Edinburgh. He was a political radical, an atheist, and a revolutionary Francophile. He admired Darwin's grandfather, Erasmus, and had sought Charles out and taken him under his wing shortly after he had arrived in the Scottish capital. The two quickly became close and worked together on sea sponges, as Lamarck had done.[12] Of course, Darwin had subsequently studied at Cambridge, which was still a bastion of Anglican orthodoxy and conservatism even though the university had returned its first Whig MP by a very narrow margin in 1829. Darwin had witnessed the trouble that an association with atheism and materialism might bring during Robert Taylor's visit to the town. Taylor, an avowed atheist and republican, was a graduate of St. John's who had taken holy orders. Having lost his faith, he had returned to his alma mater and publicly laid down a challenge to the presumptive faith of the university. Dubbed the "Devil's Chaplain" and calling himself an "infidel missionary," he invited debate about the foundations of Christian belief but had been run out of town by a mob for his troubles.[13] Whatever else Darwin might have taken from the contrasting influences of these two university towns, he certainly came away with a full appreciation of the breadth of political and religious controversy that might surround transmutationist ideas.

Making full sense of all that he had seen on the voyage would take Darwin much of the rest of his life, but it is evident that he had approached the question of the origin of new species with the presumption that there was much about man that allied him with other animals. As the historian Robert Young has pointed out, this in itself was not a particularly unusual position, but of particular significance was the fact that to Darwin man appeared to be almost infinitely malleable.[14] The effects of environment appeared to play a large part in this. At least this was so in the early days of Darwin's thinking on the subject. But always important too were the interactions between organisms.

In this, man was no exception. What was evident in the latter case, though, was that human interactions were influenced in large part by the moral character of the individuals concerned and by the politics of society.

Darwin was not alone in framing his perceptions of human nature in the context of ongoing debate about moral character and political economy—two areas of study that were considered to be intimately related. Nor did the conclusions he drew go uncontested. But it is significant that it was his impressions of the native peoples he encountered as he circumnavigated the globe that first led him to think seriously about this issue, and it was in this regard, as he later acknowledged, that "the voyage of the Beagle has been by far the most important event in my life, and has determined my whole career."[15]

When Darwin first set eyes upon the natives of Tierra del Fuego, he thought the naked and wild savages who followed *Beagle* along the coast as much animal as man. However, when, in the company of a landing party, he came face to face with them, he could not but reflect upon the similarities as much as the differences between these painted savages and the Englishmen who made up the ship's company. It is likely, too, that the thoughts that dawned on Darwin were shared to some extent by at least some of his shipmates. After all, traveling with them had been three Fuegians, each of whom had had the benefit of three years of the best education that the Church Missionary Society could provide. Newly civilized and dressed accordingly, they had even been invited to St. James palace by King William IV and there were presented to Queen Adelaide.[16] Now, FitzRoy's intention was to see them repatriated in the company of a young missionary named Matthews, reestablished in their native land as a spiritual beacon on a Godless shore.

En route from England, Darwin had become particularly friendly with the Yámana Fuegian who had been christened Jemmy Button, and thus when he first encountered the Fuegians in their native condition he was deeply impressed by the difference that environment and education could make in a man—even in only a few years. "It was without exception the most curious and interesting spectacle I had ever beheld," he later wrote. "I could not have believed how wide was the difference, between savage and civilized man. It is greater than between a wild and domesticated animal, in as much as in man there is a greater power of improvement."[17]

The analogy between wild and domestic animals, as well as that between man and animal, would prove telling, and although the historian Camille Limoges has argued that Darwin only arrived at this analogy after he had essentially worked out his theory in 1842 and that even then he adopted it as a rhetorical device in order to convince his readers, like Jean Gayon I think otherwise. Limoges's argument rests on his detailed analysis of Darwin's notebooks and the fact that Darwin does not use the term "natural selection" to describe his own ideas until 1842—long after he had developed the core of his theory. This may be so, but as Jean Gayon has pointed out in his own account of the development of Darwin's theory, Darwin's notebooks are rife with references to and reflections upon heredity taken from his interactions with domestic breeders, and it was in this context that he worked out his theory.[18] Further, and as is evident here, Darwin was also clearly applying analogies from domestic to natural breeding to make sense of and articulate his conception of the differences between Fuegians and Englishmen.

Clearly already on Darwin's mind were thoughts about what had facilitated the comparative rise of Englishmen above the dreadful state of the barely human savages he now saw before him—or what had held the Fuegians back. Darwin was led to speculate not only upon the role of education in this improvement of mankind, but also on the action of the environment and of habituated behavior in shaping a man's life and character. "What a scale of improvement is comprehended between the faculties of a Fuegian savage & a Sir Isaac Newton," he mused.[19] Darwin was struck not only by the savage nature of the Fuegians, but by the harsh and inhospitable nature of the environment in which they made their home. "The climate is certainly wretched," Darwin noted.[20] "Their country is a broken mass of wild rocks, lofty hills & useless forests, & these are viewed through mists & endless storms."[21] This was truly a desolate landscape, the weather making the place "thoroughly detested . . . by all who know her."[22] The people were only too clearly accommodated to their surroundings. "I never saw more miserable creatures," he confided to his diary. They are "the most abject and miserable creatures I any where beheld."[23] His horror turned to fascination, and his reflections spilled over into a lengthy footnote in his published account of the voyage, in which he compared the degraded South Sea islanders, the Esquimaux, and even the Aboriginal Australians, favorably to the degraded beings he saw before him. His conclusions were stark: "I believe, in this extreme part of South America, man exists in a lower state of improvement than in any other part of the world." "Viewing such men," he continued, "one can hardly make oneself believe they are fellow-creatures"—and yet, despite himself, and with Jemmy as compelling evidence, he did.[24]

The resource-scarce environment subjected the Fuegians to frequent famine and hardship and fueled perpetual intertribal conflict. Indeed, Jemmy had said enough to give the crew the impression that the Fuegians turned to cannibalism in the times of greatest dearth. Darwin could not help but speculate upon the fearful sense of dread that must have crept up on the old women of the tribe as the grip of hunger pressed in upon the community. FitzRoy saw fit to include an account of what he could only comprehend as an outrage to human decency in his own account of the *Beagle*'s second voyage. Testament to the truth of the tale, he said, was that it came from several of the Fuegians on board and had been repeated to different persons on a number of occasions. In gruesome detail, the Captain related the ghastly consequences of scarcity: "Almost always at war with adjoining tribes, they seldom meet but a hostile encounter is the result; and then those who are vanquished and taken, if not already dead, are killed and eaten by the conquerors. The arms and breast are eaten by the women; the men eat the legs; and the trunk is thrown into the sea. During a severe winter, when hard frost and deep snow prevent their obtaining food as usual, and famine is staring them in the face, extreme hunger impels them to lay violent hands on the oldest woman of their party, hold her head over a thick smoke, made by burning green wood, and pinching her throat, choke her. They then devour every particle of the flesh, not excepting the trunk, as in the former case. Jemmy Button, in telling this horrible story as a great secret, seemed to be much ashamed of his countrymen, and said, he never would do so—he would rather eat his own hands. When asked why the dogs were not eaten, he said 'Dog catch iappo' (iappo means otter). York told me that they always eat enemies whom they killed in battle; and I have no doubt that he told me the truth."[25]

Much to FitzRoy's chagrin, Darwin's book proved by far the more successful of the volumes, although it is perhaps significant that Darwin was only later to give his own account of this story when he was revising it for publication as a travel narrative.[26] Beyond the macabre fascination of this story, however, Darwin turned once more to consider the role such circumstances might have had in keeping the Fuegians in their savage state. The harshness of the environment was certainly significant—and sufficiently bleak to curtail the development of the intellectual powers that might otherwise have facilitated their rise from such a desperately primitive condition. "How little can the higher powers of the mind be brought into play!" he reflected. "What is there for imagination to picture, for reason to compare, for judgment to decide upon? To knock a limpet from the rock does not even require cunning, that lowest power of the mind."[27] Branching out, he followed the consequences of this to their logical and perhaps already evolutionary conclusions: "Nature by making habit omnipotent, and its effects hereditary, has fitted the Fuegian to the climate and the productions of his country."[28] Darwin's transmutationist train of thought betrayed the influence of his Edinburgh tutor, the Lamarckian evolutionist Robert Grant, although he would later try to distance his own ideas from the views of both Grant and the French naturalist. Grant had one day "burst forth in high admiration of Lamarck and his views on evolution," and although it may well be that Darwin "listened in silent astonishment," it is hard to believe that this was indeed "without any effect on my mind," as Darwin later claimed in his autobiography. Whatever Grant's influence, though, it is certain that Darwin's speculations on the Fuegians echoed the transmutationism of his grandfather, Dr. Erasmus Darwin, even though Darwin again was less than forthcoming about his grandfather's influence upon his thinking.[29]

Erasmus Darwin, Charles's paternal grandfather, had become famous in his own lifetime. A successful physician, he was also a prominent Enlightenment scholar, humanitarian, poet, and abolitionist.[30] A member of the Lunar Society, the famed intellectual circle of industrialists, philosophers, and radicals in Birmingham, he later moved northeast to the provincial town of Derby, where he founded the Derby Philosophical Society, which itself became a formidable powerhouse of Enlightenment ideas in the English Midlands.[31] A friend of the utilitarian materialist philosopher and dissenting minister Joseph Priestly, and admirer of the radical journalist, author, and political philosopher William Godwin, Erasmus Darwin was believed by many of his contemporaries to harbor deeply Jacobin sympathies. Both Priestly and Godwin are credited with founding utilitarianism, and here, too, Erasmus was in their debt. Godwin was England's most famous radical. An ardent Francophile, he had written *Enquiry Concerning Political Justice* in 1793 in response to the French Revolution. In it, Godwin criticized the institutions of government for their role in undermining the reason and autonomy of the citizenry, a view that immediately made him popular among radicals even though the price of his work meant that it would never be as popular as Thomas Paine's work. Like the French Enlightenment philosopher Nicolas de Condorcet, he had an unbending faith in the power of reason to effect a continual improvement in mankind. It was these views that were to bring Godwin into conflict with Malthus and that prompted Malthus to write his essay on population. Godwin's views were politically controversial, but after publishing a rather too candid biography of his late wife, Mary Wollstonecraft, whose 1793 work, *A Vindication of the Rights of Woman*, had long been read as threatening to turn women's traditional role topsy-turvy, they also became tainted with the odor of sexual impropriety. If such associations were not enough to keep Erasmus Darwin talked about, the fact that he also wrote racy poems that celebrated sex and transmutation, and fathered a large number of children as a result of liaisons with several women, ensured that this was the case.[32]

FIGURE 1.1. Erasmus Darwin, 1731–1802. (Frontispiece to Erasmus Darwin, *The Botanic Garden*; courtesy History of Science Collections, University of Oklahoma Libraries)

Darwin scholars have long noted the evolutionary themes of Erasmus Darwin's poetry, and despite the younger Darwin's claims to the contrary, they were significant for his thoughts on evolution. However, also important were the radical politics at the heart of this poetry and the connections that Erasmus Darwin drew between his hopes for humanity and his transmutationism. Erasmus Darwin wrote his poetry as a contribution to the ongoing debate that fueled the Enlightenment and had occupied both the Scottish moral philosophers and the English political economists regarding the formation of character and the origin of the moral sense. The question that motivated Enlightenment scholars and philosophers was how best mankind might organize society to improve their fellow men.

If there was one theme that unified radicals of all stripes, it was their sense of optimism that industry and reason would transform the world—and the people in it—for the better. As Greta Jones and Robert Richards have each pointed out, in the wake of the American and the French Revolutions, in Britain this was a question that was largely addressed through utilitarian and "sensationalist," or "associationalist," philosophies of one shade or another.[33] Associationalists, Erasmus Darwin among them, took the view that human character was formed through experience. Incorporating the utilitarianism of Godwin, Priestly, and, later, of Jeremy Bentham, they believed that circumstantial sensations of pleasure and pain were the phenomena that prompted the determination of preferences. Once these preferences had become internalized, and thus habitual, they formed the raw material of human character. What differentiated human character from that of mere animal life was the human capacity for reason that allowed man to reflect upon the past, anticipate the future, and to weigh these against the immediacy of the present.

Just as Godwin had done, Erasmus Darwin argued that the character of mankind had its origin in the relationship between the external environment and the free play of each man's individual reason. At the center of human character was the moral sense, which in light of Adam Smith's *Theory of Moral Sentiments* (1759) was often described in terms of "sympathy"—the quality that all agreed was the cornerstone of ethical behaviour.[34] Godwin's views were typical of the optimism that enlivened Enlightenment thought: give reason free rein and the conclusion was inevitable—"Man is perfectible . . . susceptible of perpetual improvement."[35] Erasmus Darwin clearly agreed, and in *The Temple of Nature* he gave a lyric account not only of the rise from monad to man, but of the origins and evolution of society from sympathy. Echoing Smith's *Theory of Moral Sentiments*, Darwin suggested that sympathy for other humans would result from physical sensation, association, sociability, and imitation. Significantly, and in a departure from

Godwin's work, Darwin ultimately tied the origins of society to the relations between the sexes. He suggested that one important consequence of sex was the deep passion of one human being for another that it inspired. Sexual passion would provide the initial "golden chains" that made society possible.[36]

As the historians Margaret Schabas and Donald Winch have each made clear, the debate about moral sentiment and sympathy was common currency for moral philosophers and political economists alike.[37] The distinction that we would draw today with these labels would have appeared a strange one to make in the eighteenth and early nineteenth centuries. Adam Smith is just one case in point. His *Wealth of Nations* and *Theory of Moral Sentiments* each transcend the terrain of these now distinct fields of inquiry. As Winch argues, despite the efforts of some historians to stress the amorality of nineteenth-century political economy, we should see Malthus's work in this light as well.[38] How to organize society so as to facilitate the development of character and sympathy was central to political economy, and as the historian J. D. Y. Peel has noted, how to turn this philosophical aspiration into political reality was also, therefore, a chief occupation of middle-class radicals. Influenced by French utopian theorists like Saint-Simon and Nicholas de Condorcet as much as by the Scottish moralists Smith and David Hume, radical writers focused upon education as well as social and economic relations in their quest to theorize the best means of improving mankind.[39] This was certainly the case with Erasmus Darwin. Like Godwin, he was firmly convinced that arrangements in political economy, including social and sexual relationships, were a crucial aspect in the molding of character.

Politics had been a significant topic of discussion in the Darwin household throughout Charles Darwin's youth, so many of his family members had interests in that direction.[40] Not only had his grandfather been deeply political, but so too had his uncle. Josiah Wedgwood, Darwin's "Uncle Jos," was not only the owner of the Wedgwood Pottery, he was also member of Parliament for Stoke on Trent, and so discussion of political economy could hardly be avoided. Darwin's father, Robert, although a physician by profession, had also invested in housing and was a major stockholder in the Trent and Mersey Canal.[41] Political economy, natural economy, and moral economy were familiar talking points in the Darwin household, and thus when as a young man Charles Darwin was prompted to think about the moral differences between savage and civilized man, he could not help but see connections between the political, natural, and moral economies. As a result, he not only viewed the physical environment as an important factor in determining the Fuegians condition and character, but the prevailing political economy as well. Further, and here echoing his grandfather's concerns, he also reflected on the importance of sexual politics and the apparently unsympathetic nature of the relations between the sexes that seemed to characterize Fuegian society. As Darwin noted in his diary, he saw nothing that he could recognize as domestic affection among them; the relations between man and wife were akin to those between master and slave—and infanticide was far from unheard of, he wrote.[42] He would later rationalize this latter as an evolved survival strategy, but for now he could only think it a barbaric practice that reflected the low moral nature of the Fuegians and their undeveloped sympathetic sentiments. Further—and what ultimately seems to have become the determining factor in Darwin's mind in explaining the Fuegian's pitiful condition—was the fact that they also appeared to lack even the vaguest notion of private property.[43] Darwin had noted that of all the savage races he had encountered, it was only the Fuegians who apparently lacked any semblance of government or any appreciation of private property.

Where he diverged from his grandfather's views, however, was over the extent to which he embraced the principle ideas of contemporary philosophical radicalism and the Malthusian politics that had become central to them. Whereas Erasmus Darwin had been a political radical of the old school, the 1832 Reform Act split the radical movement along class lines, dramatically changing the character of English politics. Liberal middle-class industrialists—including families like the Wedgwoods and the Darwins—had been brought into the Whig political fold largely by the efforts of Henry, Lord Brougham. Championing the moral virtues of self-reliance, hard work, and financial independence, they embraced the dramatic social and political changes of the 1830s as a progressive development. The reforms of 1832 ushered representatives of these self-made men into Parliament as liberal Whigs. Progress, industry, and free trade became the hallmarks of their politics, but so too did other progressive social causes. Indeed, it was largely Lord Brougham's own outspoken antislavery politics that had attracted them to the Whig Party in the first place. Significantly, they also embraced the neo-Malthusian politics that Brougham advocated as a part of his campaign to reform the poor laws. He had encouraged the popular author Harriet Martineau to take up this cause, and she had done so with great effect in a series of pamphlets issued under the title *Poor Laws and Paupers Illustrated* (1834). Martineau had come to national prominence as a persuasive writer on such subjects following the success of her *Illustrations of Political Economy* (1832), a series of short, didactic tales in which she made political economy accessible to the general reader.[44] Martineau fit the liberal Whig ideas pertaining to moral improvement and social progress to a revised version of the Malthusian population principle. Any such embrace of Malthus was a far cry from the radicalism that had united masters and men in the cause of political reform prior to 1832. Where Godwin had decried Malthus for naturalizing the struggle for existence, middle-class Whigs were now embracing Malthusian struggle as an endorsement of the competition of the marketplace. This was an ideology that would have been anathema to radicals only a few decades earlier, and it was this view of life that colored Darwin's view of the Fuegians as he wrote up his account of the voyage. Echoing the worldview of this rising industrial class, Darwin suggested that the only hope of the Fuegians ever raising themselves lay in their adoption of private property as an incentive for individual advancement. "The perfect equality among the individuals composing these tribes, must for a long time retard their civilization," he concluded. As long as they persisted in living in a state of primitive communism, so long would their mere animal existence continue to be dominated by a cycle of misery and famine. "Until some chief shall arise with power sufficient to secure any acquired advantages, such as the domesticated animals or other valuable presents, it seems scarcely possible that the political state of the country can be improved," he wrote. "At present, even a piece of cloth is torn into shreds and distributed; and no one individual becomes richer than another. On the other hand, it is difficult to understand how a chief can arise till there is property of some sort by which he might manifest and still increase his authority."[45]

Environment, habit, and education were the mechanisms on Darwin's mind in his consideration of the rise from savage to civilized man, but clearly, so too were the Whig politics of property ownership. Thus, when Darwin approached the question of the origin of society and of social behavior, he was well aware not only of the fact that his observations in the field of natural history were a significant contribution to this debate, but also that there were ready resources in political economy with which he could work as well—and that, among these, Malthus's famous *Essay on the Principle of Population* was the most talked about when Darwin returned to England in 1836.

Historians have long noted the importance of Malthus for the development of Darwin's ideas, but exactly what Darwin took from Malthus remains a point of contention. As Darwin recalled his Malthusian moment from the perspective of 1876, the year in which he wrote his autobiography: "In October 1838,[46] that is, fifteen months after I had begun my systematic enquiry, I happened to read for amusement Malthus on *Population*, and being well prepared to appreciate the struggle for existence which everywhere goes on from long continued observation of the habits of animals and plants, it at once struck me that under these circumstances favorable variations would tend to be preserved, and unfavorable ones to be destroyed. The result of this would be the formation of new species. Here, then, I had at last got a theory by which to work."[47] This much from Darwin's autobiography, then. However, given the significance that Darwin attributed to Malthus, it is unsurprising that historians and philosophers of science have raised a number of questions about this event, and that following the rediscovery of Darwin's notebooks in the 1950s—and of further excised pages in 1961, and again in 1967—that historians have had significant resources to work from in reconstructing Darwin's encounter with Malthus's work.

In the decades that followed the initial rediscovery of the notebooks, debate focused upon the extent to which Malthus served merely to complete an evolutionary theory that Darwin had already all but grasped, or on whether the insights he took from Malthus represented a significant break with his earlier thoughts on evolution.[48] Ernst Mayr was one of several significant voices in this debate, and as he recalls in his *One Long Argument*, so too were Gavin De Beer, Sydney Smith, Howard Gruber, David Kohn, and Camille Limoges. Mayr, De Beer, Smith, and Gruber argued that Malthus provided no more than "a little nudge that pushed Darwin across a threshold he had already reached"; Limoges and Kohn argued that Malthus was more significant—"a rather drastic break, almost equivalent to a religious conversion."[49] In tackling this question it became pertinent to ask not only what Darwin actually took from Malthus, but how he came to read the book in the first place. Sandra Herbert has argued that far from reading Malthus "for amusement," as if he did so either out of idle curiosity or merely to keep up with dinner table conversation, Darwin picked it up in full expectation that it would contribute to his thinking on species. Herbert points out that in these early years Darwin was hard at work on his geology and zoology—tasks that he quite rightly considered his real work. An aspiring geologist and the recently elected secretary to the Geological Society of London, it was imperative that Darwin write up his notes and realize the potential that his mentors had recognized in him and done so much to nurture. Even so, Herbert argues, he could not keep from transmutation. While Darwin's earliest transmutation notes appear in his Red Notebook alongside his geological observations, once he had filled its pages he consciously separated out these two areas of study. Thus, in mid-1837 he opened Notebook A, in which he focused on geology, and a month or two later, most likely in July, he opened Notebook B, which he dedicated exclusively to his thoughts on transmutation.[50] The similarity in the binding of the Red Notebook and Notebook A supports Herbert's contention that Darwin did indeed now recognize that geology was now his central occupation.[51] The existence and dating of Notebook B also substantiates her point that he could not help but turn to the species question in every spare moment, even as he confessed guilt at the time he "frittered away" in doing so.[52] Transmutation, it seems, was Darwin's "amusement" in these busy days filled with geology, and thus, while he might have been turned on to some significant passage in Malthus across the dinner table, it is clear that he picked up the book with the full expectation that it would shed light on transmutation.

Silvan Schweber has argued that Darwin's encounter with Malthus was a much more calculated affair. Indeed, he argues that Darwin actively sought out Malthus after having read a review that David Brewster had written for the *Edinburgh Review* of Comte's *Cours de philosophie positive.* This, in turn, led Darwin to Adolphe Quetelet's *Sur l'homme et le développement de ses facultés,* which had been reviewed in the *Atheneum,* a publication in which Malthus's work had been discussed at length. Comte had made the point that not only should good scientific argument be predictive, it should also have a quantitative element, and this, Schweber contends, is what inspired Darwin to seek out Malthus.[53] Perhaps so.

Whether or not Darwin sought a quantitative element for his theory in Malthus, this was certainly what he found there. By the time he recorded his views on Malthus, Darwin had thought and scribbled his way through Notebooks B and C and was on to Notebook D in his transmutation sequence. "Population in increase at geometrical ratio in FAR SHORTER time than 25 years— yet until the one sentence of Malthus no one clearly perceived the great check amongst men," he wrote.[54] As David Kohn notes, the one sentence that Darwin singles out was most likely either, "It may safely be pronounced, therefore, that population, when unchecked, goes on doubling itself every twenty-five years, or increases in a geometric ratio," or one from this passage: "A thousand millions are just as easily doubled every twenty-five years by the power of population as a thousand. But the food to support the increase from the greater number will by no means be obtained with the same facility. Man is necessarily confined in room."[55] Thus, while Darwin would not have needed a political economist to point out the fact of struggle in nature, Malthus's explicit discussion of the exponential rate of increase gave him a new appreciation of the intensity of that struggle. When it came to writing *Origin,* Darwin cited both the geologist Charles Lyell and the naturalist Augustin de Candolle as having acknowledged the struggle for existence in nature. Lyell, who was a pillar of British science, had cited de Candolle in his own work as having described "all the plants of a given country" as "at war with one another."[56] However, as Darwin recorded in Notebook D, "Even the energetic language of ~~Malthus~~ Decandolle does not convey the warring of the species as inference from Malthus."[57] The intensity of the struggle was something new then, but so, too, if Depew and Weber are correct, was the significance of individual variation to the outcome of this struggle. Nature was not witness to a war between one species and another, but rather to a bitter war between individuals, a war that was at its most intense within, and not between, species.[58] Implicit in this reading of what Darwin took from Malthus is the recognition that Darwin was using the word "species" in a radically different way than did the vast majority of contemporary naturalists—at least those who were a part of the establishment, the dons and men of science at the English universities. The significant distinction is between the notion that species were "real," in the sense that they existed in the world as natural types, and the Lamarckian assumption that the word "species" was a nominal convenience that systematists—those naturalists who are particularly interested in classification—used to categorize groups of similar-looking organisms. Frank Sulloway and John Beatty have separately argued that Darwin exploited the ambiguities in the word "species" in a deliberately strategic manner in order to win over his audience,[59] and his articulation of varieties as "incipient species" and of species as "well marked varieties" suggests that this was in fact the case. As Sulloway further notes, "In addition to providing him with a mechanism of evolutionary change, Darwin's reading of Malthus completed the revolutionary shift in his thinking about species (in terms of populations rather than 'types') that had begun with his conversion to a transmutationist position in March 1837 and resolved the mystery of extinctions." Sulloway goes on to argue that we need to pay attention to the context of Darwin's reading of Malthus. Throughout the spring and summer of 1838, Darwin had been studying animals under domestication as well as breeder's techniques, which had convinced him that the improvements in the forms of domesticated species were effected by the "picking" of desired individuals. It was his reading of Malthus, Sulloway contends, that enabled Darwin to see how this principle might be applied to the state of nature.[60] However, there was a lot at stake in the definition of "species," which had not only scientific, but political ramifications, and given the obvious importance of this to what was more generally termed "the species question," it merits further explanation.

Orthodoxy on species followed the definition advocated by the French comparative anatomist George Cuvier and his English equivalent, the Hunterian Professor at the Royal College of Surgeons, Richard Owen.[61] According to Cuvier, individual organisms varied, but for all their peculiarities each organism was but an approximation of an ideal type. Thus, throughout the early nineteenth century naturalists and systematists were more interested in individual "typical forms" than they were in either the exceptions or in the extent to which exceptions might have been indicative of variation as a topic worthy of study in itself. That there were natural limits to the extent of variation, though, was an assumption that was central to Cuvier's philosophy of anatomy, just as it was for Owen.

The notion of ideal types goes back at least to Aristotle, and Aristotelian philosophy provided much of the background for Cuvier's thinking on organisms, types, and species, and these in turn informed his views on classification. Thus, in pursuit of his comparative anatomy, Cuvier owed an immense debt to Aristotle's view that organisms must be understood as functional wholes, each of their various parts serving a particular purpose and thus being dependent upon the other parts in order to work effectively. As he put it in his 1817 work in which he laid out his approach to comparative anatomy, *Le règne animal distribute d'après son organisation, pour server de base à l'histoire naturelle des animaux et d'introduction à l'anatomie comparée*: "As nothing may exist which does not include the conditions which made its existence possible, the different parts of each creature must be coordinated in such a way as to make possible the whole organism, not only in itself but in its relationship to those which surround it."[62] Putting Aristotelian philosophy in the context of nineteenth-century natural science, Cuvier believed that organisms were subject to certain structural "general laws," and it was these laws that he held up as the basis for the science of anatomy. He expressed these basic laws in terms of "combination" and "subordination of characters" and what he referred to as an organism's "conditions of existence." Because each aspect of an organism must serve a useful function and at the same time be intimately related to the function of its other aspects, there were clearly relational limits, or "conditions," set upon the possible structure of the parts of an organism in and of itself, and of course in its interactions with the environment. Cuvier also referred to this necessary relationship between different aspects of an organism in terms of the "correlation of parts" as well as the "conditions of existence."[63]

It was from this set of beliefs that Cuvier made his famous anatomical reconstructions. Given even just one tooth or bone of an animal, he was able to deduce the likely other parts to get a picture of the whole. As Michael Ruse points out, though, this particular talent was clearly as much a function of his extensive knowledge and experience of comparative anatomy as it was of his theoretical commitments. Nevertheless, it was these theoretical commitments that meant that Cuvier could not but be an anti-evolutionist—and certainly an opponent of any hypothesis of a gradual transition from one species to another. To Cuvier's mind, Lamarck's transmutationism was simply not possible given that it relied on the gradual development or loss of particular organs or characters in the particular organism in question—a hypothesis that would breach his law of the "correlation of parts" and thus the "conditions of existence" that made it a viable subject of a life. Accordingly, the transitional stages that Lamarck hypothesized simply could not exist and, as Cuvier was keen to point out, neither had any such transitional stages been found in the fossil record.[64]

Cuvier's focus upon the gaps between fossil organisms rather than on their progressive development in the strata further prevented him from accepting the possibility of evolution. Indeed, Cuvier was adamantly opposed to the concept of a hierarchical or progressive "Great Chain of Being" at all—the one Aristotelian idea that he did not take up. Instead, he argued that since each organism had its own conditions of existence, it made no sense to talk of one organism being any more perfect than another—with the possible exception of humanity, of course. Refusing to see progression in the fossil record, Cuvier marshaled that record and his Aristotelian conception of species to counter Lamarck's argument for evolution as well as to support his own arguments for a general law of extinction.[65]

Although Cuvier kept his science largely separated from his theology, he did suggest that the primary cause of the extinctions that he believed to have occurred on a regular basis throughout earth's history were the result of catastrophic floods resulting from either significant elevation in sea level or the elevation and subsidence of the land. The floods that would result from these "revolutions" would not only explain the discovery of fossil shells at the tops of mountains, but might also explain how so many large land-based vertebrates might have gone extinct. Although Cuvier made little explicit reference to supernatural causes in his work, at least in Great Britain his emphasis upon floods as the mechanism by which such great changes had been effected was interpreted as suggesting as much. This was largely owing to the way in which Robert Jameson—the same Jameson who had taught Darwin geology in Edinburgh—prefaced his translations of Cuvier's work, not only reintroducing the idea of a progressive scale of nature but also framing Cuvier's theory in such a way as to allow the last great "revolution," which Cuvier had estimated as having taken place roughly six thousand years previously, to be easily interpreted as the Noachian Flood of Genesis.[66]

Cuvier thus led conventional thought on species across Europe, and in addition to Jameson—and of even greater consequence, William Whewell—ensured this was the case in England as well. Whewell, who had written one of the more telling of the *Bridgewater Treatises*, had been appointed Master of Trinity College in 1841 and held offices in several of the most prominent and influential scientific societies of the day, including periods as the president of the British Association for the Advancement of Science (of which, along with

Babbage and Herschel, he had been a founding member), and as variously president and vice president of the Geological Society of London. As such, he was chief among those who sought a place for science in the university curriculum, where hitherto mathematics and classics had reigned supreme. Increasingly conservative politically, Whewell saw science as having a rightful place at the heart of the establishment, but he was also well aware of the radical Lamarckians who sought to make science a means of tearing down the ramparts of long-established truth and tradition.[67] This was reflected in Whewell's philosophy of science, and in his attitude toward transmutation in particular, but it also raises the question of whether Darwin was a pioneer in thinking of "species" as merely a nominal category or whether this was in fact already a widely accepted assumption among those British naturalists who were members of nonconformist and radical communities.[68]

To Whewell's mind, central to policing the boundaries of what constituted good science was "the recognition of final causes in physiology." This had been Whewell's main concern in his 1845 *Indications of the Creator*, a claim he there repeated from his earlier, three-volume *History of the Inductive Sciences* (1837). Despite its title, Whewell wrote *Indications* in response to what he saw as the sloppy and speculative approach to grave scientific questions—and the equally flawed conclusions—of the then recently published *Vestiges of the Natural History of Creation* (1844). Whewell's philosophy of science guarded the gates of the establishment, but it was as much about keeping the second-rate, amateur, and speculative transmutationist claims of radical Lamarckians at bay.[69] Whewell quoted Cuvier on causation as well as on species—his endorsement the ultimate stamp of authority in British science: "Indefinite divergence from the original type is not possible; and the extreme limit of possible variation may usually be reached in a short period of time: in short, species have a real existence in nature and a transmutation from one to another does not exist."[70]

My point here is not that Darwin arrived at a non-typological conception of species from his reading of Malthus, but rather that it was his reading of Malthus that enabled him to see the importance of individual variation, not just as factual entities that filled in gaps between species, but as the raw material that drove speciation in the context of an environment with limited resources that thus demanded competition. The systematist Ernst Mayr, who has also given Darwin's theory building serious consideration, agrees with Depew and Weber that this was the case, and Mayr was the first to term Darwin's recognition of the significance of individual variation in a population a transition to "population thinking." However, where Depew and Weber, echoing Sulloway, see this as a direct consequence of Darwin having read Malthus, Mayr argues that it was more a consequence of his commitment to Lyellian gradualism.[71] This is why Mayr sees Malthus as merely pushing Darwin over the edge of a threshold he was already looking over rather than precipitating a more revolutionary event. However, it seems to me that Depew and Weber and other "revolutionists" have the best of this argument. The *significance* of individual variation only becomes apparent in light of a struggle for existence that is at its most intense between the slightest variations between individuals of the same species. And thus even if Darwin's commitment to gradualism had laid groundwork that allowed Darwin to appreciate Malthus in the way that he did, it seems to me that the intellectual work that Malthus did for Darwin makes the event much more than a mere step over a threshold. After all, it was only in light of reading Malthus, and again recorded in Notebook D in the hasty handwriting that betrayed his excitement, that Darwin arrived at the "wedging" metaphor that he believed best conveyed the true face of nature and which led him to the centerpiece of his theory. "One may say there is a force like a hundred thousand wedges trying force ~~into~~ every kind of adapted structure into the gaps ~~of~~ in the œconomy of Nature, or rather forming gaps by thrusting out weaker ones," he wrote. Following this passage, he inserted the logical conclusion that "the final cause of all this wedging, must be to sort out proper structure & adapt it to changes."[72] This was Darwin's first formulation of natural selection.

For all that I see Darwin's encounter with Malthus as a more significant breakthrough than Mayr does, I am not sure that it was an event I would describe as something akin to a religious conversion.[73] Importantly, though, and whatever the veracity of Darwin's claims about his motivation for reading Malthus and whatever he may have taken from doing so, it is certain that he was not naive to the political significance of adopting Malthus in the way he did. Schweber is not alone in pointing out that "it would have been difficult for anyone reading the *Edinburgh Review* or the *Quarterly Review*, as Darwin did, not to be familiar with Malthus' thesis" and the political and economic ends that various commentators appealed to it to support.[74] Malthus allowed Darwin to present evolutionary ideas as having very different political associations than they had previously enjoyed. As Adrian Desmond and Paul Elliott have each shown, the Lamarckian mechanisms that Darwin had been entertaining up to this point—his emphasis upon an organism's adaptation to environment and climate—had been very much the food and drink of political radicals and revolutionaries.[75] His adoption of Malthus turned evolution to very different political ends. Darwin was well aware that

Malthus was central to the new Whig politics that were reforging the nation. This being the case, it is important that we revisit the moral and political meaning of Malthus in the middle years of the nineteenth century.

Thomas Robert Malthus had been the first professor of political economy at the East India College at Haileybury, near Hertford Heath, and his *Essay on the Principle of Population*, which he had written as a young man of thirty, was an outright rejection of the Enlightenment views of both William Godwin and Nicolas de Condorcet on the improvement of mankind. The natural incommensurability of the ratio between population increase and food production meant that mankind would always struggle to feed their growing numbers. "Population, when unchecked, increases in a geometric ratio. Subsistence increases only in an arithmetic ratio," he wrote. "A slight acquaintance with numbers will shew the immensity of the first power in comparison to the second." First published in 1798, the essay went through six editions before 1826, and throughout this period and beyond, it received by turns adulation and abuse.[76] The political significance of Malthus was inescapable. Indeed, as Robert Young has noted, Malthus's biographer, Patricia James, has pointed out that in the popular press "it rained refutations of Malthus for thirty years."[77] No one had ever been indifferent to Malthus.[78] "An apostle of the rich," Robert Southey had called him, both amazed and affronted at the "stupid ignorance of the man"; the essay was "Adam Smith's book in code, a confession of faith in this system; a tedious and hardhearted book."[79] William Hazlitt, the radical journalist who was later to find little else to agree with Southey about, had also rained pages of abuse upon a man who would naturalize the oppression of the poor, rejecting Malthus's misguided economic assumptions. The book was nonsense. Surely corn and cattle were geometrically increasing populations too.[80] It was through human injustice that people starved, not through any inevitable law of nature. Malthus's essay was "the most complete specimen of illogical, crude and contradictory reasoning that perhaps was ever offered to the notice of the public."[81] As if this wasn't enough, to William Cobbett, the inveterate radical journalist, "Parson Malthus" was a "monster" who would make marriage a matter of money rather than of love.[82] Many of those who had been so vocal in their abhorrence of Malthus were the same radicals who naturalized their own political claims by adopting the evolutionary ideas of the French naturalists Jean Baptiste Lamarck and Étienne Geoffroy Saint-Hilaire, and, of course, Erasmus Darwin. This had been especially so in the medical schools of Edinburgh, and from the 1820s, in London, where dissolute medical students incensed at the injustices metered out by "Old Corruption" lapped up the science of revolution, as well as in the provincial towns that were expanding in the wake of the Industrial Revolution.[83]

However, by the 1830s times had changed. The 1832 reform ushered in a new class of industrialists. As I have suggested above, like Martineau and Brougham, this new class saw in Malthus the potential for quite a different politics than he had intended when he wrote his essay. While Malthus remained the bête noire of those within the radical movement who held true to the tradition of Godwin, Cobbett, and Hazlitt—those who had been excluded from the 1832 settlement—to a new generation of liberal Whigs it was clear that Malthus could be appropriated to their own ends.

Malthus was still controversial in the late 1820s and 1830s when middle-class, self-styled "philosophical radicals" like Martineau embraced Malthusian economics to rally against the state-mandated obligations for charitable provision under the poor laws. In collaboration with Brougham, Martineau had written a series of pamphlets under the title *Poor Laws and Paupers Illustrated* (1834), in which she promoted a Benthamite approach to tackling poverty based upon a neo-Malthusian moral economy. It was Jeremy Bentham, the radical utilitarian philosopher and politician, who in the late 1790s had made the case for moving poor relief into a system of workhouses, in an essay entitled "Pauper Management Improved." Martineau combined Bentham's belief that people responded best to the ministrations of pleasure and pain with Malthus's population principle to prepare the ground for the Poor Law Amendment Act on Brougham's behalf. Under the amendments that Brougham proposed, those seeking relief would indeed have to subject themselves to being separated from their families and enter a workhouse system. Anyone who could possibly prevent this by his or her own efforts would do so, the argument went, thus ensuring that only those who were really incapable of self-help received state aid. This was tough love. The mere provision of aid alone would do nothing to end the cycle of dependence: Malthus had demonstrated that the poor would continue to reproduce even more hungry mouths. Malthus had acknowledged that while there were some who might recognize the need for moral restraint, the majority would not, as they had neither the foresight nor the moral character to do so. Martineau, however, rearticulated Bentham's ideas to suggest that the severity of the workhouse would spur the poor to recognize the need both to labor and to limit their own fecundity since their own improvement and the improvement of society as a whole depended upon it. Thus, whereas Godwinite radicals had attacked Malthus for attempting to naturalize social injustice, philosophical radicals like Martineau and John Stuart Mill embraced him. However, they did so with a significantly different emphasis. Malthus had argued that the limits that nature imposed upon mankind meant that the progressive utopian ideal that Godwin and Condorcet sought was illusory; Martineau and Mill, among others, used Malthus to naturalize the very means—competitive free trade, hard work, and moral restraint—by which human improvement and progress could be effected. Mill later recalled in his autobiography that "Malthus' population principle was quite as much a banner, and point of union among us, as any opinion specially belonging to Bentham. This great doctrine, originally brought forward as an argument against the indefinite improvability of human affairs, we took up with ardent zeal in the contrary sense."[84] The difference of opinion on Malthus that existed between liberal Whig reformers like Martineau on the one hand and the more-revolutionary radicals on the other, was indicative of the opening class divide that defined the "condition of England" in the second half of the nineteenth century.

This political context thus gives us important information about what Malthus meant to Darwin. Darwin's sisters had sent him Martineau's popular pamphlets as *Beagle* ran for home, but discussion of Martineau's views among the officers on board could not prepare him for the reality of the political changes that England had undergone since he had left the Devonport quayside in the winter of 1831. By 1836, the Whigs were in the ascendant, and the Wedgwood's were in the thick of it. For over forty years, the family had been involved in the campaign to abolish slavery, and with the passing of the 1833 Abolition Act their goal was now close across the empire. Plus, now that Uncle Jos had won a seat in Parliament, the Darwins and the Wedgwoods were fully aware of the politics of the day.

England was in a whirlwind of change, and things had changed for Darwin too. No longer merely the promising student, he was now a salted and proven naturalist and he dedicated himself to a flurry of scientific work. Collections needed describing, notes needed writing up, and his newfound commitments to scientific societies also pressed upon his time. He would later count the next two years and three months as among the busiest of his life.[85] To secure his scientific reputation Darwin needed to ensure that the collections he had sent home over the preceding years were described by reputable men who could be relied upon to get the job done. He was lucky: John Stevens Henslow, his mentor from his student days, had been talking him up in his absence, and as a result he had little difficulty in getting some of the best men to take on the work. He settled into Cambridge to oversee what would eventually become the *Zoology of the Beagle*, to prepare his journal for publication, and to write up his geological notes. Consisting of five parts in nineteen numbers, *Zoology* was published in installments between

February 1838 and October 1843. Testament to Darwin's growing stature in the world of science, the Duke of Somerset (who was then president of the Linnean Society), the Earl of Derby, and William Whewell each exerted their influence to secure him a £1,000 grant from the Lords Commissioners of the Treasury to see the project through to completion.[86]

Happy that this work was underway, in March 1837 Darwin moved to London, staying for a week with his brother Ras (who was named after their grandfather Erasmus) at 43 Great Marlborough Street while he sought lodgings of his own. He found somewhere suitable just a few doors down the road at No. 36 and stayed there until after his marriage to Emma in 1839.[87] London was a bustle of excitement compared to Cambridge, and Darwin found himself thrust into Ras's busy social circle. Harriet Martineau dominated the group, which was made up of some of London's most-prominent intellectuals and political radicals. Martineau and Ras were close enough for Darwin's father to fear a scandal, and her prominence in Ras's life, and thus in his own, makes it unsurprising that Darwin read his brother's copy of Malthus's famous essay. It was the 1826 sixth edition, and he read it between September 28 and October 12 of 1838.[88] Malthus, political economy, and reform were not only high on the list of topics of conversation that enlivened the intellectual set that Erasmus ran with, but as the British economy faltered, Chartists rioted. Scarcity did indeed appear to breed struggle, and those who felt themselves left out of the 1832 compromise were demanding justice.[89] Darwin gulped down such heady notions: materialism, positivism, political economy, and the poor law filled his head—and his notebooks. Darwin clearly thought deeply about Malthusian politics, but, and perhaps in consideration of his grandfather's political preferences, he was also prompted to reconsider Godwin's side of the argument. In light of discussions with Martineau, Ras, and others, Darwin turned back to Godwin. He was clearly anxious to give the arguments that had so impressed his grandfather a fair hearing. In his reading list he noted that "Shelley says [Godwin] is victorious and decisive." Shelley, though, had family as well as atheist radicalism in common with Godwin, and while Darwin evidently thought Shelley's views worth recording, they ultimately did not sway him.[90] Over the following months and years, Darwin read a number of Godwin's works for good measure, but on the face of it, it seems that Malthus won out.[91] Indeed, although Malthus had eventually capitulated to Godwin's repeated criticism that the possibility of moral restraint was more significant than Malthus had allowed in the first edition of his essay in 1798, when it came to writing *Origin*, and once he had made the decision to exclude man from the book, Darwin could see no opening for moral restraint in nature at all, nor, it seems, among the inhabitants of

Tierra del Fuego. As a result, when he came to reflect upon Malthus in relation to the species question, as both Robert Young and Howard Gruber have pointed out, he stripped away any vestige of Godwinian moral restraint. As he distilled his thoughts into a coherent argument, the Malthus that Darwin invoked in *Origin* was that of 1798, not 1826.[92]

Darwin noted in his autobiography that he was "anxious to avoid prejudice" and so did not publish his ideas; he was still vexed by the difficulty of explaining the manner in which new species appeared to diverge in character from the parental form.[93] In addition, Lamarckism was still politically controversial. In the same year as he wrote out his long essay, 1844, the anonymous publication of the *Vestiges of the Natural History of Creation* proved as much. The reviews reminded Darwin that evolutionary ideas were bad science, largely because of the undesirable politics with which they had become associated. Thus, I contend that Darwin's repeated insistence in *Origin* that his theory was "the doctrine of Malthus, applied to the whole animal and vegetable kingdoms," that it was "the doctrine of Malthus applied with manifold force," was a clear attempt to dissociate evolution from radicalism and to signal to his readers that *Origin* was on the right side of the fence politically.[94] Darwin also went to great lengths in his introduction to attempt to distance himself from both the author of *Vestiges* and from Lamarck, and he also made it clear that his findings were not the result of hasty speculation.[95] It seems that with such an outspoken appeal to Malthus, Darwin hoped to succeed where the anonymous author of *Vestiges* had failed. He hoped to make evolution acceptable science rather than a sensation.

Lyell, whom Darwin sought most to convert to his views, had been one of Lamarck's most outspoken critics, and so it was to him that Darwin wrote of his rejection of the radical attacks on Malthus. "What a discouraging example Malthus is to show during what long years the plainest case may be misrepresented and misunderstood," he wrote.[96] His point was not that the radicals were mistaken in thinking that Malthus's work had significant political implications, but that they were mistaken in their perception of what the political implications were. Just as Martineau and John Stuart Mill appropriated Malthus to their own middle-class program of political transformation, so too did Darwin—and he was keen to advertise this fact. Desmond and Moore do not overstate the case when they say that "Darwin's biological initiative matched advanced Whig social thinking," describing natural selection as "a mechanism that was compatible with the competitive free-trading ideals of the ultra-Whigs." Darwin had indeed "broken with the radical hooligans who loathed Malthus."[97] But then so too had the rest of his class. The 1832 reform had seen to that.

Darwin's contemporaries certainly read *Origin* as having exactly these deeply political implications. Karl Marx rightly saw it as an attempt to ground individualism, capitalism, and laissez-faire in nature, famously writing to his colleague and comrade Friedrich Engels: "It is remarkable how Darwin rediscovers, among the beasts and plants, the society of England with its division of labor, competition, opening up of new markets, 'inventions' and Malthusian struggle for existence. It is Hobbes' *bellum omnium contra omnes.*"[98] Of course, Hobbes's "war of each against all" was a presumption common to the British tradition of political economy.

It wasn't only critics of capitalism who read Darwin in this way. In fact, Marx only echoed the point that Thomas Huxley had already made in his enthusiastic review of *Origin,* which had appeared in the *Westminster Review* early in 1860. Darwin had written a "veritable Whitworth gun in the armoury of liberalism," Huxley wrote with clear approval.[99] The metaphor was telling. It was the hexagonal rifling of the Whitworth that gave it its superior range; against conventional weaponry, it was quite literally unanswerable.[100]

Marx and Huxley were not the only ones to have read *Origin* as a tool in the service of an aggressively liberal ideology either. Indeed, in the immediate aftermath of publication, Darwin had noted with some amusement to Lyell that he had "received in a Manchester Newspaper rather a good squib, showing that I have proved 'might is right,' & therefore that Napoleon is right & every cheating Tradesman is also right."[101] The paper in question was the *Manchester Guardian,* the article "National and Individual Rapacity Vindicated by the Law of Nature."[102] The *Manchester Guardian* had been founded in 1821 by a group of nonconformist mill owners. Uncompromising in its editorial policy, the paper spoke for the interests of what in 1846 Disraeli had christened the "Manchester School" of Cobden, Bright, and the Free Trade Hall.[103] Darwin's amusement was most likely as much an expression of relief as one of humor. Here in the mouthpiece of the free-traders, which had previously been attacked by the radicals as "the foul prostitute and dirty parasite of the worst portion of the mill-owners" for its outspoken opposition to the 1832 Ten Hours Bill, was proof that Darwin had succeeded in distancing transmutation from its revolutionary associations.[104] However, the fact that Darwin could claim "amusement" at such connections being drawn is indicative of the gulf that had opened up in English politics, and of what Malthus had come to represent. It is certainly doubtful whether his grandfather Erasmus, who had been branded an out-and-out Jacobin for much of his life, would have found such associations amusing.[105]

It is easy to see why the author of the *Manchester Guardian* article saw *Origin* as worthy of such parody, for many read it as a vindication of the most unremitting competition and self-interest. Having flagged up Malthus in the introduction, in the third chapter, entitled "The Struggle for Existence," Darwin had laid out the full implications of a Malthusian worldview. Having gutted the book of any aspect of Godwinian moral restraint, even the brightest aspects of nature betrayed a dark underbelly: "We behold the face of nature bright with gladness, we often see superabundance of food; we do not see, or we forget, that the birds which are idly singing round us mostly live on insects or seeds, and are thus constantly destroying life; or we forget how largely these songsters, or their eggs, or their nestlings, are destroyed by birds and beasts of prey; we do not always bear in mind, that though food may be now superabundant, it is not so at all seasons of each recurring year."[106]

Death fell mainly on the young, eggs were taken, nests destroyed. This was a far cry from the "happy world" that William Paley had described in *Natural Theology*, the book with which Darwin had become so familiar in his college days. Although Darwin was clear to state that he used the term the "struggle for existence" "in a wide and metaphorical sense, including dependence of one being on another, and including (which is more important) not only the life of the individual, but success in leaving progeny," this was not a theme that he chose to develop here. Rather, the chapter is an account of the unremitting struggle and death that Darwin believed to be the true character of nature. Only the few survive, while the many fall by the wayside: "Hence, as more individuals are produced than can possibly survive, there must in every case be a struggle for existence, either one individual with another of the same species, or with individuals of distinct species, or with the physical conditions of life. It is the doctrine of Malthus applied with manifold force to the whole animal and vegetable kingdoms; for in this case there can be no artificial increase of food, and no prudential restraint from marriage."[107]

There were no exceptions. Every organism increased at such a rate that, unless the majority were destroyed by one means or another, the issue of even a single pair would quickly overrun the earth. Darwin surmised that if nature's fecundity were left unchecked, then even such a slow-breeding beast as the elephant, which reached sexual maturity only in its thirtieth year and raised no more than three pairs of young in a lifetime, would, across the generations, be responsible for fifteen million elephants in just five hundred years. The case was even more pressing with humankind, as Malthus's work had made clear. Darwin wrote: "Even slow-breeding man has doubled in twenty-five years, and at this rate, in a few thousand years, there would literally not be standing room for his progeny."[108] In the end, however, it was a single sentence in Malthus that Darwin found most compelling—"It may safely be pronounced, therefore, that the population, when unchecked, goes on doubling itself every twenty five years, or increases in a geometrical ratio"[109]—and that allowed Darwin to conclude that "no one clearly perceived the great check amongst men."[110]

It was the natural checks to this potential population growth that Darwin was keen to point out. Seasonal variations in climate combined with the complex web of interdependence that connected the many organisms in a given environment determined the checks that gave nature the appearance of balance. The one check that really became Darwin's focus, however, was the competition for available resources. This was not the war between species that de Candolle had pointed out, but a struggle between each and every individual. Indeed, Darwin argued, it was here where competition was the most intense. "The struggle almost invariably will be most severe between individuals of the same species, for they frequent the same districts, require the same food, and are exposed to the same dangers," he wrote.[111] It was this seeming endorsement of unremitting individualism that was to prove controversial. Not only did this aspect of Darwin's description of the nature of things appeal to the Manchester school of political economy and to advocates of laissez-faire generally, but it was this that radicals and socialists found objectionable. Competition between individuals was at its most fierce between the most similar and closely situated organisms, and was central to Darwin's theory of speciation. A consequence of natural selection was not simply the perfection of organisms as they became ever more adapted to their environment, but the splitting of one species into two or more different species. This was fundamental to Darwin's argument, and he described it in relation to what he termed his "theory of divergence."

Darwin's theory of divergence was central to his understanding of the origin of new species, at once both a part of and a result of Darwin's theory of natural selection. It was the solution to a problem he had long recognized—the problem, indeed, that had made him delay publication of his ideas until he could solve it. As he recalled in his autobiography, the solution had come to him while out riding. "I can remember the very spot in the road, whilst in my carriage, when to my joy the solution occurred to me," he wrote. "The solution, as I believe, is that the modified offspring of all dominant and increasing forms tend to become adapted to many and highly diversified places in the economy of nature."[112] As Sylvan Schweber points out, Darwin had outlined his theory of divergence first in a letter to his friend Joseph Dalton Hooker and then in more detail to the American botanist Asa Gray in 1857, and he considered it so vital to his case that the one illustration that he included in *Origin* was in explanation of what he called "this rather perplexing subject."[113] In *Origin* Darwin premised his theory of divergence on what the

French naturalist Henri Milne Edwards called the "physiological division of labour." It was clearly an appropriation, in name if not in all the details, of the economic theory of the division of labor with which Adam Smith had introduced his *Inquiry into the Nature and Causes of the Wealth of Nations.* Citing Milne Edwards, Darwin noted that a certain patch of ground could accommodate a greater abundance of life if it housed many different varieties and species than if it held only one.[114]

Just as many of Darwin's contemporaries recognized the easy fit between his theory and the prevailing liberal capitalist politics of the day, so too have historians. Silvan Schweber in his 1980 article "Darwin and the Political Economists" has argued that not only was Darwin influenced by Malthus in building his theory of natural selection, but that he was inspired by Adam Smith's theory of the division of labor to arrive at his theory of divergence—and this before he read Milne Edwards's work in 1842.[115] If this is so, then liberal political economy really did run throughout his theorizing. Margaret Schabas has questioned whether this really was the case, however, and as Diane Paul and John Beatty have pointed out, she is not alone among historians of economics in doubting the full extent of Schweber's claims.[116] Scott Gordon points out that there is no evidence that Darwin ever read *Wealth of Nations*, for instance, and that neither did he cite Smith on this or any other point in *Origin.*[117] Schabas has further pointed out that this could hardly have been the result of a reluctance to quote an economist when a biologist would do, given Darwin's open acknowledgment of Malthus.[118] This much is all well and good perhaps, but as Schabas herself points out, Darwin would hardly have needed to have read *Wealth of Nations* to have gained an appreciation of the division of labor. The idea was widely enough accepted for it to be common knowledge, especially given the fact that the Darwin-Wedgwood clan was so immersed in industry, politics, and commerce. Schweber has outlined in detail the political commitments and connections of Darwin's extended family, noting that his Uncle Jos was "on familiar terms with Henry Brougham, James Mackintosh, and Sidney Smith" and that "discussion of the important literary, and philosophic works of the day were usual at the dinner parties he attended when he lived in London."[119] Indeed, Darwin had read Brougham's 1839 *Dissertations on Subjects of Science Connected with Natural Theology* in which Brougham had given detailed consideration to the division of labor among bees in the construction of their combed cells.[120] Also, as Schweber points out in his essay "Scientists as Intellectuals," one of the sources of Darwin's knowledge of the division of labor was undoubtedly Charles Babbage's *On the Economy of Machinery and Manufactures* (1832), in which Babbage extolled the virtues of Smith's theory.[121] The fact that Darwin could cite Milne Edwards on the "physiological division of labour" is merely indicative of the extent to which Smithian political economy had already permeated biology, but if Darwin was in need of a quick reminder of exactly what Smith had said on the subject, Darwin's reading notebooks show that while he may not have read *Wealth of Nations*, in addition to having read Brougham and Babbage he had also read Dugald Stewart's *Account of the Life and Writings of Adam Smith* (1793), in which the theory of the division of labor was covered more than adequately for any purposes that Darwin might have had in mind.[122] He recorded in Notebook M, "D. Stewart ~~Smith~~ lives of Adam Smith. Read, etc. worth reading as giving abstract of Smith's views."[123] As I shall suggest later, in chapter 3, if Darwin did take the theory of divergence from Smith via Dugald Stewart, this was not all he took. While in *Origin* Darwin was primarily concerned with the positive checks to population, not only had Malthus considered what he called the negative checks to population of moral restraint, but so too had Smith. These were exclusive to mankind and were the result of his foresight, conscience, and morality. Although he left man out of *Origin*, it is significant that Darwin did read Smith when he read Dugald Stewart, but it was Smith's *Theory of Moral Sentiments* that drew his attention.[124]

In *Origin* Darwin described the competition between organisms in terms of the natural selection of those individuals that were best fitted to survive in the environment in which they found themselves. As Darwin had succinctly put it: "Can we doubt (remembering that many more individuals are born than can possibly survive) that individuals having any advantage, however slight, over others, would have the best chance of surviving and of procreating their kind? On the other hand, we may feel sure that any variation in the least degree injurious would be rigidly destroyed. This preservation of favourable variations and the rejection of injurious variations, I call Natural Selection."[125]

In the theory of divergence, though, Darwin described the implications of natural selection acting across a population as a whole. It was not just that some survived while others did not. Rather, because of the fecundity that pressed upon every aspect of nature, those that might most effectively make a living in one "office"—or, as we now say, in one niche—would be in immediate competition to do so with others that were most similar to them.[126] Further, given an environment in which the members of this population might make their living, then those individuals best fitted for each of these niches would tend to survive and reproduce, passing on the characteristics that made them so well suited to survive in the process. Darwin had invoked analogies from the artificial selection of breeders many times in his argument up to this point, and here he did so again. Citing pigeon breeders as an example, Darwin noted that just as some fanciers preferred birds with a shorter beak while others preferred birds with a longer beak, and based "on the acknowledged principle that 'fanciers do not and will not admire a medium standard, but like extremes,'" so the result of generations of artificial selection had been the diverse creations to be witnessed in any pigeon fanciers' meeting.[127]

To show that this process was indeed analogous to the natural world in chapter 4, on "natural selection," Darwin presented the hypothetical case of the wolf. He did so to illustrate the process of natural selection, but it also served to illustrate the principle of divergence. The wolf was a useful example as it preys upon diverse animals at different seasons of the year and in different regions of its range. It is among the clearest illustrations of Darwin's theory in the whole of *Origin*, and is worth quoting at length: "Let us take the case of the wolf, which preys upon various animals, securing some by craft, some by strength, and some by fleetness; and let us suppose that the fleetest prey, a deer for instance, had from any change in the country, increased in numbers, or that other prey had decreased in numbers, during that season of the year when the wolf is hardest pressed for food. I can under such circumstances see no reason to doubt that the swiftest and slimmest wolves would have the best chance of surviving, and so be preserved or selected." Again appealing to evidence that would have been well within the experience of his readers, he continued, "I can see no more reason to doubt this, than that man can improve the fleetness of his greyhounds by careful and methodical selection." Given the circumstances he had described, Darwin suggested that "the wolves inhabiting a mountainous district, and those frequenting the lowlands, would naturally be forced to hunt different prey; and from the continued preservation of the individuals best fitted for the two sites, two varieties might slowly be formed." Turning away from hypothesis to what he presented as verification from the natural world, he cited a letter from one of his many North American correspondents. "I may add, that according to Mr. Pierce, there are two varieties of wolf inhabiting the Catskill Mountains in the United States, one with a light greyhound-like form, which pursues deer, and the other more bulky, with shorter legs, which more frequently attacks the shepherd's flocks."[128] Not only did nature pay close heed to the individual differences that allowed one wolf to succeed in the struggle for life where others failed, but the fact that Darwin chose a wolf—a species that was, however incorrectly, deemed a voracious and immoral beast—to illustrate the most vital part of his theory, also doubtless facilitated the reading of *Origin* as an endorsement of capitalist economics and competitive individualism.

Darwin read Malthus in 1838 but did not publish *Origin* until two decades later. A good part of Darwin's reticence to publish, as Robert J. Richards and, more recently, John van Wyhe have contended, was not so much because of the theological implications, but because he was still wrestling with a number of problems.[129] It is also clear that there was a political dimension to his delay. He knew his theory would be a hard sell, and thus despite his eagerness to show the solid Whig credentials of what he called "descent with modification," he remained wary of how his speculations on man might be perceived. "I was so anxious to avoid prejudice, that I determined not for some time to write even the briefest sketch of it," he recalled in his autobiography.[130] In the last days of 1857, he responded in a similar vein to an inquiry from the naturalist and collector Alfred Russel Wallace on the subject. "You ask whether I shall discuss 'man'; —I think I shall avoid the whole subject, as so surrounded with prejudices," he wrote, adding, "though I fully admit that it is the highest and most interesting problem for the naturalist."[131]

Darwin's notebooks are testament to the dangerously materialist conclusions that his "mental rioting" had already driven him to. "Nothing for any Purpose," he had ominously concluded on the back cover of his Red Notebook, and that had been back in 1837.[132] He was aware that the implications of his views "would make a man a predestinarian of a new kind, because he would tend to be an atheist," and thus warned himself "to avoid stating how far, I believe, in Materialism."[133] Later, rushing to finish *Origin*, he remained mute on man for fear that it would be too much all at once. He had been working hard to bring Lyell into the fold from the beginning, arguing his case even as Lyell raised objections to the draft chapters that Darwin worried him with. Only five weeks away from publication, Darwin had still been testing his argument, trying to bring the old geologist round to his way of thinking and anxious to have his blessing before going public. "I suppose that you do not doubt that the intellectual powers are as important for the welfare of each being as corporeal structure," he addressed his mentor, probing for a point of agreement from which to argue the conclusions that to his mind were inevitable. "If so," he continued, "I can see no difficulty in the most intellectual individuals of a species being continually selected; & the intellect of the new species thus improved, aided probably by effects of inherited mental exercise."[134]

Darwin went on to suggest that this might explain the development and divergence of the various races of man, not only in terms of geographic dispersal, but in intellect, "the less intellectual races being exterminated" in consequence. Recognizing that Lyell was at least swayed by his argument until he touched on the delicate subject of man, mind, and morals, he urged his friend to see that it really was an all-or-nothing argument simply by force of logical inference: "If you admit in ever so little a degree the explanation which I have given of Embryology, Homology & Classification, you will find it difficult to say thus far the explanation holds good; but no further; here we must call in 'the addition of new creative forces.'" Feeling that he had brought Lyell at least this far, he concluded, "I think you will be driven to reject all, or admit all."[135]

Lyell vacillated, and in consequence Darwin's nerve failed him. Recognizing that what was a bridge too far for Lyell might cause lesser minds to reject his theory altogether, he made the decision to leave man out, eradicating any telltale incriminating references from his manuscript, if not from his notebooks. By the time that *Origin* ran off the press at John Murray's, all that remained was that one most tantalizing sentence: "Light will be thrown on the origin of man and his history."[136] A vestige of what might have been, it was an understatement indeed.

Where Darwin remained silent, though, others did not. Marx and the Manchester school were one thing: ideological trumpeting, they were almost unthinking assessments, or at least Darwin might have consoled himself with such a conclusion.[137] He looked for a more-considered response closer to home, in any case. Huxley, ever the trusty Bulldog and who was clearly still salivating from his victory over rival anatomist Richard Owen at the British Association meetings in 1862, quickly penned *Man's Place in Nature.* Hastily pulled together from lecture notes, it was published in 1863. It was a short and accessible book that at once demonstrated Owen's error on the subject while at the same time outlining the morphological similarities between men and apes as evidence of common ancestry. The famous frontispiece—the skeleton of man proudly led a procession of his ungainly cousins across the page, ever more upright and in the ascendant—quickly became iconic. The similarities were glaring, and the public read "forebears," not "cousins," as the intended relationship. The implications, of course, went well beyond morphology. The deeply religious Duke of Argyll, George Douglas Campbell, declared it "a grim and grotesque procession."[138]

Darwin loved both the clarity and the audacity of it. Following on from his eminently popular lectures for workingmen, Huxley was once more taking evolution to the people, careless of whom he might scandalize in the process. If ever a picture spoke a thousand words, it was this one. "Hurrah the Monkey Book has come!" Darwin hallooed from Down, unable to contain his enthusiasm as an advance copy arrived in the mail.[139] By comparison, however, Darwin could not hide his disappointment with Lyell's reticence in

FIGURE 1.2. "Skeletons of the Gibbon. Orang. Chimpanzee. Gorilla. Man. Photographically reduced from diagrams of the natural size (except that of the Gibbon, which was twice as large as nature), drawn by Mr. Waterhouse from specimens in the Museum of the Royal College of Surgeons." The Duke of Argyll described this illustration from Huxley's 1863 work as "a grim and grotesque procession." (Frontispiece to T. H. Huxley, *Man's Place in Nature and Other Essays*; courtesy History of Science Collections, University of Oklahoma Libraries)

Antiquity of Man. Published in the same year, not only had Lyell remained decidedly lukewarm on selection, but, a uniformitarian gradualist in geology, he gave every appearance of assuming that the intellectual and moral faculties of man were way beyond the explanatory power of natural selection without some divine intervention to bridge the gap. "To say that such leaps constitute no interruption to the ordinary course of nature is more than we are warranted in affirming," he wrote, wavering in his conclusions.[140] Lyell, like Huxley, had sent an advance copy of his work to Down, and although Darwin was too sick and mired in work to give it detailed consideration immediately, he confessed to having turned first to the last chapter "with very great interest." Noting that Lyell castigated Owen in similar tone to Huxley, he responded, "You will, I feel sure, give the whole subject of change of species an enormous advance," adding in an anticipatory postscript, "I am impatient to begin reading it."[141] Having done so, however, he could only throw up his hands, exasperated. He immediately wrote to his close friends Asa Gray and Joseph Dalton Hooker expressing his disappointment. He told the American that he would be compelled to "grumble at his excessive caution" when the Lyells visited Down House the following week.[142] To Hooker, long his most intimate confidante, he was more forthcoming. He was "disappointed" with Lyell's "timidity," he wrote. The worst of it, though, was that "he thinks he has acted with the courage of a martyr of old." Such a noncommittal stance from the very person of influence that Darwin had been claiming as a convert for the last four years was the worst of outcomes. Clearly piqued, he confided to Hooker, "I wish to Heaven he had said not a word on the subject."[143]

By far the best effort, though, came from Wallace, and took Darwin quite by surprise. Since the joint presentation of their work to the Linnean Society on 1 July 1858, Wallace had been stalwart in his defense of natural selection. Further, with no Lyell to give him pause, and without Darwin's concern for reputation, Wallace had been far less reserved in applying selection to man. In 1864 he had published what Darwin later described as "the best paper that ever appeared in the Anth[ropological] Review!," in which he had advanced a daring evolutionary account of the origins of man and society.[144] Hooker too was "amazed at its excellence."[145] Despite the success of Huxley's work, the debate about the implications of evolution for mankind had quickly moved beyond mere morphology; Lyell's conclusions indicated as much. As Wallace wrote to Darwin, "I was led to the subject by the necessity of explaining the vast mental & cranial differences between man & the apes combined with such small structural differences in other parts of the body."[146] Having anticipated Darwin on selection, here he was again pressing the very ideas that Darwin had expressed in private to Lyell back in 1859. However, and what was new, was that although Darwin had reckoned intellect significant, he confessed he had not thought that it would become the primary target of selection. "The great leading idea is quite new to me," he confessed.[147] Wallace suggested that as humanity had become social and competent tool users, natural selection had ceased to operate on their bodies so much as upon their minds.[148] By such reasoning he sought to account for the apparent intellectual gulf between men and apes despite their morphological similarity, as well as for the evolution of the different races of mankind.

Significantly, while Huxley was bent on making an animal of man, Wallace was caught up on the differences. As Darwin had noted in *Origin*—and what he had relied upon Malthus to underline—was that natural selection in the animal world focused upon each animal as an individual. Each, Wallace now noted in his essay, "depends mainly upon their self-dependence and individual isolation." In contrast, however, man was "social and sympathetic," and contrary to Darwin's impression of the Fuegians, Wallace asserted, even "in the rudest tribes the sick are assisted, at least with food," and those below the average vigor could still find a place in a mixed economy founded upon a division of labor.[149]

In this communal economy, mental qualities would quickly outstrip physical ones in significance and replace them as "the subjects of natural selection." Those individuals that had the most developed "capacity for acting in concert, for protection and for the acquisition of food and shelter," would prevail over those that insisted upon selfishness and individualism. Further, those with well marked "sympathy, which leads all in turn to assist each other," combined with a more developed "sense of right, which checks depredations upon our fellows," and who combined a "decrease of the combative and destructive propensities" with an increase in the capacity for "self restraint" would find themselves favored by selection, and would have an "increasing influence on the well-being of the race" as a result.[150]

Although Wallace did not go into details on this point, it is clear that he saw that as a result of the social nature of mankind, selection would also operate on the community as a whole rather than only upon the individuals that made up the community. "Tribes in which such mental and moral qualities were predominant, would therefore have an advantage in the struggle for existence over other tribes in which they were less developed, would live and maintain their numbers, while the others would decrease and finally succumb," he wrote.[151] It was this process that "has raised the very lowest races of man so far above the brutes (although differing so little from some of them in physical structure)," and which had also resulted in the development of "the wonderful intellect of the Germanic races."[152] Darwin was delighted that

Wallace sought to explain the void between man and brute, having clearly worried about this for some time himself, and was enthused too by Wallace's intention to show that all men, regardless of race, were brothers.

As Wallace noted, this had been an issue that had long divided anthropologists, and most of the men who made up Wallace's audience at the Anthropological Society were "polygenists"—men who argued that the different races of men had separate origins and thus might actually be classified as different species. Darwin, along with Huxley and fellow members of the Ethnological Society, so named not only to distinguish but to distance themselves from the Anthropological Society membership, were ardent "monogenists"—firmly of the belief that all races shared common ancestry. This had moral significance beyond mere classification, of course.[153] Wallace's stated intention was to address the pressing question of whether man was "of one or many species." He attempted to present an account of human natural history that might reconcile the two camps, arguing that there was truth and error on both sides. He proposed that, given the vast antiquity of man that all serious scientific men now accepted, it was credible to believe that mankind in his most primitive state had formed "a single homogeneous race," although this was certainly "at a period so remote in his history, that he had not yet acquired that wonderfully developed brain . . . at a period when he had the form, but hardly the nature of man." Prior to the development of language, the fine quality of intellect, and the moral characteristics that were deemed so characteristic of civilized humanity, natural selection would have operated primarily upon human physiological characters, and thus, as the race had spread across the globe, distinct morphological types would have formed through adaptation to their local conditions and as a result of the correlation of growth. This would explain the physiological differences between the races of mankind, he argued. However, and as Wallace went on to suggest, as the social, mental, and moral qualities of these races developed, as they acquired language and developed tools, so natural selection would cease to operate on the physiology of race, but would instead work to refine the mental and moral faculties. "This action would rapidly give the ascendency to mind: speech would probably now be first developed, leading to a still further advance of the mental faculties, and from that moment man as regards his physical form would remain almost stationary. The art of making weapons, division of labour, and anticipation of the future, restraint of appetites, moral social and sympathetic feelings, would now have a preponderating influence on his well being and would therefore be that part of his nature on which 'natural selection' would most powerfully act."[154]

He went on: "We should thus have explained that wonderful persistence of mere physical characteristics, which is the stumbling-block of those who advocate the unity of mankind." In light of this, Wallace argued, there was truth to both the polygenist and the mongenist perspective. Man in his earliest stages of development had indeed formed "one homogeneous race"; however, it was also true that significant physiological differences later developed that distinguished one race from another. It was only after this point in human history that the moral and intellectual qualities became predominant. "If, therefore, we are of opinion that he was not really man till these faculties were developed, we may fairly assert that there were many originally distinct races of men; while, if we think that a being like us in form and structure, but with mental faculties scarcely raised above the brute, must still be considered to have been human, we are fully entitled to maintain the common origin of mankind," he concluded.[155]

Wallace thus sought to establish an uneasy middle ground between the prevailing beliefs of those in the Anthropological Society and those in the Ethnological Society. His account of the origin and evolution of the human races did allow that even the lowest of the races of mankind, were, as a result of the development of their moral and intellectual characteristics, raised "far above the highest brutes." But it still allowed for the prevailing hierarchies of race to go largely unquestioned. "Is it not the fact," he asked his audience, "that in all ages, and in every quarter of the globe, the inhabitants of temperate have been superior to those of tropical countries?"[156] Few, if any, men of science of the day, even from among the ethnologicals, would have disagreed with such a typically nineteenth-century presumption. However, the concession to these presumptions that Wallace offered proved insufficient to convince his audience at the Anthropological Society. The ensuing debate, which was recorded in the pages of the journal following the article, indicates as much.[157]

There was a lot in this paper that Darwin admired. Wallace's determination to prove the common ancestry of man as well as his account of the evolution of mind and morals through the process of natural selection clearly appealed. Darwin did confess some reticence to Hooker, however, writing to his friend, "I am not sure that I fully agree with his views about man."[158] As I shall show in chapter 3, Darwin thought that sexual selection played a significant role in the evolution of race, morals, and sociability, and that Wallace was wrong to appeal only to adaptation to environment and the correlation of growth to explain the differences between the races, but for now he was more than happy with Wallace's contribution. "There is no doubt, in my opinion, of the remarkable genius shown by the paper," he concluded.[159] Wallace had shown how the vast gap between man and ape might be bridged.

Significantly, Darwin did not remark upon Wallace's shift away from Malthusian individualism in his account of man, nor did he comment upon the fact that Wallace's account of tribal life differed so radically from his own account of the Fuegians. Indeed, Wallace clearly had selection operating upon entire communities in which cooperation and mutualism were the norm rather than competition. As a result, Wallace had argued that "tribes in which such mental and moral qualities were predominant, would therefore have an advantage in the struggle for existence over other tribes in which they were less well developed, would live and maintain their numbers, while the others would decrease and finally succumb."[160] Darwin would certainly not have let this pass without comment if he had disagreed, nor would he have given the paper the praise he did. Indeed, while Darwin had focused upon the intensity of the struggle for life between individuals in his account of natural selection, he had also given significant consideration to the evolution of the social insects and the instincts that drove them to cooperate in the way that they did. He was well aware that humans were in very many respects a social species, as Wallace had made clear in print, and this would be his starting point when he did eventually return to consider human evolution himself in *Descent of Man*.

Wallace had ended his essay by pointing out the implications of his own account "for the future of the human race." In this, he offered a glimpse of the utopian future that would later lead him toward socialism but which was in fact quite representative of the radical aspirations that Erasmus Darwin had held dear, and with which Darwin also clearly felt some sympathy. He wrote:

> If my conclusions are just, it must inevitably follow that the higher—the more intellectual and moral—must displace the lower and more degraded races; and the power of "natural selection," still acting on his mental organisation, must ever lead to the more perfect adaptation of man's higher faculties to the conditions of surrounding nature, and to the exigencies of the social state. . . . Refined and ennobled by the highest intellectual faculties and sympathetic emotions, his mental constitution may continue to advance and improve till the world is again inhabited by a single homogenous race, no individual of which will be inferior to the noblest specimens of existing humanity. Each one will then work out his own happiness in relation to that of his fellows; perfect freedom of action will be maintained, since the well balanced moral faculties will never permit any one to transgress on the equal freedom of others; restrictive laws will not be wanted, for each man will be guided by the

> best laws; a thorough appreciation of the rights, and a perfect sympathy with the feelings, of all about him; compulsory government will have died away as unnecessary (for everyman will know how to govern himself), and will be replaced by voluntary associations for all beneficial public purposes.[161]

Deeply impressed, Darwin could only ruefully divulge to Hooker, "I wish he had written Lyell's last chapter on Man."[162]

Darwin was thus understandably dismayed when, in light of Wallace's conversion to spiritualism following his marriage to Annie Mitten in 1866, he reversed his earlier stated commitment to a naturalistic explanation of the intellectual abyss between man and ape.[163] In what looked like a total turnaround from the paper which Darwin had thought "most striking & original & forcible," Wallace now argued that there was much about man—in particular, the origin of his moral sense, of his sense of aesthetics, and of his conscience—that could not be accounted for by natural selection alone. The gap between man and even the highest simian primate was just too large.

In a letter to Darwin, Wallace alluded to the fact that in his forthcoming review essay of Lyell's *Elements of Geology* and the latest edition of *Principles of Geology* for the *Quarterly Review*, he would go public with his growing doubts. "I venture for the first time on some limitations to the power of natural selection," he wrote.[164] "I shall be intensely Curious to read the Quarterly," Darwin responded, clearly in some trepidation as to what Wallace might have said on this score. "I hope you have not murdered too completely your own and my child."[165]

Upon publication, it became clear that Darwin's fears had not been misplaced. Wallace had declared that while natural selection certainly had had a bearing upon the natural history of humanity, there was much of significance that he now believed it could not explain. In contrast to his earlier position, he suggested that "the moral and higher intellectual nature of man is as unique a phenomenon as was conscious life on its first appearance in the world, and the one is almost as difficult to conceive as originating by any law of evolution as the other."[166] In the margin of his personal copy, Darwin placed an emphatic "*No!*" alongside the offending paragraph, underlining it three times.[167] To make matters worse, Wallace went on to publish an essay on "The Limits of Natural Selection as Applied to Man," as the final contribution to his otherwise orthodox *Contributions to the Study of Natural Selection.* Darwin's closest ally in the selectionist cause had abandoned him when he needed him most. "The moral and higher intellectual nature of man," Wallace now maintained, was "utterly inconceivable as having been produced through the action of a law which looks only, and can look only, to the immediate material welfare of the individual or the race."[168] Backtracking further, he argued that in light of this, there were in fact even certain physiological points, "the brain, the organs of speech, the hand, and the external form of man [that] offer[ed] some special difficulties in this respect."[169] Reading these lines, Darwin could only respond, "If you had not told me I sh[d] have thought that they had been added by someone else. As you expected I differ grievously from you, & I am very sorry for it." Echoing the point he had tried to make to Lyell, he continued, "I can see no necessity for calling in an additional and proximate cause in regard to Man."[170] Eight months later he was moved to write once more, again bemoaning his colleague's disaffection. "I groan over Man," he lamented. "You write like a metamorphosed (in retrograde direction) naturalist." He signed off, "Eheu Eheu Eheu" [Alas alas alas], "Your miserable friend."[171]

What made this *volte face* bitter irony was the fact that it had been Wallace who in 1866 had suggested to Darwin that he adopt Herbert Spencer's phrase "the survival of the fittest" in order to fend off the worrying trend of even some of Darwin's supporters to read theistic teleology into natural selection. The very term "natural selection," Wallace had suggested, "requires the constant watching of an intelligent 'chooser' like man's selection to which you so often compare it." It was no wonder people were reading God back into his argument; Spencer's was a much less misleading term.[172] Darwin clearly concurred, and although he could not bring himself to rename his theory—the analogy to the artificial selection of breeders was too good to lose—beginning with the fifth edition of *Origin*, published in 1869, he did append Spencer's phrase to his own throughout the text, as well as in the title of chapter 4, which now read "Natural Selection; or the Survival of the Fittest."[173] Publication coincided with Wallace's change of heart.

In May 1864 Darwin had offered Wallace his own notes on man, but now he was glad that Wallace had refused them.[174] Whereas in the early 1860s men like Charles Kingsley and Asa Gray had invoked a theistic interpretation of natural selection as a way to reconcile their evolutionary views with orthodoxy—God created through natural law, and natural selection was just one of these laws—by the end of the decade even his supporters were invoking God to explain the things that they thought selection could not accomplish. Not only had Lyell hinted at such a view, but so too had Herschel—and now Wallace. Later, Huxley's former student, the talented anatomist St. George Jackson Mivart, would also make similar claims, and clearly with malicious intent.[175] Huxley would respond with typically savage wit to both Mivart and Wallace in an essay refuting "Mr. Darwin's Critics," but what was needed was a detailed and thorough account from the sage at Down himself. With the page proofs of his book on *Variations* finally off to Murray, and feeling in better health than he had for a long time, Darwin recognized that finally the time had come to state his own views on man. He sat down to begin the project in February 1868.[176] He had notes on the subject in his *Beagle* diary as well as throughout his transmutation notebooks. Not only could natural selection account for the evolution of man as an animal and unite all races under one ancestor, but—and as Wallace had intimated in the 1864 paper that Darwin had so admired—so too could it account for the evolution of those attributes, like conscience, morality, and ethics, that made man most human, most liberal, and, by nature, a Whig.

2

A Very Social Darwinist: Herbert Spencer's Lamarckian Radicalism

> If moral systems are adopted or condemned, because of their consistency or inconsistency, with what we know of men and things, then it is taken for granted that men and things will ever be as they are. . . . It is a trite enough remark that change is the law of all things.
>
> HERBERT SPENCER

That evolution has been read as having significant political implications is nothing new. However, the majority of the scholarship on this subject has tended to focus on what has become known as "Social Darwinism"—exactly the sort of reading of Darwin's ideas that fueled the Manchester economists' enthusiastic response to *Origin*. As I have suggested in chapter 1, though, while there was much in *Origin* that might have supported just such an interpretation of the implications of natural selection for humanity, this was not the message that Darwin had intended his readers to take. Having stepped back from his original intention of including his views on human evolution in *Origin*, Darwin said only that "light will be thrown on the origin of man and his history."[1] In doing so, he inadvertently invited others to speculate on the question, and many arrived at conclusions quite different to those he had in mind. Rather, and as I shall show in chapter 3, Darwin was keen to ground a liberal communitarian ethic as the outcome of human evolution, and he embraced Malthusian political economy in order to do so. Although he told Lyell that he was amused to have been taken to have written an endorsement of "every cheating Tradesman," when he saw the extent to which such misconceptions had spread, he was moved to put pen to paper himself.[2]

As I have already noted, Darwin's theory of natural selection was not by any means the only transmutationist theory in nineteenth-century natural history. Scholars such as Adrian Desmond and Paul Elliott have documented the extent to which the ideas of the French naturalists Jean-Baptiste Lamarck and Étienne Geoffroy Saint-Hilaire, and those of Darwin's own grandfather, Erasmus Darwin, were widespread across England in the first half of the century—and that they were most popular among the political radical community in particular.[3] Thus, in this chapter I want to bring into the narrative of *Political Descent* the man who has been most prominently—and in many ways, most erroneously—associated with evolutionary politics, Herbert Spencer, for although he has become associated with what has been termed "Social Darwinism," he drew more from these alternative evolutionary traditions than he drew from Darwin.

Spencer has been written into history as an ardent defender of laissez-faire economics and an advocate of the unremitting "struggle for existence," and as a result he has been cast as representative of all that might be and might have been wrong with laissez-faire, either as it was manifest in the context of the liberal calls for free trade in the English Industrial Revolution or at any point since. Richard Hofstadter's work has done the most to establish this view of Spencer, and even though subsequent scholars have tried to ameliorate this picture, none has matched the popularity of Hofstadter's *Social Darwinism in American Thought.* If Hofstadter is to be believed, Spencer shared the views of the Manchester economists that Darwin legitimized: rampant capitalist exploitation. While Hofstadter marshaled Spencer in support of his own critique of American capitalism, others, such as Gertrude Himmelfarb, were much more dismissive of Spencer's contribution. Spencer was far from being the most important philosopher of the age, as some of his contemporaries thought him. Indeed, Himmelfarb derides the multivolume *Synthetic Philosophy* that was Spencer's lifelong endeavor as a mere "parody of philosophy."[4]

There have been scholars who have reached different conclusions about Spencer, however—who have attempted to view Spencer and his work in the context of their times. Perhaps most prominent among those who have called for a reevaluation of Spencer is the historian and philosopher of science Robert J. Richards. As long ago as 1987, Richards argued that there was much more to Spencer than Hofstadter's rather limited view allowed, and that in consequence Spencer was a much more significant contributor to the history of evolutionary biology than had hitherto been acknowledged.[5] Michael Ruse agreed, and although he has found little else on which to agree with Richards since, he concurred that not only was Spencer important, he was more so than Darwin, in terms of both his respective influence upon nineteenth-century biology as well as upon nineteenth-century sociology.[6] Other historians and philosophers have also argued that we need to take Spencer on his own terms rather than on Hofstadter's, but in the present-day public mind, at least, it is still Hofstadter's view that prevails.[7]

Of primary concern to Richards, and clearly important here too, is the fact that Spencer's lifelong motivation throughout a very productive career was "how the natural processes of evolution could produce a moral society."[8]

This is clearly significant, and in stark contrast to the notion that Spencer sought only to justify merciless competition. To understand Spencer's motivations and conclusions we need to consider the intellectual and social context in which he developed his ideas. Here I wish to do more than simply recapitulate Spencer's views though. We have seen that Darwin grew up in the radical tradition of English politics: his grandfather and uncle were both deeply sympathetic to William Godwin's views, and it was not without some soul-searching, and careful reading, that in 1838 Darwin finally concluded that Malthus had the best of the argument. Subsequently, he had made every effort to show that his evolutionary ideas were not indebted to Lamarck, the author of *Vestiges of the Natural History of Creation*, or even to his own grandfather, but rather were inspired by Malthus. By emphasizing this connection in *Origin*, Darwin was able to show that evolution was thus quite compatible with the new Whig politics of the philosophical radicals John Stuart Mill, Harriet Martineau, and their circle, and had nothing to do with the radicalism with which transmutation had hitherto been associated. Herbert Spencer grew up in the same tradition as Darwin's forebears: his family was deeply involved in Derbyshire radicalism, in which Erasmus Darwin and Lamarck were familiar names. Familiar, too, was the name of William Godwin, and what I wish to demonstrate here is that, whereas Darwin rejected Godwin for Malthus, Spencer remained true to the Godwinian cause. For the vast majority of his life, Spencer articulated a Lamarckian evolutionism that was deeply Godwinian in hue. Without an appreciation of this fact, we misread Spencer's intentions, his politics, and thus the diversity of his appeal. Further, given that many of Darwin's and Spencer's contemporaries came at Darwin's ideas through an appreciation of Spencer, this view of Spencer gives a new perspective upon many who subsequently called themselves "Darwinists" or "Darwinians."

Although Spencer had humble beginnings as a none-too-successful radical journalist in the Midlands in the 1840s, following the publication of Darwin's *Origin of Species* and the popularity it generated for evolutionary theories in general, Spencer quickly rose to be among England's foremost commentators on the subject. Pigeons were all well and good—Darwin had made them central to his first chapter—but people were more interested in the light that Darwin's work might throw on "the origin of man and his history," and this is what Spencer gave them and more. Spencer's lifelong and multivolume investigation into the political and philosophical implications of cosmic evolution for the progressive development of mankind and society was a best-seller, gaining him an international reputation in the process. When his opinion was sought, which it frequently was, he obliged, and in addition to becoming a household name in England, Spencer quickly became the nation's most significant intellectual export to the United States as well.[9]

Spencer's rise to fame was a long time in the making, but when it came, his fame spread quickly. However, although he hoped to be counted among the most significant contributors to science—alongside even Newton perhaps—by the 1870s his ideas were already being eclipsed by reform. By the 1880s, his adherence to laissez-faire appeared old-fashioned and the Lamarckian biology that underpinned his ideas was under attack. Thereafter, his fall from grace was precipitous, at least in England.[10] This was not the case in the United States, where free-market beliefs as well as Lamarckism enjoyed a more-enduring popularity.[11] Even there, however, the move toward a greater role for government in the early decades of the twentieth century, which reached its peak in the New Deal of the 1930s, undermined the perceived relevance of Spencer's philosophy. Even champions of laissez-faire economics were reticent to promote him after the Second World War—for a time, few among the Allies felt comfortable publicly embracing biological arguments for the survival of the fittest. As I have suggested, however, even though the view of Spencer as an unapologetic advocate of laissez-faire who thought it necessary that the weak perish to further the advance of society is an accurate reflection of how many American entrepreneurs chose to interpret Spencer's work, it is at odds with Spencer's self-perception. In fact, Spencer was clearly as ambivalent about the ways in which his own theory was being appropriated as Darwin had been about the uses to which Manchester economists put natural selection.[12] When Spencer did eventually tour America, in 1882, he was duly feted and feasted by American captains of industry, who hailed him as one of their own. According to Andrew Carnegie, Spencer had given them "the truth of evolution" in support of industry and unrestrained commerce. In replying to such a hearty toast, however, Spencer's departing remarks were to the effect that, for all that he had enjoyed their hospitality, they had misunderstood him.[13]

Like the young Darwin, Spencer was caught at a crossroads in the development of English radical politics. On the one hand were the Enlightenment views of William Godwin, while on the other lay the neo-Malthusian views of Harriet Martineau, Henry Brougham, and John Stuart Mill. Their philosophical radicalism demanded that Malthus be taken seriously, even if they had effectively turned his argument to conclusions other than those he had intended. Where Darwin ultimately rejected Godwin in favor of the Malthusian dynamic that had given him natural selection, Spencer remained true to the Godwinian radicalism of his upbringing, to which Malthus was anathema.[14] Although Spencer did not deny the struggle for existence, of course, like Godwin, Spencer believed that the antagonism between population and resources would ultimately resolve itself as humanity evolved toward a stable socialist utopia. Like many who shared his nonconformist upbringing, Spencer was critical of the role of the state; "Old Corruption" did little to serve the interests of the people, he believed.[15] However, where many of his contemporaries looked to reform government, in light of what he took from Godwin and the transmutationism he learned from Lamarck and Erasmus Darwin, Spencer rejected the state entirely. If the individual was infinitely malleable, as Spencer believed, and was so most through the exercise of his or her own agency, then clearly any state intervention was to be avoided. Not only would it undermine the development of the self-reliance he hoped to see develop as one of mankind's most significant natural characters, but it would thwart the development of human sociability as well. It was only through the exercise of the moral faculties that they might be maintained and improved, he argued. Otherwise, the non-state socialist future he anticipated would remain a futile dream. Spencer's concerns were biological as well as political. Indeed, he considered the two synonymous.

Spencer developed his ideas about the *longue durée* of human history across his multivolume *Synthetic Philosophy*, and while he did foresee that nations and races would war with one another, as they had done throughout history, far from reveling in the prospect of such eventualities, he looked forward to an eventual epoch of peace. Spencer only drew back from his initial socialist conclusions in the late 1880s as the English socialist movement grew, strengthening their demands for revolution and—to Spencer, even more disturbing—for the centralization of the state. This was a radically different vision of what socialism might entail than Spencer had in mind.[16]

Herbert Spencer grew up in world of change. In the early decades of the nineteenth century industry and commerce were transforming the nation, and nowhere was this more apparent than in the provinces—in the coalfields and textile manufacturing regions of the North and Midlands, in particular. The Industrial Revolution not only made England the "workshop of the world," it gave new vigor and a fresh sense of urgency to the Enlightenment ideals of progress and improvement. Spencer's father was honorary secretary of the Derby Philosophical Society, an organization that kept alight the ideas and ideals of its founder, Erasmus Darwin. Darwin had established the Society intent upon transplanting a seed from the thriving radical intellectual culture of the Lunar Society he had left behind in Birmingham—and in this his endeavors had been successful.[17] By the 1830s, provincial radicalism was an intellectual as well as a political force to be reckoned with and it molded the young Spencer's outlook on the world, on man, and on society. Spencer

FIGURE 2.1. "Herbert Spencer when 38." (Frontispiece to Spencer, *An Autobiography*; courtesy History of Science Collections, University of Oklahoma Libraries)

retained the impress of this heritage throughout a literary career that would see him rise from relative obscurity—his one claim to fame was that his uncle, Thomas Spencer, was a successful political pamphleteer—to become one of the most highly regarded philosophers of the age. In Erasmus Darwin's wake, Derby had become a town with rich and deep radical traditions and, as Paul Elliott has documented, evolutionary ideas were part and parcel of Derbyshire radicalism long before Darwin's now more famous younger son put pen to paper.[18]

It was in Derby that Herbert Spencer was born in the spring of 1820. Spencer's parents, William George Spencer and his wife Harriet, were Wesleyan Methodists, although his father later developed Quaker sympathies. As a result, from about the age of ten until he was thirteen, the young Spencer would accompany his father to the local Friends Meeting House on Sunday mornings and his mother to the Methodist chapel in the evenings.[19] By Spencer's own account, though, his own disapprobation of authority was the offspring more of lax parental discipline on his father's part than of the lasting influence of his brief association with the Friends. "I do not know that any marked effect on me followed," he recalled, although both of Spencer's biographers, J. D. Y. Peel and Mark Francis, argue that Spencer's religious upbringing colored his worldview significantly.[20]

Educated at his father's teaching academy, Spencer's schooling was far from run-of-the-mill.[21] Erasmus Darwin's own children had been educated there by Matthew Spencer, Herbert's grandfather, and as Michael Taylor notes, the pedagogy of the school was based on that of the contemporary Swiss educational reformer Johann Heinrich Pestalozzi. Pestalozzi believed that children should play an active part in their own education; all aspects of their life, he argued, played a significant role in the formation and development of their character.[22] Pestalozzi's views resonated with the prevailing associationist philosophy that informed much of contemporary radicalism and fit well with Lamarckian ideas of heredity. In 1833, at the age of thirteen, the young Spencer was sent to live with his uncle, Thomas Spencer, in order to continue his education. His uncle lived in the town of Hinton Charterhouse, in the southwest of the country. It seems that this was something of a surprise to Spencer, however, who had been led to believe that this was to be a short vacation rather than a long-term arrangement.[23]

Thomas Spencer was an evangelical Church of England clergyman who harbored somewhat unorthodox views on a number of issues, including church reform, the revision of the prayer book, and the separation of church and state.[24] Of greater significance for his nephew's later life, though, was the fact that he too was active in local radical politics. Indeed, it seems likely that it was these credentials that had been the motivating factor in the decision to relocate the young man. At the hands of his uncle, in addition to studying Euclid and Latin, Spencer also imbibed the politics of the Anti-Corn Law League, and he was required to read Harriet Martineau's pamphlets on political economy to his uncle in the evenings. Martineau's works were well known to him from his earlier childhood, although he confessed that while a child he had read her *Tales in Political Economy* mainly for the stories she employed to teach her philosophy. Despite this confession, it seems that her methods had the desired effect, for upon examination he found that her underlying message had stuck with him.[25]

Thomas Spencer was also Chairman of the Board of the Guardians of the Bath workhouse and oversaw severe cuts in the local parish poor rates, and his influence upon his nephew is evident in Herbert Spencer's choice of topic for his first publication—a series of letters in the *Bath and West of England Magazine.* Writing on the Poor Law, the young Spencer argued that aid given too readily to the poor only undermined their capacity for self reliance.[26] As Greta Jones notes, Spencer's politics were pretty typical of many of his background and upbringing, with the exception of a growing religious skepticism. Spencer increasingly saw the world through Deist eyes, appealing to the workings of Divine Law rather than to the careful oversight of an intervening personal God.[27] It is likely that this influenced his decision not to follow his uncle's example of matriculating into Cambridge University, which would have required that he assent to the Thirty-Nine Articles of the Anglican faith. Rather, on the basis of his scientific training, at the age of seventeen he instead took work as a civil engineer on the London and Birmingham Railway before moving to the Birmingham and Gloucester Railway in the following year.[28]

The historian Iain McCalman has recently argued that the fact that Darwin, Joseph Dalton Hooker, Thomas Huxley, and Alfred Russel Wallace each had the experience of an ocean-going voyage forged a bond between them that was important in the campaign to promote and defend evolution in the post-*Origin* years.[29] To the extent that this is so, it is surely also significant that Spencer shared in the experience of bringing about the railway revolution with Alfred Russel Wallace and the physicist John Tyndall, each of whom did time on the country's expanding network of railway cuttings. Both would later come to admire and promote Spencer's views. Indeed, Wallace admired Spencer to such an extent that he named his first child, who was born in 1867, Herbert Spencer Wallace.[30]

It was in light of the fossils that were unearthed in the process of extending the lines across the county that Spencer was first induced to read Lyell's

Principles of Geology, something he recalled as "a fact of considerable significance," for it was here that he first read of the transmutationist ideas of Lamarck. Uncannily anticipating exactly the response that Darwin would later have to Lyell's rejection of Lamarck, Spencer found himself developing "a decided leaning" toward the ideas of the French naturalist. Substantiating Elliott's view that transmutationist ideas were more widespread in English radical circles than historians have recognized, Spencer acknowledged that even prior to his Lyellian encounter with Lamarck, he was already familiar with transmutation. "I had during previous years been cognizant of the hypothesis that the human race has been developed from some lower race," he recalled. The family connections to the Derby Philosophical Society were doubtless the source of these views.[31]

Spencer believed that adaptation to environment accounted for the full range of observed variation that individuals of a species exhibited—the result of "an adjustment of structure to function."[32] Clearly a gradualist and under strong Lamarckian influence, Spencer also thought that variation would not only fill in the gaps between one recognized species and another, but—and of much greater radical import—would also expose the category of "species" to be a nominal convenience employed by systematists. As I have already suggested in chapter 1, this was in stark contrast to the prevailing view among naturalists that "species" was a "real" category denoting essential types that existed as discreet and immutable entities.[33]

However much Spencer's views on species might have resonated with what radical naturalists might believe, they were far from orthodox science among the dons and natural theologians who controlled science at the English universities. As I have made clear in the previous chapter, in those circles it was an established convention to follow the renowned comparative anatomist and Lamarck's long-time rival, Georges Cuvier, on species, and no less a man than William Whewell took it upon himself to enforce this point. To many, the conservative political implications of Cuvier's anti-evolutionary science were as attractive as his impressive feats of comparative anatomy.[34] The fixity of species in a naturally ordered hierarchy provided a basis from which the existing social hierarchy could be claimed to be of natural origin, and thus equally immutable to change. Indeed, it was doubtless significant for the reception of Cuvier's work in England that in addition to his obvious skill as an anatomist he was critical of the French Revolution and of transmutation. As Martin Rudwick has pointed out, Cuvier was feted as a national scientific hero and eventually received many honors and decorations by the French: Perpetual Secretary of the French Institute and Professor and Administrator of the Museum of Natural History in Paris were but two of the titles bestowed upon him. Cuvier was born in the town of Montbéliard, a small and largely Protestant territory belonging to the duchy of Württemberg. Thus, despite the fact that he ostensibly became a Frenchman when Montbéliard was annexed to France in the course of the Revolution, in 1793, "his cultural affinities were with the small Protestant minority in France, rather than with the dominant Catholic culture."[35] English science remained a deeply Protestant affair throughout the nineteenth century, and while historians have tended to emphasize the religious issues at stake in this period of the history of science, the political implications were equally contested. Indeed, it was Richard Owen's anti-Lamarckian stance in the 1840s that had seen him acknowledged as "a worthy successor of Cuvier and an honour to his country and its science" by the *Literary Gazette.* As Desmond has pointed out, the *Gazette* trumpeted the moral and political implications of Owen's science as much as his skill as an anatomist. His denial of transmutationism was "essential to science," the publication claimed.[36]

In light of having read Lyell's account of Lamarck, and prompted by the fossils he was digging up, Spencer certainly had the opportunity to explore and discuss his evolutionary ideas further. In between spells on the railway, he returned to Derby, spending much of his time in the library of the Philosophical Society. He also deepened his commitment to the radical community. He had written a number of articles for provincial radical and nonconformist journals in the late 1830s, but in the early 1840s he put pen to paper on a series of letters to the recently founded *Nonconformist.* The *Nonconformist* was a paper then under the control of its founding editor, Edward Miall, and it was quickly becoming one of the most significant voices in the nation to advocate religious disestablishment.[37] Indicative of the level of Spencer's commitment and of his involvement with the radical politics of Derbyshire is the fact that in 1843 he published these letters together as the pamphlet *On the Proper Sphere of Government* at his own expense. Although the venture was far from a commercial success—political pamphleteering rarely was—in the company of a letter of recommendation from his uncle they were sufficient to secure him the post of subeditor of the free-trade journal *The Economist.* This was in 1848, the year that revolutions swept Europe and Chartists took to the streets of Britain in numbers.[38]

It was while at the *Economist* that Spencer met with the other great formative influence upon his political development, Thomas Hodgskin. An ardent Godwinian, Hodgskin reinforced Spencer's already significant beliefs in the importance of the natural laws of development, the importance of environmental circumstances in the molding of moral character, and in the evils of the state.[39] In the free time left to him around his fairly minor editorial commitments at the *Economist*, Spencer brought his ideas together in his first book, which he entitled *Social Statics: or, The Conditions Essential to Human Happiness Specified, and the First of Them Developed.* John Chapman was enthusiastic about the project, and once he was assured that Spencer had the financial wherewithal to insure against potential loss, he agreed to publish the work. A small print-run appeared in 1851.[40] This was the same John Chapman who had also recently bought the radical *Westminster Review*, and even while still working for the *Economist*, the offices of which were located further down the Strand, Spencer was soon mixing with Chapman's circle, which included some of the most outspoken and progressive liberal intellectuals in London, including Harriet Martineau. As Spencer would later reflect, in those days "the Westminster Review had been an organ of genuine Liberalism," by which he meant "the Liberalism which seeks to extend men's liberties," not that which sought to impose taxation and unnecessary regulation.[41]

Books about Spencer usually stress the significance of *Social Statics* as being the work that first brought him to the attention of the radical literary set that wrote for and identified themselves with the *Westminster Review*—and quite rightly, for this was a transformative moment in Spencer's young career if ever there was one. What I want to emphasize here, though, is that it was in *Social Statics* that Spencer first outlined an evolutionary account of a radical and Godwinian theory of social ethics that not only built upon the political convictions he had earlier expressed in *On the Proper Sphere of Government*, but that laid the foundations for his essay "A Theory of Population, Deduced from the General Law of Animal Fertility," which Chapman would publish in the *Westminster Review* the following year. The opinions that informed these works—with only one or two exceptions—remained the basis of the progressive and laissez-faire evolutionism that Spencer developed across his multivolume *Synthetic Philosophy*, a project that would take him almost forty years to complete.

Social Statics is a document that was in many ways characteristic of the most advanced radical politics of the 1850s, but in others looked back to Godwin's 1793 *An Enquiry Concerning Political Justice.* In its pages, Spencer portrayed a dynamic and progressive conception of human history. Free trade and a diminution of the role of government were necessary for the development of both individual and social morality, and the nationalization of the land and graduated death duties were prerequisite to the establishment of the conditions of equality that were held in abeyance by the corrupt politics of landed interests. Grounded in a combination of Godwinian radicalism and Lamarckian social evolution, however, Spencer anticipated that the "disequilibrium" of present-day social conflict would ultimately find resolu-

FIGURE 2.2. William Godwin, 1756–1836. Painted by James Northcote. Godwin was a significant influence on Herbert Spencer. (NPG 1236; © National Portrait Gallery, London)

tion in a utopian-socialist future—a point at which all men would live in harmony, obeying their evolved and ultimately perfected social inclinations. This would be the realization of the "social statics" of his title.

Significantly, Spencer prefaced his work with a rejection of the utilitarian school of rationalist thought whose leading light at that time was still the patrician Jeremy Bentham—although Spencer noted too that the Cambridge natural theologian William Paley was also still a revered name in utilitarian philosophy.[42] Spencer targeted the expediency of utilitarianism not because it was so bad as a theory of morals, but rather because it was so good. Indeed, Spencer was deeply sympathetic to utilitarianism and Godwin was a utilitarian as well, of course. However, the one fault he found with utilitarian philosophy he believed to be a major one: ultimately, none of its advocates could offer any solid ground for following it other than their own opinion.[43] Spencer sought a more compelling ground for normative claims and arrived at that of natural necessity. He developed a science of ethics by uncovering the natural laws to which man should adhere in order to adapt to the conditions of life in society. This would only be possible, Spencer claimed, if it became generally recognized that morals are not of rational origin but rather have been derived from instinctive desires.[44]

Although he was closer to Godwin in his conclusions, true to the political economy he had learned from Martineau, Spencer did not deny the truths that Malthus had pointed out in the world. Hunger and thirst each existed as an ever-vigilant but unconscious "punctual monitor" to ensure that we fulfill our physiological needs, but so too, Spencer believed that there were equally necessary ideal social and mental conditions under which humanity would flourish. As Spencer put it—and in the process driving a wedge between these instinctive compulsions and the reasoned calculations of Benthamite utility: "The longings for food, for sleep, for warmth, are irresistible; and thus quite independent of foreseen advantages. The continuance of the race is secured by others [other longings, such as the sexual passions, that were] equally strong, whose dictates are followed, not in obedience to reason, but often in defiance of it."[45]

This was clearly no denial of the utility of such compelling urges, but it was certainly a radically different motivation than Bentham had had in mind. Indeed, if reasoned utilitarian calculations were to be the necessary dictate of "all other requirements of our nature," including "knowledge, property, freedom, reputation, [and] friends, . . . then would our investigations be so perpetual, our estimates so complex, our decisions so difficult, that life would be wholly occupied in the collection of evidence, and the balancing of probabilities." Given his conclusions about the instinctive nature of the utilities that served our physiological well-being, and refusing the distinction that others would later draw between the physiological and the mental properties of mankind, he continued, "May we not then reasonably expect to find a like instrumentality employed in impelling us to that line of conduct, in the due observance of which consists what we call *morality*?"[46]

It is notable that the "like instrumentality" that Spencer believed to be the motivating force that drove the development of morality—of a "system of regulating our conduct to our fellows"—was similarly driven by instinctive desires. Thus, and despite Spencer's oft-repeated emphasis upon independence of mind and action, first and foremost among the human instincts was the desire for sociability and the company of others. It was the social instinct that had facilitated the division of labor upon which civilization had been built, but also instrumental in this process had been our instinctive care for reputation. Indeed, it was in order that "we may behave in the public sight in a most agreeable manner, [that] we possess a love of praise."[47]

As man had become ever more civilized, Spencer contended, so too public opinion had usurped the arbitrary imposition of one man's will upon another as the dominant arbiter of human social life. "Thus, as civilization advances, does Government decay." To Spencer's mind this was another compelling indictment of utilitarianism: it "implies the eternity of government." In contrast, and loudly echoing the Godwinism he had learned from Hodgskin, Spencer stated as a bare fact that "it is a mistake to assume that government must necessarily last forever." Certainly, the institutions of government had marked "a certain stage of civilization" and indeed were "natural to a particular phase of human development," but they were "not essential but incidental" to human social life. Indeed, he anticipated a future in which government would become unnecessary—replaced at first by the sanction of public opinion, but as individual actions and desires became ever more closely correlated to life in society, society would be less and less a sanction placed upon individual aspirations and more and more the ground for their fulfillment. Turning to anthropology for evidence, Spencer argued that, just as "among the Bushmen we find a state antecedent to government; so may there be one in which it shall have become extinct." Legislative government might be necessary to the extent to which society was made up of selfish villains, but in a society of civilized men government was only so much restraint upon individual liberties. "Restraint is for the savage, the rapacious, the violent; not for the just, the gentle, the benevolent," he intoned. Spencer charted a historical correlation between the growth of freedom and the decline of government. Feudalism, serfdom, and the tyranny of slavery had each in turn fallen by the wayside as constitutionalism, democracy, and liberty had spread. "The triumph of the Anti-Corn-law League is simply the most marked instance yet, of the new style of government—that of opinion, overcoming the old style—that of force"[48]

As a result, in answer to those who would insist upon "certain notions of what man *is*, and what society *must* be," Spencer argued that surely it was obvious that human nature was no fixed quality. "If moral systems are adopted or condemned, because of their consistency or inconsistency, with what we know of men and things, then it is taken for granted that men and things will ever be as they are." But this was patently not the case. "It is a trite enough remark that change is the law of all things: true equally of a single object, and of the universe," so it was true of man as for the rest of the natural world.[49] In what can only be read as a mixture of Lyell, Lamarck, and the still anonymous *Vestiges of the Natural History of Creation*, which he had read appreciatively if critically shortly after its publication, Spencer saw the evidence as indicative of a long history of the inexorable operation of a law of progressive development.[50] From deep in the earth's many-layered crust, testimony to this fact could be read from a past without beginning extending into the future without end: "As we turn over the leaves of the earth's primeval history . . . we find this same ever-beginning, never ceasing change. We see it alike in the organic and the inorganic—in the decompositions and recombinations of matter, and in the constantly-varying forms of animal and vegetable life. Old formations are worn down; new ones are deposited. Forests and bogs become coal basins; and the now igneous rock was once sedimentary. With an altering atmosphere, and a decreasing temperature, land and sea perpetually bring forth fresh races of insects, plants, and animals. All things are metamorphosed. . . . Thus also is it with systems as well as worlds. Orbits vary in their forms, axes in their inclinations, suns in their brightness."[51]

Having left his reader to ponder "satellites" that "sweep forever onward into unexplored infinity," Spencer prompted the inevitable conclusion: "Strange indeed would it be, if, in the midst of this universal mutation, man alone were constant, unchangeable. But it is not so." If change really was the law of all things, then this much was obvious: man, like all else, "also obeys the law of indefinite variation." In a passage that demonstrates the extent of his evolutionary thinking on man even by 1851, Spencer uttered his total disbelief that it was still possible to hold a non-evolutionary account of man in light of the evidence: "Every age, every nation, every climate, exhibits a modified form of humanity; and in all times, and amongst all peoples, a greater or less amount of change is going on. There cannot indeed be a more astounding instance of the tenacity with which men will cling to an opinion in spite of an overwhelming mass of adverse evidence, than is shown in this prevalent belief that human nature is uniform. One would have thought it impossible to use eyes or ears without learning that mankind vary indefinitely, in instincts, in morals, in opinions, in tastes, in rationality, in everything."[52]

Contrary to the Cuvierian view of species, Spencer presumed Lamarckian adaptation to circumstance was the origin of a heritable variation that knew no bounds. Regarding mankind, he argued, "his circumstances are ever altering; and he is ever adapting himself to them. Between the naked houseless savage, and the Shakespeares and Newtons of a civilized state, lie unnumbered degrees of difference," each of which could thus be accounted for. That man might evolve ever onward toward a perfect adaptation with the environment followed as a logical conclusion.[53]

Spencer's radical Lamarckism thus put him at a far remove from the orthodoxy of British science, and his connections at the *Economist* and now at the *Westminster Review* meant that he was in just the right company if he wanted to debate the connections between physiology and radical politics—for among those who received Chapman's patronage were the science popularizer, naturalist, reviewer, and literary editor of the *Leader*, George Henry Lewes, and the Scottish publisher Robert Chambers.

Spencer and Lewes had begun their friendship in the spring of 1850 while Spencer was still writing for the *Economist*. Resident in the same lodgings, the two had hit it off as they shared the walk along the Strand to their respective workplaces one morning, engaging in an opportune conversation about what was now coming to be referred to as "the development hypothesis." This discussion had garnered Spencer an invitation to one of Chapman's infamous soirées that evening, at which the two had continued their conversation. Unsurprisingly, this included a deeper discussion of the still anonymous *Vestiges of the Natural History of Creation*. Spencer recalled later that he had boldly stated his differences from the author of *Vestiges* on the means although not the fact of development—and that Lewes had been pleasantly surprised to find a fellow philosophical thinker who had original ideas on the subject.[54] It was the start of a lifelong friendship, and Lewes ensured that Spencer found favor among the inner circle of friends who made Chapman's house on the Strand and the *Westminster* their intellectual home. Within the year, Spencer was unknowingly rubbing shoulders with the "Vestiginarian" himself at Chapman's soirées, and he wrote a number of pieces for the *Leader*, too.

The author of *Vestiges* brought to the table not only his own considerable ability to synthesize the work of others into a readable, slim, and highly provocative volume, but the desire to see scientific knowledge made available to all. As coeditor, with his brother William, of *Chambers's Edinburgh Journal*, and as a contributor, Robert Chambers—for he was the author of *Vestiges*—had a ready-to-hand reviewer's knowledge of many of the latest developments in a wide range of the sciences. He was also well-versed in William B. Carpenter's popular *Principles of General and Comparative Physiology*, that work having been clear inspiration for a number of the chapters in *Vestiges*. Physiology, phrenology, and politics were each and all hotly debated at Chapman's soirées. Nothing was taboo in this circle, and transmutation was par for the course.[55]

Carpenter influenced Spencer too. He had at one time been Robert Grant's student, but where Grant had found himself quickly marginalized by the scientific establishment for mixing his physiology and his radical politics too freely, as a result of his evident talent, his gentlemanly manner, and ability to communicate his ideas effectively, Carpenter had managed to carve a niche for himself at the heart of British science. Carpenter's *Physiology* was widely used in teaching and went through a number of editions; indeed, it was in part his ability to give a clear and accessible exposition of even complex subjects in physiology that had led to his appointment as Fullerian Professor at the Royal Institution, a position he held until 1848.[56]

Carpenter was thus no renegade, speculator, or quack theorist—all charges that had been leveled against the author of *Vestiges.* Widely regarded as a "gentleman of science," he had been elected Fellow of the Royal Society and would receive the Royal Medal in 1861. From 1847, he served as editor of the *British and Foreign Medical Review*, and like Chambers he was deeply interested in presenting science to a broad public audience. Despite the public controversy that surrounded *Vestiges*, Carpenter found the book deeply compelling, and in stark contrast to Whewell, he wrote a favorable account of it for the *British and Foreign Medical Review*, as well as an extensive serial commentary (that stretched to some eighteen parts) for the Unitarian *Inquirer.* As was the fashion, though, and doubtless with some forethought on the matter, the commentaries were published unsigned.[57]

Carpenter exemplifies the political tensions in science in the middle years of the century. An acknowledged gentleman of science, he harbored sympathies for both Lamarckian transmutationism and Geoffroyan anatomy—both of which carried the odor of Republican radicalism.[58] He rejected Whewell's "Doctrine of Final Causes in Physiology," which he had outlined in his *History of the Inductive Sciences.* There, Whewell had restated Cuvier's functionalist comparative anatomy in support of his claims, but to Carpenter's mind, and as he wrote to his friend John Herschel, Whewell's "comprehensive mind had failed to appreciate the true import of the data."[59] Carpenter was careful not to oppose Whewell in public, however, nor to openly state his own radical sympathies. He could see what doing so had done for Robert Grant's reputation and he had also witnessed the vilification of the author of *Vestiges.* Carpenter, who had been one of Grant's students, played his cards carefully—he offered qualified support for *Vestiges* in unsigned reviews and muted and often only implied endorsement of such views in his own physiological works, all the while decrying the "inflammatory political trash" that issued from the radical presses when pressed for his opinion at society soirées and meetings of the British Association.[60] Even so, this did not stop him from helping revise the text of the offending book, and as James Secord notes, he made a handsome £35 from doing so, slipping discreetly into the publisher's office to do the work. This was easy enough in practice and did not attract undue attention since Chambers had had an intermediary convince John Churchill to publish the book. By happy coincidence, Churchill also published Carpenter's work.[61]

Carpenter was not alone in enduring the kind of schizophrenia that the politics of science could force upon its practitioners in these tumultuous years as he sought entry to the scientific establishment while harboring distinctly anti-establishment views. Neither was he a lone hand in the ghost-editing of *Vestiges.* Edwin Lankester, whose son would later also make his mark as a science popularizer and a diehard Darwinian, would also put his hand in the publishers purse, receiving thirteen guineas for his part in correcting the text.[62] These revisions and corrections were important. Not only did they show a concern among science popularizers that the science consumed by the public should be accurate and earnest, but in strengthening the science in *Vestiges* they considerably strengthened the argument as well—and thereby the radical claims that were associated with it.

While *Vestiges* was lambasted by those within the only recently erected walls of the scientific establishment—and in public, at least by those who sought access—the book succeeded in presenting a compelling, law-bound argument for a science of creation. As Spencer had argued in that first conversation with Lewes, the author may not have all the facts straight, but that was of little consequence; others, like Carpenter and Lankester—and like Spencer—could correct that. Spencer would not go sneaking into publisher's offices to edit someone else's work, however; he would write his own multivolume work giving a painstakingly detailed treatment of the very issues that *Vestiges* had raised in outline.

Spencer's start on this quest came when Chapman asked him to review a new edition of Carpenter's *Physiology.* In his autobiography, written much later, Spencer acknowledged that he had given the book "such perusal as was needed to give an account of its contents," but in truth he took a lot from Carpenter. He recalled that it was here that he first came across Karl Ernst von Baer's work on embryological development, and he acknowledged Carpenter's account of this to be the source of his own belief that all life tended to begin in a state of homogeneity and develop to become ever more heterogeneous.[63] Chambers had utilized Von Baer in *Vestiges,* of course, passages that Carpenter had corrected as necessary—the two books even have common illustrations. Even if Spencer had recalled Chambers's use of Von Baer in *Vestiges,* it was Carpenter, whose work was infinitely more respectable, that he was willing to acknowledge.

Spencer's belief that species were not fixed but might vary indefinitely placed him well into radical territory, and well beyond the pale of the scientific establishment. His anticipations for the future of mankind reflected a transition in English radical thought that hinged upon attempts to reframe the moral meaning of Malthus. Indeed, given the prevailing debate about Malthus in radical circles, as well as the contemporary obsession with population—the first census had been taken in 1801—it is unsurprising that Spencer would have had something to say on the matter.

In doing so, he would try to resolve the tensions between the two major influences upon his own intellectual development: the Godwinian anarchism of Hodgskin and the Malthusian political economy of Harriet Martineau. The Godwinian tradition was grounded in a total rejection of Malthus. Godwin had argued—as later Owenites and socialists would too—that Malthusian claims were ideologically motivated and as a consequence were not legitimate descriptions of the natural order at all. Spencer had grown up with an ingrained appreciation of Malthusian Poor Law economics, but these were tempered by Godwinian ideas about the plasticity of human character and an ideal of future social harmony that he learned from Hodgskin.

Even though Spencer's early works were far from a commercial success, the philosophical and literary group of writers who organized themselves around Chapman's *Westminster Review* had a growing faith that here was "a very remarkable man" possessed of a mind of great compass. Others certainly thought so too.[64] Spencer's closest associates among the *Westminster* crowd were Lewes and the woman who would come to play a significant role in both of their lives—Marian Evans, the author who would later be famous as George Eliot. However, to Spencer's mind both Harriet Martineau and John Stuart Mill remained the central figures in John Chapman's salon at the *Westminster*, even though they were rarely present. Martineau was perhaps the most influential personality, for even though she was absent from London more often than not, she and Chapman were close friends as well as business associates, and she took a personal interest in the success of the *Review*. She was convinced that "the cause of free thought and free speech was under great obligation to Mr. Chapman" and supported the *Review* financially for a number of years.[65]

Both Mill and Martineau embraced Malthus but drew from his work quite contrary conclusions to those he had intended. Where Malthus had been prompted by what he had seen as the overzealous Enlightenment aspirations of Godwin and Condorcet to point out what he believed to be natural constraints upon human improvement, both Martineau and Mill read him as having revealed the very motor of social progress, that scarcity would prompt the morality and labor that would drive progress forward.[66]

While Godwin and the Owenites had rejected Malthus, under the influence of Chapman's circle, Spencer did not do so out of hand. He clearly believed that Malthus had been correct to highlight the conditions of struggle in nature; however, he believed that Malthus had been wrong in his statement that the relationship between population and resources would always remain as he had described them. As he had argued in *Social Statics*, and went on to repeat in "A Theory of Population," an essay he wrote for the *Westminster* on the subject, to the extent to which humanity had become civilized so had the Malthusian struggle for existence diminished.

In this article, Spencer set out his stall on the population question and what he took to be the moral meaning of Malthus. He chose as his point of departure the anti-Malthusian claims that the Newcastle-born political reformer and secretary of the Northern Political Union of Whigs and Radicals, Thomas Doubleday, had made in his *True Law of Population Shown to be Connected with the Food of the People* (1842). In making his case, Spencer considered evidence and argument taken from a plethora of recently published works in physiology, generation, and parthenogenesis.

Doubleday had argued that there was a necessary correlation between the amount of food available to an organism and its fertility. "Overfeeding checks increase," he had asserted, "whilst on the other hand, a limited or deficient nutriment stimulates and adds to it."[67] Doubleday gave each of these states names: an overabundance of food he called the "plethoric" state; dearth he named the "deplethoric" state. The relationship between food and fertility oscillated around a point of balance as organisms regulated their fertility in response to changes in their environment, thereby maintaining the existence of all species. Spencer drew out the ready associations that his readers might make between Doubleday's views on this point and the natural theology of William Paley: in both, everything tended toward a beneficent outcome in service of an overarching utility.[68] As the title of Doubleday's work suggests, the implications of this for man were his primary concern, and Spencer summarized Doubleday's conclusions for the benefit of his readers before moving on to show the weakness of Doubleday's argument compared with his own: "And hence, applying the law to mankind, he infers that there is a state of body intermediate between the plethoric and the deplethoric, under which the rate of increase will not be greater than needful; and that a sufficient supply of good food to all, is the chief condition to the attainment of such a state."[69]

This much, however, seemed against all observed phenomena and against all logic, and Spencer went on to question the reasoning that had led Doubleday to conclude that despite the fact that man was currently in a deplethoric state in which the level of fertility exceeded available resources, that the subsequent generation would produce either more resources or fewer children. In place of such lax thinking, and here doubtless indebted to Carpenter, Chambers, and to what little of the positivist philosophy of Auguste Comte he had read or otherwise imbibed (Harriet Martineau's translation of Comte was popular with the *Westminster* crowd), Spencer sought a more general cause than the particulars of Doubleday's argument admitted. In this, he turned instead to deeper laws of nature.[70]

He suggested that competition would be the obvious result of an increase in population in the face of scarce resources, and did so with such clarity that Darwin later wrote to Spencer conceding that he had "put the case for selection in your Pamphlet on population in a very striking & clear manner" (the article had also circulated as a pamphlet).[71] As Spencer acknowledged, though, his was an overgenerous assessment of what he had been thinking in 1851 when he wrote the article. Spencer argued that any species was subject to two antagonistic forces, the one destructive, the other preservative of the species. Among the former he ranked death from age, enemies, dearth of food, and an adverse climate; among the latter, the strength, swiftness, and sagacity of its individual members as well as their fertility. The shifting balance between these forces would oscillate around a point of equilibrium, Spencer suggested, the number of individuals making up the species diminishing in the face of enemies and scarcity, increasing again when either their enemies died off from the insufficiency of their prey or when the species in question had declined in number to the extent to which the available food was sufficient to sustain them.[72] Unlike Doubleday's model of natural balance, in Spencer's scheme of things it was the harsh realities of the external environment—both climate and the existence of predators—that checked the expansion of a given population.

While equilibrium was the overall outcome of this natural condition, there were no guarantees of any given species survival. Cuvier, and his own knowledge of fossils, had assured him of the reality of extinctions, and it was quite possible that "should the destroying forces be of a kind that cannot be thus met (as great change of climate), the race, by becoming extinct is removed from the category." He concluded, "Hence this is necessarily the *law of maintenance* of all races; seeing that when they cease to conform to it they cease to be."[73]

These were ideas that Spencer had already applied to mankind in *Social Statics*. The context had been the historical development of civilized society and human morals out of an earlier stage of human life that was dominated by savagery and barbarism. The rise of the human social and moral sentiments, and the civilization that they facilitated, were indicative of an ongoing progressive development. Humanity had yet to emerge from the struggle that was the result of their maladaptation to the environment and so had yet to reach the social equilibrium that he thought would be the end-point of human social evolution. Mankind was thus still subject to natural tendencies toward strife and conflict, and was still enthralled to the particularly evil propensity to delight, however occasionally, in the sufferings of other men. Mercifully, this was becoming a less-dominant feature of mankind, Spencer argued. Once civil society had been established, the industrial, intellectual, and moral progress that accompanied civilized and social life had taken off, leaving the savage races—and the worst of man's ancestral instincts—far behind. Spencer looked forward to the day, at some point in the future of the species, that would see the end of man's inhumanity to man.[74]

For previous long millennia savage races had been dominated by passions that led to warfare. This was only to be expected, Spencer argued, given that their environment had remained fairly constant. While limited resources had led to conflict, the selfish inclinations that these circumstances bred had consistently undermined any hope of moral advance. Indeed, given the persistence of almost constant intertribal warfare, it was also true that these sentiments had not been without their uses.[75]

Highlighting the importance of the environment, Spencer stated that "only when a revolution in circumstances is at once both marked and permanent, does a decisive alteration in character follow." However, as it was the case that "the warfare between man and the creatures at enmity with him has continued up to the present time, and over a large portion of the globe is going on now," then so would the propensity for conflict and self-assertion be maintained. In savage life this "old predatory disposition" dominated day-to-day existence. "The desires of the savage acting . . . indiscriminately, necessarily lead him to perpetual trespass against his fellows, and, consequently, to endless antagonisms—to quarrels of individuals, to fightings of tribes, to feuds of clan with clan, to wars of nations." Subject to such contrary tendencies—the social and the antagonistic—the general character of man had thus remained fairly constant throughout the ages. However, among those tribes who had either enjoyed a period of respite from conflict or for whom, as a result of the division of labor, warfare had become "the employment of but a portion of the people, the effects of living in the associated state have become greater than the effects of barbarizing antagonisms, and progress has resulted."[76]

As warring nations exterminated each other, and the more cohesive societies had cleared the earth of both their animal and their human antagonists, so the way had been cleared for the proliferation of the social instincts. "Just as the savage has taken the place of the lower creatures, so must he, if he remained too long a savage, give place to his superior. . . . From the very beginning, the conquest of one people over another has been, in the main, the conquest of the social man over the anti social man; or strictly speaking, of the more adapted over the less adapted."[77]

However Darwinian this may have sounded, Spencer's theory of population had distinctly un-Darwinian characteristics. Certainly, Spencer envisaged that the struggle for existence would continue to go on between savage races, and even among those who had advanced to the "industrial" stage of development, but he believed it would eventually cease to be a factor among civilized peoples—and not because of technological, agricultural, or other interventions per se. Technology and industry were important, of course, but—and true to his Godwinian heritage—struggle would diminish because of a biological diminution of the sexual passions corresponding to an increased stimulation of the nervous and intellectual faculties.

As Adam Smith had made clear, the combination of an increasingly specialized division of labor and the application of technology to industry certainly had the potential to increase the amount of resources available to humankind significantly. However, at the same time and in light of the connection he drew between intellectual and reproductive energy, Spencer believed that as man exercised his mind in the performance of these tasks, so, consequently, he would sire fewer offspring. It was this solution to the problem that Malthus had thought so intractable that made possible the utopia that Spencer so eagerly anticipated, and even as he wrote his autobiography at the turn of the century, he was still describing the possibility of a Godwinian future full of leisure, culture, and play: "The progress of mankind is, under one aspect, a means of liberating more and more life from mere toil and leaving more and more life available for relaxation—for pleasurable culture, for aesthetic gratification, for travels, for games." Realizing that this was so far out of the way of the thinking of the majority of his contemporaries, he added, "So little is this truth recognised that the assertion of it will seem a paradox."[78] Certainly, this statement does seem quite paradoxical coming from the same man who coined the phrase "survival of the fittest," and it was this element of his thought that Spencer would later point out to the American industrialists who hosted him when he visited the United States later in his career.

The relations between the sexes was an ongoing question throughout the second half of the nineteenth century, and it is unsurprising that any theory of evolutionary social development would include some level of engagement with the sexual division of labor. As Evelleen Richards has noted, "There was scarcely an evolutionist who did not take up and pronounce upon the woman question."[79] Spencer was no exception. As we might expect given his concern with population and the nature of the challenge that Malthus had presented to all utopians, sex was to prove fundamental to Spencer's conception of the overarching progressive development of human evolution. By Spencer's reckoning, the relations between the sexes were the expression of innate differences even to the most fundamental cellular level. The different characters of the sexes—the more assertive and intellectual nature of the male over and above the nurturing and emotional qualities of the female—were representative of deeper, fundamental differences between the male sperm cell and the female germ cell. "We must infer that the sperm-cell and germ cell respectively consist of coordinating matter and matter to be coordinated," he wrote, labeling them "neurine" and "nutriment," respectively.[80] Appealing to Carpenter's *Physiology*, Spencer maintained that there was a connection between the coordinating material—the neurine—and the expenditure of nervous and intellectual energy. He provided charts to show the correlation between the amount of phosphorous, which Spencer took to be the chemical compound of neurine, and the stage of intellectual development across the lifespan of the human male. As the intellectual power of man increased, so too did the proportion of phosphorous matter in the brain; in idiots the amount of this substance was below that in infants.[81]

Spencer's reading of the connection between the intellectual and the reproductive forces had significant implications for his views of the proper relationship between the sexes and, consequently, upon the role that women might be expected to play in society. While he firmly believed that there were innate physiological differences between the brains of men and women and thus that women were not the intellectual equals of men, Spencer's Lamarckian beliefs about heredity led him to think that the intellectual qualities of women were open to development and change just as were those of men. Spencer was an ardent and lifelong advocate of education: the exercise of the intellect would not just benefit a man, but through the inheritance of acquired characters might also benefit his offspring. In doing so, of course, the increased intellectual activity that education involved would consume neurine, answering Malthus. This was all well and good in men, Spencer thought, but he was concerned about the ultimate effects that education might have on women, particularly under the prevailing system of education. This was not a subject that Spencer developed fully until he wrote a series of articles that were brought together and published as *Education: Intellectual, Moral, and Physical*, in 1860. In it, he acknowledged that he thought that female education was certainly important. The effect of it might become hereditary and thus be exacerbated across the generations to her offspring of both sexes. However, given the relatively poor state of the female intellect compared to that of her male counterpart, Spencer believed it would be sheer folly to presume that they might benefit from the same rigorous intellectual stimulation as their brothers. To subject them to such training would sap them of their femininity and all that made them attractive. "In the pale, angular, flat-chested young ladies, so abundant in London drawing-rooms, we see the effect of [such] merciless application," he wrote. Excessive intellectual stimulation would drain their physical health as well as their beauty to the extent that it would either prevent them from having children or harm the children they had if they did conceive. Education was the means by which society could tame the dreadful fecundity that Malthus thought would bring an end to social progress, but to take it too far threatened evolutionary degeneration.[82]

Spencer took the connections between sex, intellect, and society seriously, and one might perhaps not unreasonably speculate about the conclusions he drew on the subject and their relation to Spencer's own immersion in his work and refusal to burden himself with a wife and family. Certainly, his friends seemed aware of the possibility, and told him so. As Spencer recalled in a rare moment of humor, his good friend Huxley had recommended a full treatment of "gynoeopathy" to alleviate the mental strain he put himself under. Even Huxley was forced to admit, however, that "the remedy had the serious inconvenience that it could not be left off if it proved unsuitable."[83]

Unsurprisingly, given the presumptions involved, the results that Spencer relied upon in his estimation of the relationship between the sexual passions and the intellect were found to be consistent with contemporary estimations of the different brain capacities not only of the sexes but of different races as well. Just as the female was frequently presented as being a less-developed form of the male, and the female brain a lesser-developed form of the male brain, so too different races were presumed to occupy different levels of development on a hierarchical ladder. The mental characteristics—and thus brain capacities—of black males were often compared to these characteristics in white women or children, and the various non-white races were also ranked accordingly.[84]

Perhaps anticipating objections to such analogies, Spencer argued that the idea that "an enlargement of the nervous centres of mankind is going on, is an ascertained fact. Not alone from a general survey of human progress—not alone from the greater power of self-preservation shown by civilised races, are we left to infer such enlargement; it is proved by actual measurement." He utilized figures from a lecture that had been delivered to the Zoological Society earlier in 1851 as evidence. Owen, the "English Cuvier," was the most renowned comparative anatomist in the country, and although Spencer later confessed that he found some of Owen's ideas "anything but logical," on this subject he considered the opinion of the Hunterian Professor at the Royal College of Surgeons solid testimony.[85] Utilizing Owen's data, Spencer wrote: "The mean capacities of the crania in the leading divisions of the species have been found to be—

In the Australian	75	cubic inches
" African	82	"
" Malayan	86	"
" Englishman	96	"

The results indicated "an increase in the course of the advance from the savage state to our present phase of civilization, amounting to nearly 30 per cent. on the original size." To Spencer, this was clear confirmation of his presumptions.[86]

If it were needed, further evidence for Spencer's theory that fertility would diminish as intellect increased might be taken from the relative fertility of simple, as compared to complex, organisms—between those with primitive and those with more developed "nervous centres." Whereas at the bottom of the scale the most simple organisms were almost invariably infinitely fecund, those higher up the scale of vertebrate life not only reached sexual maturity much later in life, they also produced proportionally far fewer offspring. From this Spencer deduced that there was a "marked antagonism" between "the nervous and generative systems." He continued, "The fact [is] that intense mental application, involving great waste of the nervous tissues, and a corresponding consumption of nervous matter for their repair, is accompanied by a cessation in the production of sperm cells."[87] Implicit in this conclusion was not only the fact that further progress in man's intellectual development would result from the continuation of this trend, but so too would a diminishment in the power of increase that had led Malthus to such pessimistic conclusions.

What Malthus had missed was the fact that it was "*the excess of fertility itself*" that would spur the evolution toward social equilibrium. As Malthus had himself noted: "The first great awakeners of the mind seem to be the wants of the body. They are the first stimulants that rouse the brain of infant man into sentient activity. . . . The savage would slumber for ever under his tree, unless he were roused from his torpor by the cravings of hunger, or the pinchings of cold."[88] As Spencer had made clear, as the brain of infant man was stimulated, so his neurine would be sapped and, in consequence, so too would his fertility. In refutation of the idea that those less-civilized, and thus more-fecund, races would then gain the upper hand, Spencer pointed out that to the extent to which they excelled in fertility, so too would they lack intellect, and their society, cohesion. The division of labor facilitated by social harmony and the technological advances of civilization meant that any such contest would, in effect, be no contest at all. Here was a Godwinian biology to answer Malthus. Of course, the flip side of Spencer's theory was that the "undue production of sperm-cells" could have catastrophic consequences. "The first result of a morbid excess in this direction is headache, which may be taken to indicate that the brain is out of repair; this is followed by stupidity; should the disorder continue, imbecility supervenes, ending occasionally in insanity"[89]—a stern warning indeed for any would-be masturbator! Darwin would later describe this part of Spencer's theory as "such dreadful hypothetical rubbish."[90]

By 1852 Spencer had thus developed his own theory of evolution totally independent of the work that Darwin was quietly busying himself with in the Kent countryside.[91] There would remain significant differences between them even after Spencer read *Origin*, and although it is clear that Spencer admitted a role for natural selection in his grand scheme once he had read Darwin's work, it would always be only of secondary importance to his hopes for what might ensue from the Lamarckian idea of the inheritance of acquired characters. However, when it came to the consideration of human evolution—which Darwin would not write about explicitly until 1871—there would also be similarities, which, when pressed on the issue, Darwin acknowledged. In the sixth edition of *Origin* Darwin expanded upon his teasing comment about the implications of his work for mankind: "In the future I see open fields for far more important researches. Psychology will be securely based on the foundation already well laid by Mr. Herbert Spencer, that of the necessary acquirement of each mental power and capacity by gradation. Much light will be thrown on the origin of man and his history."[92]

Spencer's account of human mental evolution was of perhaps the greatest significance to his overall evolutionary scheme and the political implications that he believed to follow from it. He had certainly recognized early on that an evolutionary account of humankind could not content itself with a mere account of morphology. This was not so much from any conviction that mind was the crucible of mankind—the holy grail of what it means to be human—but rather was simply the logical outcome of the conviction that if the laws of nature were indeed universal, then they would encompass psychology as surely as they did physiology, even if they might be more difficult to discern. Spencer had made this point in his earliest essays for the *Nonconformist* and later stated the case in no uncertain terms in the essays that were reprinted as *On the Proper Sphere of Government*: "Mind has its laws as well as matter."[93]

In 1852 Spencer had taken this further, in an essay provocatively entitled "The Development Hypothesis," which appeared in the *Leader*, a six-penny weekly folio newspaper. Under George Henry Lewes's editorship the *Leader* promised "free utterance to the most advanced opinion," and this was no false advertisement.[94] Extending the argument that Whewell had made in treating the material world as the outcome of law rather than "divine interposition," Spencer argued that a naturalistic explanation of the origin and development of man simply made more sense and had a lot more to recommend it than had any supernatural alternative.[95] Spencer had long held to the belief that "supernaturalism, in whatever form, had never commended itself."[96]

As Darwin was later to do in *Descent of Man*, Spencer gave an account of a gradual and historical development of mind from matter. Echoing the developmental hypothesis of *Vestiges*, which also and for obvious reasons bore a close resemblance to Carpenter's work, Spencer went on to explain that the origin of consciousness lay in the nervous centers of the most-primitive organisms. His account of how this had occurred displays the extent to which Spencer had been influenced not only by Carpenter but by the radical transmutationist thinking, picked up from his Derbyshire radical connections, of both Erasmus Darwin and Jean-Baptiste Lamarck. Certainly, in his essay on population he had acknowledged a role for competition and conflict, but as he set out to explain his views on the development of mind in detail, it is clear that Spencer set much store by his belief in the existence of an inherently progressive law of development as the motive force of evolution.

Thus, Spencer followed Carpenter in the belief that the development of the nervous system was prompted by compound sensations, which in turn promoted the development of further nervous complexity. Intelligence had thus arisen as an outgrowth of feeling and was therefore not some novel faculty of mind that necessitated an unusual or novel explanation beyond that which might account for physiological adaptations to the environment. As Spencer put it, the inner relations of the organism, including nervous and physiological structures, would increasingly come into correspondence with the outer relations of prevailing circumstance. The primary mechanism of this adaptation "long before set forth by Adam Smith, [is] that from the sympathetic excitement of pleasurable and painful feelings in ourselves, there originate the actions commonly grouped as benevolent."[97] As Paul Elliott has indicated, Spencer was also clearly guided on this point by the prevailing "Darwinism" of the Derby philosophers.[98] Erasmus Darwin, like Lamarck, had recognized at least some measure of nervous interaction with the environment in even the most simple homogenous life forms. This point of view had two significant implications: first, it meant that neither scheme was subject to a simple environmental determinism, allowing instead for an increasingly complex level of interaction between organisms and their environment; and second, it undermined any sense of a boundary between the sentient reason of man and the sensations attributed to beings lower down the organic scale.

From such a position it might not be unreasonable to believe that something akin to memory, at first only in a physiological sense—a muscle memory—but later in a cognitive sense too, might have developed as a result of an organism with a sufficiently developed nervous system repeating actions that more closely correlated its internal to the prevailing external conditions and which thus proved both beneficial and, through association, "pleasurable," at least in some minimally meaningful sense of the word, as implied in Spencer's comment above. To Spencer's mind, it was but a small step from these incipient faculties of memory and feeling to the development of intelligence and, ultimately, of reason. These were ideas that Spencer developed further in *Principles of Psychology* (1855). Although the first edition, like *Social Statics*, also failed as a commercial venture, selling only two hundred copies in its first year, Spencer deemed it a most significant work.[99] He confessed in private to his father that he hoped to see it achieve the status of Newton's *Principia*, and although this was clearly wide of the mark, the mental investment that he poured into its creation told upon his nervous constitution ever after.[100]

Spencer wrote the bulk of *Psychology* while in France, taking lodgings in the small coastal town of Tréport, northeast of Dieppe.[101] He was motivated to write the book in order to give an adequate account of the development of mind, but also was pressed in large part to respond to—and he hoped to resolve—the ongoing epistemological debate between John Stuart Mill and William Whewell on the nature and methods of inductive science. It was here that he aspired to make a contribution of Newtonian scale and significance. Spencer sought to describe no less than the overarching and conciliate law that had governed the development of mind as well as matter, embracing both Whewell's idealism and Mill's empiricism. "I hope to show that both of these hypotheses are right in a limited sense, and both wrong in a limited sense; that they admit of reconciliation; and that the truth is expressed by their union," he told his readers.[102] As Laura Snyder has noted, both Whewell and Mill saw their respective philosophies as attempts to reform society as well as science, and this was no less the case for Spencer.[103] It is perhaps somewhat surprising, therefore, that although Spencer's politics placed him closer to Mill than to Whewell, and his diminishing Deism was at odds with Whewell's efforts to defend the established church, as Peel has pointed out, Spencer was more sympathetic to Whewell's account of inductive reasoning.[104] Whewell's test of a true cause, or a *vera causa*, came out of what in his *Philosophy of the Inductive Sciences* he had referred to as a "consilience of inductions"—the unification of a number of distinct facts or inductions under a broad overarching explanatory theory.[105] To Whewell, knowledge derived from induction in this manner was cumulative: more and more phenomena might be brought under ever greater generalization, ultimately pointing to the existence of one ultimate First Cause. Despite the fact that by the mid-1850s, influenced by the metaphysical agnosticism of the theologian and philosopher Henry Longueville Mansel, Spencer had lost the overt Deist convictions that had animated *Social Statics*, in practice his increasing tendency to talk only of the "unknowable" was not too far removed from the position that Whewell had laid out in his treatment of the possibility of gaining knowledge of the First Cause, both in his philosophy of science and his theology.

While Spencer thus found at least some common cause with Whewell, he also had reason to be skeptical of Mill's claims, for despite the fact that Mill had been very much the flag bearer of philosophical radicalism throughout Spencer's youth, the two differed tremendously on the extent and significance of heredity. True to Enlightenment ideals, Mill came to see all hereditarian philosophies as harbingers of injustice.[106] Just as Adam Smith had pointed out the innate similarities between men, "between a philosopher and a common street porter, for example," over and above their differences, so Mill stressed environment and education over heredity.[107] This of course was where Erasmus Darwin and Godwin had diverged—and where Erasmus Darwin led, Spencer followed. What was instinct if not unconscious memory inherited from the experience of previous generations? The experiences of the philosopher and of the street porter were very different; so too would be the effects of these experiences—and these effects were hereditary. On this point at least, Spencer differed from both Smith and Mill, as well as, of course, from Whewell.

This adaptation of an organism to its environment was the core of Spencer's evolutionary theory, his commitment to an overarching "unknowable" holding him back from the relativism and contingency that seemed implicit in Mill's work. Spencer had certainly come close to recognizing selection at work in his "Theory of Population," but the manner in which he subsequently developed his ideas throughout *Synthetic Philosophy* shows that his concerns lay elsewhere. Thus, and in contrast to how the historian Mike Hawkins has characterized Spencer's work, Spencer's conviction that everything was tending toward social harmony denied anything but a peripheral and primitive role for struggle in his system.[108] It is not insignificant that the Russian-born anarchist and author of *Mutual Aid*, Peter Kropotkin, later found much of Spencer's philosophy appealing, just as Alfred Russel Wallace had done. Each of them thought that the struggle that did go on in nature was predominantly that between organisms and their environment rather than between one organism and another. Certainly, for those that were less well suited to their surroundings, the struggle to accommodate themselves to such surroundings would be more intense—and extinction as a result of failure perhaps more likely—but the reward of this struggle was adaptation and survival. The state of nature was not the brutal internecine war of each against all that Darwin, and even Wallace, would later see, in which the efforts of the less well adapted were tragic in their futility. The 1850s and 1860s, at least, were years of optimism for Spencer; there was a need to struggle, certainly, but whatever the outcome for the individual—and he clearly thought progress to be as likely as failure and death—overall, the universal system would continue along its path of progressive development, characterized by an ever greater adaptation of organisms to their environment. As they did so, the struggle for existence would gradually and inexorably diminish.

As in the more primitive organisms, so it was in man. Individuals had developed through ever greater adaptation to their environment—and in man, of course, just as in any other social species, other men were also a part of that environment. Adaptation to others had been implicit in Spencer's "Theory of Population," but in *Principles of Psychology* he gave a clear outline of his understanding of the process and character of human mental development. Spencer had explained how men might combine to mutual advantage and thus divide their labor for greater efficiency, but this was not the result of individual rational action per se, but rather the expression of the general laws that characterized all phenomena—a gradual adaptation toward an ever more complex and heterogeneous state of being. Spencer was keen to point out that what contemporary philosophers lauded as the "Will" was an evolved character just like any other and thus had been "necessitated by the same conditions" that had determined the development of any other aspect of human mental or physiological evolution.[109]

The materialist implications of this were evident to Spencer's readers, and even though only a few copies were sold, he received negative reviews in the nonconformist press. R. H. Hutton wrote a stinging review of *Psychology* that appeared in the *National Review* under the title "Modern Atheism," which prompted Spencer to defend his views in the pages of the *Nonconformist.* "A review so entitled was of course damaging," Spencer later recalled, "and the more so because it gave the cue to some other reviewers."[110] Spencer declared himself a follower of William Hamilton, who had defended his own religious convictions with reference to the limits of what it was possible for humans to know. Citing Hamilton's "Philosophy of the Unconditioned," Spencer claimed that his conclusion that "a knowledge of the absolute is impossible to man" was thus far from atheism.[111] For all his deep discussion of the "Unknowable," however, his account of the development of man and mind remained thoroughly naturalistic, and as James Secord has documented in *Victorian Sensation*, even a decade after the publication of *Vestiges of the Natural History of Creation*, in certain locales and in particular communities, such materialist accounts of morals and mind were still beyond the pale.[112] The Scottish Presbyterian David Brewster had clearly spoken for many when he branded *Vestiges* "prophetic of infidel times."[113]

While *Principles of Psychology* drew the ire of a number of reviewers in the nonconformist press, Spencer's views fit well with those of his colleagues at the *Westminster Review*, many of whom saw themselves as a part of a "New Reformation" that sought reconciliation of religion and natural science in a deistic worldview.[114] In his discussion of what these naturalistic mechanisms of development might be, Spencer acknowledged that there was variability across individuals to the extent to which their various faculties were developed. This much was uncontroversial. So too was his belief that those organisms whose associational and sensational faculties more readily responded to stimulation would find reward in adapting themselves to their environment. However, whereas Spencer might have gained conservative allies in his rejection of reasoned utilitarianism as an inadequate account of ethics, he lost them just as quickly by suggesting that a biologically adaptive utility was up to the task. Organisms responded in a positive manner to the pleasurable sensations that were increasingly aroused the more they came into equilibrium with their environment; having already made the connection between feeling, sensation, mind, memory, and consciousness, there seemed nothing that could not be accounted for. Deeply involved in his work, Spencer was clearly more concerned to distance himself from phrenology than to avoid the accusation of materialism that many would read into his writing. Although phrenology had been important to him in his youth, he was aware that many of his contemporaries had concluded that it was little more than quackery. In spite of this, however, he urged his readers not to throw the baby out with the bathwater and to recognize the truth of the existence of "the feeling which phrenologists have named love of approbation," as well as the sentiments of admiration, respect, and reverence.[115] The interactions between men were thus fundamental to the development of their psychology as well as of their morals—and, by extension, of the politics to which they were by nature adapted. Just as the interactions between all men were dictated by both physiological and psychological adaptations to their environment, so too, clearly, were the interactions between men and women.

In any theory of inheritance, reproduction was always going to be a central concern, but when it came to the matter of sex, Spencer was no Erasmus Darwin. Indeed, attitudes to sex throughout history were significant for all who considered the future condition of mankind. Erasmus Darwin had reveled in the subject; Malthus had resignedly acknowledged sexual desire as a natural instinct; and Martineau and Mill thought it something that needed to be controlled. Godwin was the only one who suggested that as man became ever less animalistic, ever more reasoned, so the natural urges that tied us to the animal world would diminish. It was from Godwin that Spencer took his lead on the subject of sex, although as we have seen, this did not leave him without deep concerns about the potentially negative effects upon the species of female education.

Given that Spencer believed a redirection of neurine from the reproductive to the mental energies was a necessary outcome of the development of human society, he thought it only reasonable to assume that as the purely reproductive energies were diminished, the affections and emotions, which Spencer saw as of-a-kind with the intellectual faculties, would be enhanced. He argued that these affections and feelings would develop from the sentiment of the "love of approbation," which would find their expression and fulfillment in the company of a member of the opposite sex. The corresponding development of a sensibility to "beauty," which would attract one particular individual, he continued, would be one of the most significant factors in the history of human development. Love of approbation would find its ultimate realization in these attachments: "To be preferred above all the world, and by one admired beyond all others, is to have the love of approbation gratified" to its fullest extent, a fact that would only be exacerbated by being witnessed by "unconcerned persons."[116]

Such sentiments were closely connected with the self-esteem of both the individuals in question, but, and more to the point, Spencer believed that they would also serve to break down the egoistic individualism that had hitherto dominated the sense of self of each organism. Without this connection, each organism saw its interests as being defined only in terms of its own adaptation—of its own pleasures, pains, and well-being. The sexual passions and emotions, however, fractured the bounds of such individualism: "In this case the barriers are thrown down; the freedom of another's individuality is conceded; and thus the love of unrestrained activity is gratified." Each might find property in the other, and the sympathies and sentiments thereby enjoyed "are doubled by being shared." Spencer held nothing back from this analysis. Even that most-treasured of Christian sentiments, the love of one person for another, had evolved through these same psycho-physiological processes: "Thus, round the physical feeling framing the nucleus of the whole, there are gathered the feelings produced by personal beauty, that constituting simple attachment, those of reverence, of love of approbation, of self esteem, of property, of love of freedom, of sympathy. All these, each excited in the highest degree, and severally tending to reflect this excitement on each other, form the composite physical state we call love."[117]

While, through aesthetic discrimination and the appreciation of beauty, the sexual passions were particular—Spencer was not about to predict a future in which "marriage," to use that respectable Victorian euphemism, was indiscriminate and general among the population—it is clear that he saw the emotions, and the faculties that facilitated them, as having more general implications in the adaptation of an individual to his or her environment. Man as a species was becoming ever more social and ever less egoistic as platonic love became general throughout the population. "Solitude . . . leads by and by to great misery," he wrote, and perhaps saying more about himself than he realized, added, "The entire absence of marks of approval from those around us, causes a state of consciousness difficult to bear; and persons accustomed to positive applause feel unhappy when it is not given."[118] Clearly, the negative reviews had stung. The implications were clear: as humans grew more adapted to the social conditions of their environment, so they would grow increasingly sensitive to the opinions of their fellows, taking pleasure in their praise and finding pain in their scorn.

Which actions would be praiseworthy; which would least provoke disapproval? To Spencer the answer was obvious: those actions on the part of the individual that resulted in the better adaptation of the social group to its environmental circumstances. What Spencer referred to as the "knightly character"—a chivalrous willingness to put the commonweal before the interests of the self—might thereby develop as each individual in society adapted to the opinions of his peers, and as society, in turn, adapted to prevailing environmental circumstances.

This conception of society as a collective entity was the subject of "The Social Organism," another *Westminster* essay, this one published in January 1860. Although the essay was published after *Origin* had appeared, Spencer had had no chance to take Darwin's work into account in preparing it. Darwin had instructed Murray to send an advance copy to Spencer, "*amongst the first* distributed, in November" 1859, but it had gone astray in the post and was not recovered until the following February.[119] Spencer went to lengths in this essay to explain the analogous relationship that might legitimately be drawn between biology and society: "That they gradually increase in mass; that they become little by little more complex; that at the same time their parts grow more mutually dependent; and that they continue to live and grow as wholes, while successive generations of their units appear and disappear; are broad peculiarities which bodies politic display in common with all living bodies; and in which they and living bodies differ from everything else."[120]

Having read *Origin*, Spencer took natural selection on board in his later works, although as I have already indicated, it would never supplant the Lamarckism that informed the synthetic system he had already outlined. In *Principles of Sociology*, the three volumes of which were published and expanded across the last three decades of the century, he returned to the idea of the evolution of society as a collective entity, referring to the social evolution involved as "super-organic evolution." Citing the social insects as exemplary of a primitive stage in the development of this phenomenon, he noted that "the processes carried on by these show us cooperation, with, in some cases, considerable division of labor; as well as products of a size and complexity far beyond any that would be possible in the absence of united efforts."[121] Again analogizing between animal and human societies, he went on to outline his conception of the evolution of human society. Once more, Lamarckian mechanisms of the inheritance of acquired characters were to the fore, and in this instance it is likely that he had read *Descent of Man*, published in 1871, in which Darwin had tackled exactly this subject, and—as I shall show in the next chapter—invoked just such a mechanism of inheritance in doing so.

Although the extent to which Spencer was influenced by Darwin's *Descent of Man* is unclear—there is certainly a good deal of superficial similarity between the two books—the ground for Spencer's *Sociology*, and, even later, his *Ethics* had been laid much earlier. In any case, any influence was clearly a two-way street. Darwin's copy of Spencer's *Psychology* has some light annotation, largely in relation to instinct and expression, while his annotation of Spencer's *Principles of Biology* show that he paid close attention to what Spencer had to say on the subject. This is unsurprising, of course, for Darwin not only rated Spencer as among the nation's "greatest philosophers," but it was in *Biology* that Spencer had coined his memorable phrase "survival of the fittest"—and had done so specifically in relation to what he had read in *Origin*: "This survival of the fittest, which I have here sought to express in mechanical terms, is that which Mr. Darwin has called 'natural selection,' or the preservation of favoured races in the struggle for life."[122] Darwin also owned both *Principles of Sociology* and *Study of Sociology*, but neither of these is marked.[123]

In *Principles of Sociology* Spencer recounted a history that developed in ordered stages, a concept which was by now familiar to his readers. Not only were there plenty of similar accounts published in the *Anthropological Review* and like venues, but he had hinted at as much throughout the earlier volumes of *Synthetic Philosophy*. Spencer named the primitive state of society the "militant" stage. In this stage of development societies had been at war with one another to secure resources and supremacy. This had been followed by the "industrial" stage—the one in which Spencer and his contemporaries found themselves. In this stage open warfare had been replaced by economic struggle, but so too had actions become much more corporate and cooperative. Following this progressive and increasingly mutualistic tendency, Spencer anticipated the final realization of man as a social being in the "civilised" stage of social development, in which man would have adapted to the social nature of his environment, arriving at the point of equilibration that he had first outlined in *Social Statics*.

Thus, although Spencer had good ideological reasons for wanting to deny that warfare played a positive role in developing adaptive characteristics—and was quite explicit in his assertion that this was so—indirectly, it is less certain that warfare was not significant, for it was in this instance that heroism and chivalry, or what Spencer had referred to as the "knightly character," flourished. The nobility and fearlessness of the warrior would certainly be adaptive in such circumstances, but it seems that rather than this being a character that had developed as a positive response to conflict, it had developed in the would-be warrior as a response to the esteem of his fellows. In light of what Spencer had already said about the significance of the sexual passions as the ultimate expression and fulfillment of the love of approbation, it seems that Spencer was suggesting that chivalry and self-sacrifice persisted not as an adaptation to warfare but as an adaptation to the love of sociality and the desire to receive the accolade of one's fellows. Although Spencer did not develop this idea more explicitly in terms of the implications this might have had for selection and heredity—as a Lamarckian, he did not have to—as we shall see in the next chapter, this was something that Darwin would include in his own understanding of the development of such social sentiments in the context of sexual selection.[124]

Having given so much time to war and conflict, Spencer was keen to point out that he was being descriptive and not normative in his account of the social evolution of humanity. But the distinction is a fine one, and many of his contemporaries, and not a few historians, have read him as an ardent advocate of conflict as the means to social progress. Spencer was passionately opposed to war and imperialism, even if he realized that these endeavors had played only an indirect role in the evolution of the characteristics that he found laudable. However, there are passages, in *Sociology* at least, that could easily be read as an endorsement of sectarian aggression by those inclined to do so. The lesson to be learned, though, Spencer urged, was that even in the most brutal warfare it was joint actions taken in concert with one's fellows that led to success; individualism on the battlefield led more often than not to failure rather than glory: "Given two societies of which the members are all either warriors or those who supply the needs of warriors, and, other things equal, supremacy will be gained by that in which the efforts of all are most effectually combined. . . . In open warfare joint action triumphs over individual action. Military history is a history of the successes of men trained to move and fight in concert."[125]

This last should be read in the context of the program that Spencer had outlined up to this point. The tendency of human history was in the direction of an ever greater sociality. Love, fellow-feeling, and mutualism everywhere ousted the egotism and individualism that had persisted among man's forebears and which continued to hang over contemporary nineteenth-century society. Spencer was adamant that all that was required for the fulfillment of human history and the realization of his utopian-socialist ideal was that nature be left to run its course. While it might seem needful as an act of morality to intervene in these processes in order to alleviate the sufferings and wants of those who were not sufficiently adapted to the demands of their circumstances, the promise of charity—especially if it were to be provided by an outside agency like the state—would only create an artificial circumstance to which the needy would very quickly adapt themselves and upon which they would henceforward come to depend. Private charity might at least exercise the altruistic qualities among the better endowed, but even then such charitable measures should not be indiscriminate. The "undeserving poor" outnumbered the deserving poor in multitudes. Of one thing, though, Spencer was certain. No good whatsoever could come from state intervention, in this or any other matter. The laws that governed progress in nature would best teach men to improve themselves, and there was no mandate for government to presume to improve upon the rule of natural law. "By a sharp experience alone can anything be done," Spencer argued. "*Educating* must be left to the discipline of nature, and [the maladapted] allowed to bear the pains attendant on their defect of character. The only cure for imprudence is the suffering which imprudence entails. Nothing but bringing him face to face with stern necessity, and letting him feel how unbending, how unpitying, are her laws, can improve the man of ill-governed desires. . . . All interposing between humanity and the conditions of existence—cushioning-off consequences by poor-laws or the like—serves but to neutralize the remedy and prolong the evil. Let us never forget that the law is—adaptation to circumstances, be they what they may."[126]

It would be necessary to be cruel to be kind. "The inner relations are determined by the outer relations," he reminded his readers in the very last paragraph of *Psychology*. "Were the inner relations to any extent determined by some other agency, the harmony at any moment subsisting, and the advance to a higher harmony, would alike be interrupted to a proportionate extent: there would be an arrest of that grand progression which is now bearing Humanity onwards to perfection."[127]

By the 1860s, and increasingly so thereafter, this last paragraph, which might have been written as an optimistic anticipation of the new millennium, quickly became an alarm call, as Spencer's liberal colleagues—even his good friend Huxley was among them—sought an increasing role for the state in social-welfare provision, in education, and in the regulation of industry. Besides the fact that many liberals embraced what Spencer later derided as the "New Toryism" of reform, Spencer was even more alarmed by those radicals who in the 1880s took their embrace of the state to the extremes of socialism.[128] On the one hand, Marxists in the small but vocal Social Democratic Federation saw evolution as working toward not only an ever greater collectivism, but toward an ever greater centralization. They argued that the expansion of the state was itself an evolutionary development. On the other hand, anarchist-communists like the exiled Russian revolutionary Peter Kropotkin embraced Spencer's ideas as a ringing endorsement of their own political creed. Horrified, Spencer pulled back from his own socialistic conclusions, and away too from any anticipation of social equilibrium in the immediate future. Indeed, the fact that the environment would constantly change, he reluctantly acknowledged, was a telling criticism of his former static conclusions.

Spencer published the final volume of his synthetic philosophy, volume 3 of his *Principles of Sociology*, in 1897, and by this time he was now suggesting that the socialist conclusions he had pointed to in earlier volumes were not the end-point of evolution but rather had merely been a stage along the way to a further development of an even greater individuation. As he was now eager to make clear, it was the ethics of the industrial stage of society that were the relevant ones, and even here he revised his emphasis away from the implied mutualism of his earlier work and back toward a renewed emphasis upon individual responsibility: "Living and working within the restraints imposed by one another's presence, justice requires that individuals shall severally take the consequences of their conduct. . . . The superior shall have the good of his superiority; and the inferior the evil of his inferiority. A veto is therefore put on all public action which abstracts from some men part of the advantages they have earned, and awards to other men advantages they have not earned."[129]

We should pause before we accept the views that Spencer articulated in 1897 as an adequate account of his contribution to contemporary debate about the evolution of man, mind, and morals, and their implications for how he believed humanity might live. The view of Spencer that I have presented here, which is based largely on the books and essays he published in the 1850s and 1860s, is at odds with Spencer's last works. It is an undue focus upon and selective reading of these later volumes that has fed the traditional view of Spencer as an ardent Social Darwinist who saw competitive individualism as the means of social advancement. As I suggested in the introduction to this chapter, Hofstadter has done the most to popularize this view of Spencer. Motivated by his own disenchantment with the liberal capitalism of his own time, he sought to condemn it by demonizing Spencer. Subsequent historians have argued that we need to pay much closer attention to Spencer's evolutionary argument, although Mike Hawkins in his 1997 study, *Social Darwinism in European and American Thought*, has reiterated Hofstadter's view. To Hawkins, Spencer's Lamarckism in no way lessened the ruthless competition between races and nations that was implicit to Spencer's conception of adaptation. Referring to Spencer's account of the struggle for existence and the extinction of those nations and races that failed to adapt to circumstance, Hawkins concludes that "there is nothing Lamarckian about these arguments or the world view they express."[130] The historians Robert Richards, Greta Jones, and others, think otherwise, however. They suggest that detailed attention to the development of Spencer's evolutionary ideas provides us with a more-nuanced picture, and I would have to agree. Further, as Michael Taylor has pointed out in his recent study of Spencer's philosophical views, if Hofstadter and Hawkins are right, then how can we explain the popularity of Spencer's views among so many radicals, socialists, and anarchists?[131] I contend that it was certainly not as a result of a misconception of Spencer's intentions, or of a selective and misrepresentative reading of his works.

The historian Daniel Todes has argued that the Russian naturalist and geographer Peter Kropotkin was representative of those Russians who embraced *Origin.* They did so selectively, choosing to ignore the obvious bias of Darwin's English political-economic views. Todes sums up their views in the title of his book, *Darwin without Malthus.* I shall have cause to turn to Todes and to Kropotkin in more detail in later chapters of *Political Descent,* but here it is important to note that this was, in effect, potentially what Spencer offered his readers. Those who came to Darwin through the already evolutionary radical tradition could appreciate Spencer's Godwinian rejection of Malthusian conclusions. Certainly he had coined the phrase "survival of the fittest," but this had been in description of Darwin's theory, not his own. Thus, even though Spencer attempted to incorporate natural selection into his own account of evolution after 1860, it was always half-hearted. To the extent to which Malthus might be right, Spencer believed that the problem he had pointed out would diminish over time, not increase. Thus, as I have suggested here, Spencer became a major figure among those radicals who ultimately found Godwin more compelling than Malthus in their understanding of the human condition. Embracing Lamarck and Erasmus Darwin, Spencer anticipated the evolution of humanity toward an ever more social condition. As mankind became more and more adapted to their circumstances, so struggle and conflict would diminish. In the years in which Spencer was being accoladed as one of the nations greatest philosophers, he was a utopian socialist, not a social conservative. As will be seen in subsequent chapters of *Political Descent,* Spencer influenced many of those who, after 1860, called themselves "Darwinists." Indeed, as I shall show in the next chapter, even though Darwin embraced Malthus as he wrote *Origin,* he was in fact much closer to Spencer's views on the outcomes of human social evolution than he was to those who accused him of providing an endorsement in nature for the self-serving ambitions of "every cheating Tradesman."

3

A Liberal Descent: Charles Darwin and the Evolution of Ethics

> This great question has been discussed by many writers of consummate ability; and my sole excuse for touching on it, is the impossibility of here passing it over; and because as far as I know, no one has approached it exclusively from the side of natural history.
>
> CHARLES DARWIN, *Descent of Man, 1871*

Darwin may have had man in mind from the earliest days of his speculations on evolution, but as the historian Joel Schwartz has noted, a close analysis of his correspondence reveals that he was ambivalent about putting his views on the subject into print even before the publication of *Origin*. "I think I shall avoid the whole subject," he had confessed to Wallace at the end of 1857.[1] The response to *Origin* and his fluctuating health did nothing to change his mind either. Deeply impressed by Wallace's paper for the Anthropological Society, Darwin had even offered Wallace his notes on man, encouraging him to tackle the subject. Feeling overburdened with work and in ill health, he wrote "I do not suppose I shall ever use them."[2] Wallace refused Darwin's offer, however, and it was only in February of 1867, when Darwin had finally shipped the page proofs of *Variation of Animals and Plants under Domestication* off to Murray and found his health much improved, that he revisited his decision to avoid writing on man. "I have almost resolved to publish a little essay on the Origin of Mankind," he wrote to Wallace.[3] In his 1864 paper, Wallace had not only attempted to give an evolutionary account of the differences between the different human races, but had also suggested how natural selection could account for the vast gap in intellect and morality between even the highest type of ape and the lowest type of man. Society had driven the development of human intellect and of morality. The division of labor, the development of technology, and other aspects of human social life had removed the focus of selection from man's morphology to his intellect and conscience. Across many thousands of generations, mankind remained very much among the apes in terms of their morphology, but where their mind and morals were concerned, they had risen to be in a class of their own.

Between 1864 and 1867, Darwin and Wallace corresponded at some length on sexual selection. Wallace was never quite convinced, but Darwin believed that sexual selection explained aspects of the natural world that the mere struggle to survive could not. He believed, for example, that many of the differences between the human races could be explained in terms of aesthetic preference and mate selection, just as he believed that sexual selection also explained sexual dimorphism in other sexually reproducing animal species. This was to be the subject of Darwin's book on the origin of man. However, as I have shown in chapter 1, Darwin was only two chapters into his account when Wallace published his review of Lyell's works in which he retracted his earlier faith in the efficacy of natural selection to account for so many aspects of humanity. Now Wallace was arguing for a supernatural explanation. Schwartz has suggested that it was Wallace's belief in phrenology and spiritualism that drove his change of heart but that much more important in this regard was his socialism and "his inability to bridge his scientific and moral beliefs."[4] Darwin's thoroughly naturalistic account of evolution was simply not capable of accounting for the evolution of morals.

Schwartz's account of Wallace's motivations for turning to a supernatural account of the origin and development of humanity is provocative but problematic. Provocative because it suggests that Wallace read Darwin as describing a state of nature that was rife with individualism and competition from which socialism could not evolve; problematic because this was exactly what Wallace had illustrated in his 1864 paper. There, Wallace had argued that natural selection when applied to human societies could quite easily account for the evolution of ethics and morality. Indeed, Wallace had pointed out that, whereas in the early evolution of mankind human morphology had been molded by natural selection just as other animals had been, once man became a social animal, so selection began to target his mental and moral capacities: "In proportion as these physical characteristics became of less importance, mental and moral qualities will have increasing influence on the well-being of the race. Capacity for acting in concert, for protection and for the acquisition of food and shelter; sympathy, which leads all in turn to assist each other; the sense of right, which checks depredations upon our fellows; the decrease of the combative and destructive propensities; self restraint in present appetites; and that intelligent foresight which prepares for the future, are all qualities that from their earliest appearance must have been for the benefit of each community, and would therefore have become the subjects of 'natural selection.'"[5]

While Wallace had primarily been concerned to show that Darwinian theory could account for the morphological differences between the different human races, he thus also provided an explanation of how humanity might have evolved to have such better mental and moral faculties than even the most advanced apes. Wallace had noted that "tribes in which such mental and moral qualities were predominant, would therefore have an advantage in the struggle for existence over other tribes in which they were less developed, would live and maintain their numbers, while others would decrease and finally succumb."[6] Thus, far from Wallace not being able to give a naturalistic account of the development of the socialist morals he hoped to see realized, at least in his 1864 paper, he articulated exactly the means by which they might come about. This was in contrast to the majority of his contemporaries. Even though the number of men of science who embraced an evolutionary account of human morphology had increased significantly since the first publication of *Origin*, the capacity for moral sentiments had remained a point presumed to be beyond the explanatory power of natural selection. Not only were human mind and morals traditionally thought of as products of the soul and thus more a fit subject for theologians than for naturalists,[7] but ethical regard for the welfare of others seemed something that would be difficult to account for as having arisen from natural selection, at least as Darwin had described it in *Origin*. Although Darwin did acknowledge that selection worked upon communities of social insects, the overriding emphasis in *Origin* had been on the competition between individual organisms in a merciless struggle for existence. It seemed reasonable to doubt that a compassionate regard for others could possibly result from a process that appeared only to reward self-interest. In a struggle for life in which even the smallest advantage could make the difference between life and death, surely, on the average, the selfish would prevail and the selfless would be driven to extinction—this much seemed evident for the way in which Darwin had applied Malthus. The positivist philosopher Auguste Comte had coined the word "altruism" in the 1850s to signify the "other-regarding" actions that soon became the center of this debate.[8]

Thus, while Darwin had some reservations about what Wallace had had to say on man—Darwin thought that the morphological differences between the races were the result of sexual selection rather than adaptation to climate and correlated to growth, as Wallace had suggested—he remained full of enthusiasm for Wallace's account of the evolution of mind and morals, and it was this that made Wallace's subsequent defection such a disappointment. Wallace, after all, had been among the most ardent defenders of natural selection; he had seemed a lot closer to Darwin's own position than even Huxley. However, Wallace had become increasingly interested in spiritualism following his return to England in 1861, and he attended a number of séances as a result. Although initially he did so with a critical intent, by 1865 he had become convinced that there was a spiritual dimension to the world. He now claimed that "the moral and higher intellectual life of man" was "difficult to conceive as originating by any law of evolution,"[9] and that "we must therefore admit the possibility that in the development of the human race, a Higher Intelligence has guided the same laws for nobler ends."[10]

As I have mentioned above, by this point Darwin was already well into writing the book that was to become *Descent of Man, and Selection in Relation to Sex*. As he had outlined in his introduction, his main intention had been to correct Wallace's 1864 argument about the origins of man and his evolutionary history by arguing the significance of sexual selection in human evolution, thus establishing that it was a force that operated throughout the natural world. His aim, he wrote, was to shed light upon three questions: "firstly, whether man, like every other species, is descended from some pre-existing form; secondly, the manner of his development; and thirdly, the value of the differences between the so-called races of man."[11] The historians Adrian Desmond and James Moore have drawn attention to this passage to argue that Darwin was motivated in his study of evolution to demonstrate the common ancestry of the different races of mankind.[12] While the claim that Darwin was driven to attempt to solve the problem of the origin of new species as a contribution to the antislavery campaign is an overstatement, it is indisputable that Darwin was passionate about the unity of man and that he wrote *Descent* with the intention of explaining the role of sexual selection in the evolution of the human races. However, as soon as Wallace published his turn to the supernatural, Darwin changed tack to reassert the conclusions that Wallace had originally drawn and to focus on substantiating a naturalistic account of the evolution of mind and morals. As Robert J. Richards has demonstrated, Darwin had already written the first two chapters of *Descent* by the spring of 1869 and there is little to suggest that he planned to give an extensive treatment to the evolution of mind or morality before Wallace's change of heart, even though he had filled notebooks with speculation on the subject decades earlier.[13] Indeed, in the opening chapters, Darwin had recapitulated much that was old news, laying the groundwork for sexual selection with a summary of evidence for human evolution from comparative anatomy and embryology. As Darwin himself pointed out, by this time, "the conclusion that man is the co-descendent with other species of some ancient, lower, and extinct form, is not in any degree new."[14] In outlining the evolution of language and intellect, he was merely expanding on Wallace's 1864 argument. It was in April 1869 that Darwin first read the review of Lyell's work in which Wallace had retracted much of what he had earlier believed, and as Richards notes, this was only the beginning of a deluge of criticism of the efficacy of natural selection. The following year, Wallace sent Darwin his *Contributions to the Theory of Natural Selection*, which contained the essay "Limits of Natural Selection as Applied to Man" as well as a new version of his 1864 paper, revised to reflect Wallace's newfound recourse to supernatural explanation.

By January 1870, Darwin had also read the first two of Walter Bagehot's essays on "Physics and Politics." Bagehot, a liberal essayist, economist, and banker, was also the editor of the *Economist*. In this series of what would grow to be five essays by January 1872, Bagehot discussed the historical and evolutionary development of the social organization of animal and human groups. Darwin was impressed with what he had had to say on both the social and the evolutionary function that deference of the individual to the group would play. Bagehot argued that the suppression of self-interest would ultimately strengthen one group in relation to others that were less coherent, and that as a result the more socially coherent groups would win out in competition with others that were more individualistic. Darwin agreed with Bagehot's suggestion on this point that over time this process could quite readily account for the development of the ethic of self-sacrifice and selflessness.[15] Further to his consideration of Bagehot's views, though, Darwin had also given serious thought to William Rathbone Greg's 1868 article for *Fraser's Magazine*, "The Failure of Natural Selection in the Case of Man." Significantly, Greg's paper was an attack on the 1864 version of Wallace's *Anthropological Review* article, which Darwin had so admired. Greg, a mill owner and an outspoken advocate of free trade, had argued that in arriving at such socialistic conclusions Wallace had not considered all of the relevant factors. Certainly, natural selection might have favored the most-cohesive and cooperative societies, but any society that preserved its unfit members at the expense of those who were more useful would actually weaken and not strengthen its position relative to other societies.[16] This was an objection that was echoed by Darwin's own cousin, Francis Galton, who had argued that the evolution of other-regarding ethics would be unsustainable.[17] Darwin found consolation in the installments of Spencer's revised and expanded *Principles of Psychology*, which he had begun receiving beginning in December 1869.At least Spencer had something positive to say on the evolution of other-regarding feelings.

Richards is surely right to say that whatever Darwin's initial intentions for *Descent*, after 1869 he could hardly avoid engaging with mind and morals.[18] Already two chapters in, Darwin turned immediately in chapter 3 to include an account of the evolution of the moral sense, going on to defend much of what Wallace had had to say about the evolution of mind and morals in 1864.[19] Not only as a response to Wallace's newfound supernaturalism, but ultimately also in response to Greg and Galton, Darwin marshaled the argument that mind and morals had developed from man's social condition. Richards has given the best account of the main thrust of Darwin's explanation of the evolution of the moral sentiments, but I differ from him in a few important respects. My main argument here is that Darwin not only utilized sexual selection to account for the evolution of morphological differences between the races, as he had originally intended, but also to account for the evolution and persistence of other-regarding morals, about which Greg and Galton had raised doubts. In the process, he was determined to demonstrate that natural selection could account for the evolution of genuinely other-regarding "altruistic" moral sentiments—that evolution did not endorse the moral standards of "every cheating Tradesman." He was reluctant to allow that "the most noble part of our nature," our moral conscience, had its origin in what he called "the base principle of selfishness."[20]

Darwin's interest in this question has been noted by a number of scholars, among whom Richards, Elliott Sober, and, most recently, E. O. Wilson, are arguably the most prominent.[21] While Wallace had argued that sociality would prompt the development of mutual sympathy, this was not something that Darwin had not already given extensive consideration. From 1837 his notebooks were filled with reflections on exactly by what means truly other-regarding sentiments might have arisen. Richards and Edward Manier, both of whom are historians, have emphasized the influence that Sir James Mackintosh and Harriet Martineau had upon Darwin in this regard.[22] Here, though, although I acknowledge that both Mackintosh and Martineau were certainly important, I argue that other thinkers were equally so. I emphasize Adam Smith and John Stuart Mill in particular, but also David Hume. Darwin studied works by each of them in the years following his return to England from the *Beagle* voyage, and even though he differed from each of them, they each contributed significantly to his thinking on the evolution and development of other-regarding sentiments. In addition to considering the role these authors played in Darwin's theory of moral evolution, I also engage with some of the contemporary commentators that Darwin read and responded to—not only Greg and Galton, whom Darwin engaged with as he wrote *Descent*, but also two notable reviewers, the ardent feminist Frances Power Cobbe, with whom Darwin had discussed the role of heredity in human development as he was in the process of writing the book, and the liberal statesman and viscount, and then editor of the *Fortnightly Review*, John Morley. Despite her admiration for *Origin* and of Darwin, Cobbe felt that in *Descent* he had pushed things too far. She thought that his attempt to ground the highest attributes of mankind in material explanation was beyond the pale. When they had discussed the moral sentiments in 1868, she had recommended that he read Kant, only to see him turn the categorical imperative and even the Golden Rule into but contingent outcomes of pragmatism and instinct. She shuddered at the thought and attacked Darwin in an article in the *Theological Review* entitled "Darwinism in Morals." Morley's review, which appeared in the *Pall Mall Gazette*, was more favorable. The review had been published anonymously, and it was only later that Morley revealed his identity to Darwin. In his review, Morley, who had been caricatured as "Mill's representative on earth" by contemporaries, had suggested that Darwin need not have attempted to distance himself from utilitarian philosophy to the extent that he had, and that in fact he might find much in the work of John Stuart Mill that would resonate with his argument in *Descent*. The review led the two men to correspond, and in light of Morley's comments, Darwin revisited Mill's work, even enlisting his son William to the task.[23] The juxtaposition of these two reviews by Cobbe and Morley, Darwin later wryly observed, "affords an amusing contrast."[24]

Darwin read widely on contemporary understandings of the origin of morals, noting the significance of the work of the Scottish philosopher and psychologist Alexander Bain in particular, but the subject was one he had long pondered. As Darwin noted in *Descent*, Bain's *Mental and Moral Science* (1868) summarized the latest work in the field of some twenty-six British authors. He observed too that the physiologist Sir Benjamin Brodie had recognized the relevance of human sociability in his discussion of the origin of morals, and it is clear that he had also considered what John Stuart Mill had to say on the matter. Darwin read Mill's *On Liberty* as he wrote *Origin*, which he thought "very good," and *Utilitarianism* shortly thereafter when it was serially published in 1861.[25] However, Darwin's notebooks reveal that he had first begun to give the moral sentiments serious consideration some three decades earlier, when he first arrived back from the *Beagle* voyage. At about the same time that he had read Malthus, Darwin also read Mackintosh's *Dissertation on the Progress of Ethical Philosophy* (1836); Martineau's *How to Observe Morals and Manners* (1838); Dugald Stewart's *Account of the Life and Works of Adam Smith* (1793); Smith's *Theory of Moral Sentiments* (1759), and David Hume's *Enquiry Concerning Human Understanding* (1772).[26]

Mackintosh was Darwin's distant relative by marriage, being the late brother-in-law of his Uncle Jos. Before his death in 1832, Mackintosh had been an occasional visitor to the Wedgwood household at Maer, and Darwin recalled meeting him there in 1827. Mackintosh had also been professor of law and politics at Haileybury at the same time that Malthus taught political economy there, and Darwin recalled that he talked extensively about law, political economy, and morals. In his dissertation, too, Mackintosh argued that the moral faculties were both innate and instinctive, rather than learned.[27] Darwin had rated Mackintosh "the best converser I ever listened to," and also gave his dissertation considerable attention, writing up a 2,500-word essay "On the Moral Sense" in May 1839 in response to what he read.[28] He opened this essay with the statement that he was "looking at man, as a naturalist would at any other Mammiferous animal," an indication of just how far and how early Darwin had sought out a fully material account of mankind.[29] As Edward Manier has pointed out, from the start of this short essay Darwin emphasized the importance of "parental, conjugal and social instincts," which is perhaps unsurprising given Darwin's familiarity with his grandfather's views on the subject.[30] The importance of sex was thus in Darwin's mind from the beginning of his consideration of the origin and development of the moral faculties, a point he would later develop in relation to sexual selection in *Descent.*

Richards has pointed out that Mackintosh's intention was to formulate an account of the moral sentiments that was not tied to the self-regarding utilitarian emphasis upon pleasure and pain as the motivation for moral action. Mackintosh recognized that this had been central to the account of the moral sentiments of the utilitarian natural theologian William Paley, as well as that of Jeremy Bentham. But Mackintosh had argued that the moral sentiments were innate rather than learned through their association with pleasure or pain, as most utilitarian philosophers believed to be the case. Men instinctively approved of virtuous acts, sought the welfare of their children, and despised cowardice, he argued. Significantly, and as Richards again points out, Mackintosh took this position not because he was unsympathetic to utilitarian philosophy—he was—but rather because he sought to point out that the instinctive motive for moral action was distinct from the learned utilitarian criterion by which it was judged. However, lacking any other explanation of this coincidence of the motive for moral action with prevailing moral standards, he could only appeal to divine action to account for how this coincidence had come about.[31] In his own essay on the subject, and as he later worked out in *Descent of Man,* Darwin took on the task of providing a naturalistic explanation of this phenomenon. Further, he would also move beyond Mackintosh's account to explain both the origin of other-regarding moral sentiments as well as how they might persist in light of the fact that natural selection appeared to promote self-interest.

Darwin was also deeply impressed by Harriet Martineau. He had read a lot of her works, and he talked with her frequently upon his return to

England. He thought her *How to Observe Morals and Manners* particularly important.[32] Martineau had traveled widely in researching her book, documenting her firsthand experiences of the different standards of morality held by different cultures as well as historical accounts of the moral norms of the cultures she encountered.[33] Darwin was clearly already thinking about how to explain the origin and development of other-regarding moral sentiments, and it was with this in mind that he read both Mackintosh and Martineau. He immediately saw the significance of Martineau's account of the widely differing ethical standards that were to be found across different cultures.[34] Even though most cultures shared a number of the same taboos, it was clear that there was no one universal standard of morality—Divine or otherwise. Darwin had noticed as much himself regarding the moral standards of the Fuegians compared to the other native peoples he had encountered on the *Beagle* voyage, of course. But Martineau helped him to generalize his views and gave him pause in thinking through what Mackintosh had said about the moral sense being instinctive.

Manier has made the case that Darwin was also influenced by the Scottish philosopher and mathematician Dugald Stewart, and Darwin certainly looked to Dugald Stewart's *Account of the Life and Writings of Adam Smith* as well as to Adam Smith's *Theory of Moral Sentiments* as he attempted to theorize his own account of the origin and development of morals.[35] Darwin's notebooks reveal that he also turned to David Hume's *Enquiry Concerning Human Understanding* to shed more light on Mackintosh's contention that the moral sense was innate and not learned. Significantly, Hume had suggested that what we call reason in man is akin to instinct in animals. As I shall go on to argue in this chapter, John Stuart Mill was also of greater importance for the development of Darwin's thought than historians have hitherto acknowledged.

Even in the nineteenth century, Adam Smith was still the starting point for consideration of the origin of the moral sentiments, added to which his political economy defined the age. In both his *Theory of Moral Sentiments* and in *Wealth of Nations*, the latter of which Dugald Stewart described in detail, Smith had characterized human morals as being grounded in reason and "sympathy." According to Smith, sympathy was essentially a "self-regarding" sentiment, based upon one man's recognition of his fellows as similarly rational beings who experienced like pleasures and pains to himself. It was this presumption of utilitarian self-regard that had become the mainstay of classical liberal political and moral economy, and it was this that informed both Paley's and Bentham's approach to the question of morals. It was this position that Mackintosh had attempted to move away from, but Darwin believed he could do better. Indeed, as Manier has noted, Darwin left his judgment of Mackintosh's attempt to account for human moral sentiments in the margins of his copy of the *Dissertation*: he thought Mackintosh's decision to fall back on divine explanation was "Trash!"[36]

Like Mackintosh, while Darwin acknowledged that self-regarding sympathy and the rational calculations of utilitarian self-interest had certainly been important in the development of human morals, he was convinced that they were not their origin: the human moral sentiments were not reducible to what he described as the "low motive of self-interest," however enlightened that self-interest might be.[37] Rather, in line with Brodie's observation, Darwin set out to show that the human moral sentiments had their origin in our sociability, and, following Hume as much as Mackintosh, he would argue that they were the product of animal instinct, not reason.[38] While Darwin could not deny that there was a significant gap between the highest animal and even the lowest type of human being, he was adamant that the difference was ultimately "one of degree and not of kind."[39]

Before jumping into the detail of Darwin's argument though, we need to appreciate the deeply political context in which he wrote and, indeed, Darwin's own political commitments to the outcomes of the question he was addressing. In chapter 1, I have made the case that Darwin feared the political associations of transmutation as much as the theological unorthodoxy. Prior to *Origin*, transmutation was viewed by the Anglican-dominated scientific community as an uncivilized and radical affair with distinctly revolutionary associations. Darwin's acknowledged embrace of Malthus had certainly been successful in distancing both himself and his theory from the revolutionary elements of the radical movement, but his emphasis upon individual competition had been readily embraced by those who took his ideas to the opposite extreme. The Manchester political economists held him up as having endorsed their own view that laissez-faire and self-interest were king and that might made right.

This was not Darwin's view at all. Certainly, he was an advocate of free trade, but he also recognized the importance of charity, compassion, and—now that the government was coming under the sway of the middle classes—of a limited amount of state regulation. As I have pointed out in chapter 1, both sides of Darwin's family, both the Wedgwoods and the Darwins, were interested and active in politics. Darwin had more than one family member in Parliament: Mackintosh had served as the member for Knaresborough until his death in 1832, and his Uncle Jos had served for Stoke-upon-Trent from 1832 to 1835. Darwin's wife Emma was also enamored by reform; she admired Henry Brougham's politics, and, like her brother-in-law Erasmus, she was also well acquainted with Harriet Martineau. Indeed, and despite Darwin's later claim not to have had much time for politics, both his son William and grandson Francis had different recollections. William recalled of his father that "he was an ardent Liberal and had a very great admiration for John Stuart Mill and Mr. Gladstone"; Francis remembered him as someone whose "interest in politics was considerable."[40] Thus, although Darwin may not have set out to write about the evolution of morals in *Descent*, when he did engage with the subject he would have recognized that it was a deeply political matter, and it is clear from what he wrote that he was determined to portray the evolution of humankind as dependent upon the gradual evolution of the liberal civil society that he and his family hoped to see as the outcome of reform. This was different morality than had presided in *Origin*. In his earlier work, although he had acknowledged the collectivism of the social insects, Darwin had emphasized the individual nature of the struggle for existence. In *Descent*, he gave pride of place to the development of liberal collectivist morals in man. As Thomas Dixon has pointed out, this development of Darwin's views in natural history echoed that which Mill had expressed in political economy. In *On Liberty*, Mill had mounted a defense of liberal individualism as the basis of civil society; in *Utilitarianism*, he seemed more concerned with the establishment of the conditions that would foster collectivism. Indeed, when it came to describing the evolution of human society, Darwin clearly saw himself as engaged in the same questions that had occupied Mill, and although he eventually concluded that there were significant differences between Mill's views and his own, he studied Mill's work in detail as he worked through his ideas on man and morals. The differences as well as the similarities between Darwin and Mill are relevant.

Back in 1860, when Darwin was waiting to see what response he would get to *Origin*, he had deeply appreciated the quite unexpected support he had received from John Stuart Mill. Henry Fawcett, the economist and statesman who had been tragically blinded in a shooting accident in 1858, had written a favorable review of *Origin* for *Macmillan's Magazine*.[41] Writing to Darwin in July 1861, he explained that, quite contrary to the opinion of many of Darwin's detractors, he had been "particularly anxious to point out that the Method of Investigation pursued was in every respect, philosophically correct." Further, and what clearly delighted Darwin the most about this communication, was the anecdote that Fawcett then went on to relate: "I was spending an evening last week with Mr. John Stuart Mill and I am sure you will be pleased to hear from such an authority that he considers your reasoning throughout is in the most exact accordance with the strict principles of Logic. He also says, the

Method of investigation you have followed is the only one proper to such a subject."[42]

Darwin was elated to have the support of someone of Mill's stature, for as well as an accomplished political philosopher, Mill was an acknowledged authority in the philosophy of science. Darwin wrote as much to Asa Gray, saying, "We in England think John Stuart Mill the highest authority on such subjects."[43] To Fawcett, Darwin replied: "You could not possibly have told me anything which would have given me more satisfaction than what you say about Mr. Mill's opinion. Until your review appeared I began to think that perhaps I did not understand at all how to reason scientifically."[44]

Mill's reputation as a philosopher of science was one thing, but when it came to Darwin's thinking through the evolution of the human moral sentiments, it was Mill's politics that were of greater interest to him. Indeed, and what doubtless made Mill's particular political project significant, was the fact that he had been deeply engaged in a project akin to that which Darwin had now taken on. When he came to write up *Descent*, it is true that Darwin noted that he ultimately differed from Mill on the importance of heredity, but it is also clear that he gave Mill's views serious consideration as he was thinking through the origin and development of morals, and that despite their differences, there are telling similarities that merit further consideration.[45] Like Darwin, Mill was determined to see beyond man as a self-interested rational actor. Both men were clearly concerned to find a way in which they might account for the development of truly "other-regarding" actions, the essence of what Mill called "virtue." Mill worked toward this goal from political economy, Darwin from natural history. Indeed, in introducing the topic in *Descent*, Darwin acknowledged that the origin and nature of the moral sentiments had been previously considered by some of the greatest moral philosophers of the age, but excused his intervention among such company on account of the novelty of his perspective. "This great question has been discussed by many writers of consummate ability; and my sole excuse for touching on it, is the impossibility of here passing it over; and because as far as I know, no one has approached it exclusively from the side of natural history," he wrote.[46] Mill had made exactly the sort of move from individualism to a more communitarian philosophy that Darwin sought to make in *Descent*; indeed, scholars have noted the striking parallels between *On Liberty* and *Origin*, and between *Utilitarianism* and *Descent* in this respect.[47]

Darwin had read *On Liberty* in 1859 as he was writing *Origin* and had considered it "very good," and he now eagerly turned to *Utilitarianism* as he wrote *Descent*.[48] Mill had argued that although man came into society as an ethical egoist, it was through the interactions and associations made possible by civil society that man had become sociable. Each had come to recognize his fellow men as similar beings to himself, subject to reason and to similar pleasures and pains. As the historian and Mill's biographer Bernard Semmel has argued, despite his familiarity with the utilitarian associationism that tied all apprehension of the world back to self-regarding sensations of pleasure and pain, Mill was ultimately dissatisfied with the utilitarian logic of all human motivations being reduced to rational calculation. Instead, he sought to theorize a move from enlightened self-interest to an account of the development of the "right morality" of other-regarding "virtue."[49] In both *On Liberty*, and especially in *Utilitarianism*, it is clear that he did so in the context of a liberal communitarianism, and in the process significantly expanded upon traditional associationist assumptions.[50] "It would be a great misunderstanding of this doctrine" of utility, he wrote in *On Liberty*, "to suppose that it is one of selfish indifference, which pretends that human beings have no business with each other's conduct in life, and that they should not concern themselves about the well-doing or well-being of one another, unless their own interest is involved."[51] Influenced by the French political theorist and historian Alexis de Tocqueville,[52] Mill had come to believe that although man was born an ethical egoist, he had, through the long history of the development of politics and society, come to live not in a Hobbesian state of nature, but rather in a civil society in which minimal rights—or, more properly, minimal standards of utility—were legislated in order to protect each citizen from the self-assertion of his neighbors. Further, Mill suggested that as a result of the pursuit of utility through social interaction, it had become the case that man's quest for individual happiness had not only come to coincide with the attainment of the greatest general happiness, but had become synonymous with it. Ultimately, Mill believed that human morals would continue to develop in this way to a point at which there would no longer be any tension between the wants of the individual and the good of society. Mill imagined a time in which it could not only be said of each individual that he was "unable to conceive of the possibility of happiness to himself, consistently with conduct opposed to the general good, but also that a direct impulse to promote the general good may be in every individual one of the habitual motives of action."[53] The doctrine of utility—according to which the measure of the morality of an action was its tendency to bring about the best outcomes for the community rather than the best outcomes for the individual—became central to contemporary debate about the direction and development of liberal politics.[54]

What excited Darwin was not only the fact that Mill clearly shared his own dissatisfaction with the attempt to ground human moral action in self-interest, but that he too did seem to recognize that there was a "natural basis of sentiment for utilitarian morality" and, more explicitly, that morality was grounded in a "powerful natural sentiment." Thus, despite the fact that Mill came at the question from the point of view of political economy, he did seem to have some cognizance of the importance of the natural-historical view—at least this was Darwin's initial impression.[55]

Mill had had three ends in mind when he started on the reformulation of utility. First, he was eager to move away from the mechanism and determinism that had characterized utilitarian political economy up to this point—as a humanist, Mill was concerned that the rise of the law-bound Newtonian universe only substituted for the theistic doctrine of predestination that had been humanism's traditional foe. Second, Mill sought to theorize a shift from the mere calculation of happiness to the recognition and pursuit of virtue. It was not a satisfactory system of ethics that sought only pleasure, he concluded, and in contrast to Bentham's famous statement on the subject, he now believed that it really did make a difference if a man preferred poetry to pushpin.[56] Third, Mill sought to take into account "the whole of human nature, not the ratiocinative faculty only." He believed "feeling" to be "at least as valuable as thought."[57] Despite being famed for his rational empiricism, Mill had come to embrace the idealism of the German transcendental philosophers as he moved further and further from a belief in the all-encompassing reasoned calculation that was central to Benthamite morality. Nevertheless, despite this determination to move away from an exclusive emphasis upon rational calculation, and however much he talked of "feelings" and "natural sentiments," Mill remained skeptical of the tendency to ascribe human action to instinct. He ardently believed that whatever was born into man through heredity could be countered by education and reason. Indeed, this was central to both his conception of what constituted moral action as well as to his belief in the possibilities of a liberal social democracy for social justice and racial and sexual equality. Across *On Liberty* and *Utilitarianism* Mill had outlined how he hoped to see the transition from self-regarding to other-regarding motives effected. Highlighting the importance of free will and agency (and, incidentally, in contrast to the older radical tradition of William Godwin), Mill was clear that people played an active role in the formation of their own character, but that they did so not as isolated individuals but in society with others. In order for this to be effective, Mill believed that it was imperative that society be governed by liberal principles of free and equal participation, and it was to this end that he argued that the prerequisites of utilitarian political economy were what have subsequently become traditional liberal freedoms: the freedom of conscience (in which he included freedom of speech and freedom of the press) and the freedom of action (which included freedom of association).

Mill was concerned that attempts to explain human nature with reference to innate characters such as instinct were in opposition to his hopes of social progress. Indeed, many of his contemporaries marshaled "nature" as grounds for their belief in persistent and immutable differences, not only between men, but between men and women and between the various races of mankind as well.[58] Further, he was adamant that self-interest was no more inherent to man than any of these other presumed natural characters. "Little is there an inherent necessity that any human being should be a selfish egoist, devoid of every feeling or care but those which centre his own miserable individuality," he contended. "Genuine private affections and a sincere interest in the public good, are possible . . . to every rightly brought up human."[59] Given his faith in the plasticity of human nature, Mill argued that the laws and institutions of society should be established in such a way as to mold the preferences of the individual so as to "place the happiness . . . or the interest . . . of the individual, as nearly as possible in harmony with the interest of the whole; and secondly, that education and opinion, which have so vast a power over human character, should so use that power as to establish in the mind of every individual an indissoluble association between his own happiness and the good of the whole."[60] The correct legislation, he argued, was that of liberal utilitarianism; only thus, through right government, might the ethical egoist be transformed into the virtuous citizen.

Of course, what Darwin found both tantalizing and frustrating in equal measure were the number of passages in Mill's writings that so very nearly either appeared to resonate with what he had already said in *Origin* or anticipated his own thinking as he prepared *Descent*, only finally to fall short of any clear statement about heredity. Darwin read Mill's *The Subjection of Women* as he was working through his material on sexual selection and was frustrated, if not entirely surprised, by the clearly anti-hereditarian stance that Mill had included there. Since the publication of *Origin*, all around him Mill had seen people eager to explain their own preferred politics as grounded in the natural instincts of mankind. As often as not, evolutionary hierarchies had been invoked to naturalize what Mill perceived to be only social inequalities and to justify prejudice and slavery. He had been an ardent supporter of the North from the very start of the American Civil War, a war he believed to be entirely about slavery—a struggle between "free and slave holding America." At stake were "the most important consequences to humanity, stretching into the remotest future," and he was deeply angered by those who claimed to be liberals who could yet support the South.[61] Mill's conclusions in *Subjection of*

Women showed the extent to which he had become jaded by those who attempted to justify slavery and injustice with reference to biology. "It is one of the characteristic prejudices of the reaction of the nineteenth century against the eighteenth," he lamented, "to accord to the unreasoning elements in human nature the infallibility which the eighteenth century is supposed to have ascribed to the reasoning elements. For the apotheosis of Reason we have substituted that of Instinct." This "idolatry" of instinct he found "infinitely more degrading."[62]

Darwin had occasion to discuss Mill's work—both his utilitarian theory of the development of moral sentiments and his work on what had become known as the "woman question"—with the ardent suffragist, reform campaigner, and anti-vivisectionist, Frances Power Cobbe. Cobbe, who had recently written a review of Mill's book, would later be quick to take up her pen in response to what she perceived to be the moral vacuity of Darwin's theory of human morals, but for a brief moment in 1869 they became almost confederates.

Cobbe first met Charles and Emma Darwin in 1868 while they visited Elizabeth Wedgwood in London, and always the keen socialite, Cobbe took advantage of their coincidentally shared choice of summer vacation in the small Welsh village of Caerdeon in 1869 to further pursue their acquaintance.[63] The Darwins stayed at Plas Caerdeon, "a beautiful house with a terraced garden," as Darwin recorded in his diary, which overlooked the Mawddach Estuary. The place was quiet, remote, and geologically distinct, and Darwin, who was enduring another bout of illness, had gone there to rest, to recuperate, and to write.[64] Emma enjoyed Cobbe's company, finding her "fresh and natural," writing of the encounter that "Miss Cobbe was very agreeable."[65] However, reading between the lines of even Cobbe's own account of things, it seems that Darwin found her ebullient personality somewhat intrusive.[66] Cobbe, like Darwin, was passionate about animal welfare and, both confident and well-read, she had a forceful personality and strong opinions.[67] As a result, both Charles and Emma found her conversation stimulating, if, in Darwin's case, perhaps a little overly so. Cobbe later reported her conversation with Darwin about Mill's new book to one of her friends. "I am glad you like Mill's book," she had written. "Mr. Charles Darwin, with whom I am enchanted, is greatly excited about it, but says that Mill could learn a thing or two from Physical Science." The conversation had evidently moved more explicitly to Darwin's work, and to the origin and nature of morals in particular, a subject that interested them both. As Cobbe continued in her letter, "He intensely agrees with what I say in my review of Mill about *inherited* qualities being more important than education, on which Mill alone insists."[68]

The source of the moral sense was a subject that was not unfamiliar to Cobbe, who had written at length on the subject herself. She too was disillusioned with the utilitarian associationist explanation of morality. In her 1855 work, *An Essay on Intuitive Morals, Being an Attempt to Popularise Ethical Science*, she had sought to bring Kantian arguments to a popular audience to counter the utilitarian associationist ideas of the English school. "We want a system that shall not degrade the Law of Eternal Right by announcing it as a mere contrivance for the production of human Happiness, or by tracing our knowledge of it to the experience of the senses, or by cajoling us into obeying it as a matter of Expediency," she had written.[69] Despite her enthusiasm at having conversed with the author of the *Origin of Species*, the religious commitments that underpinned Cobbe's critique of associationist philosophy would later color her reading of *Descent* as well.

Cobbe had prevailed upon Darwin to read Kant, and with some success, as he later read the copy of *Metaphysic of Ethics* that she sent him, despite his protests that she need not trouble herself to do so.[70] Darwin read there only a confirmation that his own views on the moral sense were far at odds with those of the German idealist, but he did see that he had to explain the development of a sense of "duty" and of the Kantian ideal that no man should treat another as merely a means to an end, as having arisen through natural selection. Wrapped up in the workings of mind, Darwin found himself reflecting upon the nature of the difference between his own thoughts on the matter and those expressed by the eminent German philosopher. He wrote to Cobbe, "It has interested me much to see how differently two men may look at the same points, . . . the one man a great philosopher looking exclusively into his own mind, the other a degraded wretch looking from the outside thro' apes & savages at the moral sense of mankind."[71] As he mulled over the insights that evolutionary history demanded of any account of morality, Darwin confessed to Cobbe that he had it in mind to "introduce some new view of the nature of the Moral Sense." In light of her views on the importance of heredity, Darwin was doubtless testing the water for the case "from the side of natural history."[72] In any case, he made sure that Cobbe received an advance copy of *Descent*—she had written asking if she might have one in order to write a review for the *Theological Review*. Darwin had instructed his publisher to send her one even before any other copies had been sent out for review, so it seems that he was at least keen for her to express her opinion, and he may yet have hoped to win her over.[73] If Darwin had hoped to persuade her with a full explication of his argument, however, he was sorely mistaken. In fact, the book had almost exactly the opposite effect upon her. *Descent*, she later confessed, "inspired me with the deadliest alarm," and she had set to work immediately to write her objections in an article for the *Theological Review,* choosing "Darwinism in Morals" for her title.[74]

Before turning to Cobbe's objections, however, it behooves us to look in further detail at what Darwin actually said in *Descent.* Darwin was frustrated with those who had grounded self-interest in associationist philosophy and who now looked to utilitarianism to do the same. "Philosophers of the derivative school of morals formerly assumed that the foundation of morality lay in a form of selfishness; but more recently in the Greatest Happiness principle," he wrote. However, he found neither position tenable in light of his appreciation of man's natural history.[75] Further, despite the similarities between what he and Mill were trying to achieve, when he came to write *Descent* Darwin reluctantly had to conclude that, notwithstanding Mill's best efforts, they disagreed over fundamentals. "It is with hesitation that I venture to differ from so profound a thinker," Darwin confessed in a footnote, but Mill had been clear that he believed that at the end of the day, "the moral feelings are not innate, but acquired."[76] Mill certainly shared Darwin's dissatisfaction with the notion that human morality could be encompassed by rational self-interest, but to Darwin's mind it remained the case that despite Mill's attempt to move beyond self-interest, the fact that his account of morals ostensibly remained grounded in reason not only left significant phenomena unaccounted for, but did little to close the presumed gap between man and the rest of the animal world. In short, Darwin had sympathy for Mill's intentions but not for his argument. Having found both Manchester and now Mill wanting, Darwin returned to the thinkers who had inspired his earliest notes on man and who had served him so well when he was preparing *Origin.* Ironically, they were the same names that had inspired the Manchester political economists he disagreed with.[77]

The social nature of man was key. Wallace might have pointed it out to him anew in his 1864 paper, and Bain had discussed it too, but it had been there in Adam Smith all along—and Darwin had known it. His notebooks are testament to as much.[78] However, Smith seemed to be caught in two minds. A number of scholars have noted that Smith's account of the motivation for sociability—which had ramifications for his conception of human nature—differed quite radically between *Wealth of Nations* and that which he had proposed in his *Theory of Moral Sentiments.* The former, it has generally been assumed, was grounded much more in economic self-interest, whereas in *Moral Sentiments* Smith gave greater emphasis to the social nature of man. To an extent, this is true—and paying attention to the context in which Smith wrote each of these books would doubtless be important in resolving why this might be the case. Nevertheless, as far as we are concerned here at least, it seems that by Darwin's reading the more significant distinction came not between the two volumes but between the account of sympathy that was derived from self-interest and the love that had its origins in sexual passion and familial attachment—a distinction that Smith had drawn in the earlier volume.[79] Darwin had clearly read keenly and thought deeply about Smith's conception of sympathy and love, respectively, and it was in the latter that he saw the basis of truly other-regarding sentiment. Far from a rational sentiment, or even one derived from the immediate pleasure or pain to be derived from experience, Smith argued that love was, "according to some ancient philosophers," one of "the passions which we share in common with the brutes," and far from being a pleasure, had "consequences [that] are often fatal and dreadful."[80] Darwin was to make this distinction central to his own account of the origin of moral sentiments that he would offer in *Descent.*

In the *Wealth of Nations* Smith had suggested that man's sociability might readily be accounted for in terms of self-interest. Self-interest alone would bring men together to cooperate on a given task, because in doing so they would gain more to themselves than those who attempted all their travails independently. Thus, while Smith's message to industrial England was that collective action was infinitely more productive than isolated individualism, the beauty of Smith's argument for liberal political economy was that this was accomplished by nothing more than self-interest. Recall that it was "not from the benevolence of the butcher, the brewer, or the baker that we expect our dinner, but from their regard to their self interest."[81] This was not the end of the story, however, and as Smith had pointed out in his *Theory of Moral Sentiments*, a perhaps unintended consequence of collective action was the mutual sympathy that would result from the association of one man with another. It was through such associations that man had initially come to recognize his fellows as beings alike to himself—capable of feeling like pleasures and pains. Smith's account of sympathy was certainly what was emphasized by contemporary liberal theorists. Citing Smith but quoting the more-recent work on "Mental and Moral Science" by Alexander Bain to make this point, Darwin noted that "Adam Smith formerly argued, as has Mr. Bain recently, that the basis of sympathy lies in our strong retentiveness of former states of pain or pleasure. Hence, 'the sight of another person enduring hunger, cold, fatigue, revives in us some recollection of these states, which are painful even in idea.' We are thus impelled to relieve the sufferings of another, in order that our own painful feelings may be at the same time relieved."[82] Thus, this formulation of sympathy ultimately suggested a self-regarding impetus to moral action. This was truly enlightened classical liberalism: a man could be self-interested with impunity, for out of his actions would spring moral outcomes.

As I have intimated above, though, Darwin found this grounding of the moral sentiments in self-interest wanting, and did so for the same reasons he also found Mill's grounding of morality in reason inadequate. The first was quite straightforward and a matter of conviction more than anything else: Darwin was simply unwilling to see "the most noble part of our nature," namely, human morality, as the outcome of nothing more than "the base principle of selfishness."[83] This was the case regardless of how enlightened that self-interest might be, or how much sympathy might result from it.[84] Also, and at odds with Darwin's fundamental understanding of the origin of man and morals, was the presumption, present in both Smith and Mill, that if morals had derived from reason, then they were necessarily exclusive to humans. If this were the case, then morality might once more be held up as evidence of human exceptionalism.

But Darwin had other objections that were arguably more substantive; there were problems that a satisfactory theory of morals might be expected to answer that an appeal to reason alone could not. How might an account of ethics grounded in reasoned self-interest account for those instantaneous acts that seemed to defy rational calculations of pleasures or pains? And what of those actions that worked to the detriment of the actor for the benefit of someone else: What of the man who rushed to rescue a child from drowning without a thought to his own peril, for instance?[85] Further, what of discernment or discrimination? If ethics had derived from the recognition of other men as reasoning, feeling creatures who felt pleasures and pains akin to oneself, then how might one account for the all-too-evident extension of moral consideration to some men in the face of a total disregard for the sufferings of others? This was something that Darwin recognized from his encounters with the various native populations of Tierra del Fuego and Patagonia on the *Beagle* voyage, as well as from the testimony of others on the native peoples of North America. While a sense of morality and loyalty had arisen among those who enjoyed a close affiliation, even the Fuegians combined in their wars against other tribes, for example—except that too often they displayed abject indifference, if not delight, in the sufferings of men not of their own tribe. A savage who offends against a member of his own tribe is branded "with everlasting infamy," Darwin noted, citing an article in the *North British Review* and those two essays by Walter Bagehot that had recently appeared in the *Fortnightly Review* under the title "Physics and Politics."[86]

As Greta Jones has pointed out in *Social Darwinism and English Thought*,

Bagehot, who had written the influential *The English Constitution* (1867), was interested in the origin, development, and social utility of political organization. While in the *English Constitution* he had given a deep analysis of the institutions of English government, in his essays on "Physics and Politics" (which he expanded into a book of the same name in 1872) he naturalized liberal constitutional government as the most advanced and therefore the "fittest" form of governance that had won out through Darwinian competition.[87] Natural selection had favored those societies that were the most deferential to authority and thus acted with the most cohesion. "Whatever may be said against the principle of 'natural selection' in other departments, there is no doubt of its preponderance in early human history," Bagehot wrote. "The strongest killed out the weak, as they could."[88] History had shown that "the first duty of society is the preservation of society" and thus that, in earlier epochs, the "yoke" and "terrible tyranny" of unquestioning adherence to "customary law" had seen one society thrive while those with less authoritarian regimes fell. However, and in response to Wallace's apparently socialist conclusions in his 1864 article, Bagehot argued that while it was true that natural selection was now not merely selecting biological types per se, it *was* selecting the social customs of one society over another. It was through intertribal conflict that nations had been formed and through war between nations that national character had emerged and social progress had been won. "Conquest is the premium given by nature to those national characters which their national customs have made most fit to win in war, and in many most material respects those winning characters are really the best characters," he wrote. This was not only a descriptive account of human history, but one with a clearly prescriptive message as well: "The characters which do win in war are the characters which we should wish to win in war."[89]

Bagehot was adamant that it was through this competition between social customs and character that social and evolutionary progress would be certain to continue, for "it is only by the competition of customs that bad customs can be eliminated and good customs multiplied." But he also argued that it was not authoritarianism alone that bred success; rather, a society's ability to adapt to new circumstances, new challenges, and new threats was also important. Instead, Bagehot believed that the kind of society that promoted independence of thought, rational discussion, and innovation would be best fitted to survive. Without the customary deference to authority that was bred by authoritarian rule, however, society was at risk of falling apart, and this is where Bagehot saw social morality playing an important role.[90] Bagehot thus explained why primitive tribes might surrender their own lives for the good of their fellows and yet have total disregard for outsiders. And this coincided with Darwin's own thinking. "A North-American Indian is well pleased with himself, and is honoured by others, when he scalps a man of another tribe; and a Dyak cuts off the head of an unoffending person and dries it as a trophy," Darwin noted. In fact, he continued, "most savages are utterly indifferent to the sufferings of strangers, or even delight in witnessing them."[91] This was something else that Darwin had discussed at length with Cobbe as being a real weakness with Mill's emphasis upon a rational account of moral consideration. "I cannot see how this view explains the fact that sympathy is excited in an immeasurably stronger degree by a beloved than by an indifferent person," he wrote.[92] Surely, if it had been as Mill and Smith suggested, then a man should recognize all men as subject to like pleasures and pains as himself, and extend equal moral consideration to them all on this basis.

This was not the end of Darwin's consideration of Smith, however, for in the very opening lines of his *Theory of Moral Sentiments* Smith had hinted at something more than the idea that morality was simply acquired in the course of our associations with others. Rather, he seemed to imply that something much deeper than this was the case. "How selfish soever man may be supposed," he wrote. "There are evidently some principles in his nature, which interest him in the fortune of others, and render their happiness necessary to him, though he derives nothing from it except the pleasure of seeing it."[93] Indeed, Smith would go on to develop an alternative source for the origin of moral sentiments, a source that was genuinely other-regarding in its origin: Love. As he argued, love, which found its original residence not in reason but in the social instincts, including the parental and filial affections, was of an entirely different character. Darwin not only appreciated this point but made it central to his own attempt to define and defend a genuinely altruistic account of morals in nature. Summing up his views on the issue, he concluded that rather than being grounded in reason, "the moral sense is fundamentally identical with the social instincts," and this being the case, they were not exclusive to humans at all, and "in the case of the lower animals it would be absurd to speak of these instincts as having been developed from selfishness, or for the happiness of the community. They have, however, certainly been developed for the general good of the community." To clarify his point, he continued, "The term, general good, may be defined as the means by which the greatest possible number of individuals can be reared in full vigour and health, with all their faculties perfect, under the conditions to which they are exposed."[94] He urged that this definition of the "good" was equally applicable in arriving at a right understanding of the development and evaluation of human morality.

While instinct was the focus, it is important to realize that Darwin did not by any means consider reason a dead-end. He fully believed that reason—even reasoned self-interest, although of limited use in explaining the evolutionary origin of genuinely other-regarding morals—might still influence their later development. While Mill rejected what he referred to as "innate differences" as insignificant in the face of the possibilities of education and environment, Darwin gave "inherited instincts" center stage. Indeed, both Darwin and Cobbe had agreed that Mill would have to start taking this aspect of human character seriously if his book was to be anything but wide of the mark. Again, Smith seemed to have more to say that resonated with Darwin's thoughts on the matter. In *Descent*, Darwin confessed that he, like Mill (and like Smith in *Wealth of Nations*), had initially doubted that the mental differences between one man and another could really be all that significant, but in light of Francis Galton's work he was now convinced of this point, and he now saw an evolutionary significance to the point that Smith had made in *Moral Sentiments* that the "original passions of human nature" were universally felt by all men, although in varying degrees, while "the virtuous and humane . . . may feel [virtue] with the most exquisite sensibility. The greatest ruffian, the most hardened violator of the laws of society, is not altogether without it."[95]

Darwin must have looked at Smith from a new perspective in light of his reading not only of Galton but of the argument that Wallace had made in his 1864 paper. There, Wallace had emphasized collective action as the means by which humans had subverted the Malthusian individualism that Darwin had made central in *Origin*.[96] Further, and quite possibly in light of taking note of Malthus's concession to Godwin that humans were indeed capable of lifting themselves out of the struggle for existence through moral restraint, Wallace had appealed to the social and moral nature of humanity as something that distinguished them from the rest of the animal kingdom. "'Natural selection' acts so powerfully upon animals" largely as a result of "their self-dependence and individual isolation," he argued. "There is, as a general rule, no mutual assistance between animals, which enables them to tide over a period of sickness. Neither is there any division of labour. . . . But in man, as we now behold him, this is different. He is social and sympathetic. In the rudest tribes the sick are assisted at least with food. . . . Some division of labour takes place. . . . The action of natural selection is therefore checked."[97] Wallace had read Malthus in 1844, but while he was willing to accept that individualism and self-interest might reign in the natural world, he would not allow that this was the case among men. Humans were different: they collaborated and combined, they were moral and social, and—given the right education and environment—they were socialist.

Like Darwin, Wallace refused to see the collaborative nature of man as stemming from self-interest. Rather—and again like Darwin—he believed it had arisen from an innate social tendency. It was this tendency, Wallace had argued, that, being variable, had been subject to natural selection and thus been the origin of human morals. "In proportion [that man was sociable] . . . mental and moral qualities will have an increasing influence on the well-being of the race," he had argued.[98] As I have noted earlier, Darwin seemingly read Wallace's essay almost as a point of revelation, but he could also see its limitations—and through them, some of the limitations of his own perhaps overeager readiness to jettison all of the Godwinian ideas about morality from his reading of Malthus in *Origin.* It is quite possible that Darwin was led to reflect that nature was not always quite so "red in tooth and claw" as he had originally hypothesized. Indeed, he had acknowledged that this was the case among the social insects in *Origin,* and so it would clearly also apply to man. Darwin's notebooks are testament to the fact that he believed that mankind had always been a social species, even though, and as we shall see in later chapters, others—including Huxley—thought differently. In Notebook N, and in light of having read Hume, Mackintosh, Malthus, and Lamarck, as early as July 1839 Darwin had recorded that "if this view holds good, then man, a socialist."[99]

Of course, what Darwin meant by the word "socialist" was not what Wallace would later understand by the word, and certainly not what Marx intended by it. It remains intriguing nonetheless to ponder just how radical a term this was in the context of the late 1830s when Darwin confided it to his notebooks. What is clear, though, is that even in light of what he took from Malthus as he developed his theory of natural selection, Darwin had never seen man as an out-and-out individualist. As we shall see in the coming chapters, the debate over what it meant to call oneself a socialist was no less contested throughout this period than what it meant to call oneself an evolutionist. Indeed, and certainly from the 1880s, each of these terms were not infrequently defined with reference to the other. With revolutionary socialism becoming a force to be reckoned with on the Continent, it is notable that by 1871, with the Paris Commune on the horizon, Darwin was willing to go only so far as to say that "most persons admit that man is a social being"—a far cry from his 1839 musing on man's socialist tendencies.[100]

Was this a significant change of perspective on Darwin's part? It was certainly a significant clarification. Historians are agreed that the one overarching thing that Darwin took from reading Malthus was the fact that it was individual variation in a population that mattered and that these variations would be vital in determining which organisms survived the struggle for existence and which perished. Certainly, many of Darwin's contemporaries read him to be endorsing the devil-take-the-hindmost politics of laissez-faire individualism. Although in *Origin* the focus was upon the effect of individual variation on the life-chances of the individual organism, he had discussed at some length the exception to this rule in the case of the social insects. In their case, he recognized that natural selection operated not on the effects of individual variation upon the individuals themselves but upon the society of which they were a part. Wallace, too, invoked a similar appreciation of the importance of the social group for the evolution of genuinely other-regarding behaviors, but where Darwin differed from Wallace was in his recognition that this was by no means an exclusively human attribute. Other-regarding instincts were not only present in ants and humans, but across all social species. Where in *Origin*, in his most-telling exemplar of selection, he had emphasized the importance of the variations of each individual wolf to its success in catching either sheep or deer as prey, in *Descent* he sought to emphasize the importance of cooperation and sociability, making the important, if perhaps somewhat obvious point that "wolves hunt in packs."[101] Making the logical connections across the animal kingdom, he wrote that "it can hardly be disputed that the social feelings are instinctive or innate in the lower animals; and why should they not be so in man?"[102]

It was sociability which was fundamental to the development of genuinely other-regarding morality, Darwin argued. In man's evolutionary history, sociability had preceded reason, and was not the consequence of a rational calculation of individual advantages. Mill, Bain, and all the other political philosophers of the "derivative school" had been wrong in this regard. In the opening of the chapter he dedicated to the evolution of the moral sentiments, Darwin had made it clear that he saw the development of morals as the necessary outcome of a number of conditions. As Darwin stated this quite clearly in the opening pages of his chapter on the evolution of morals, "The following proposition seems to me in a high degree probable—namely, that any animal whatever, endowed with well-marked social instincts, would inevitably acquire a moral sense or conscience, as soon as its intellectual powers had become as developed, or nearly as well developed, as in man."[103]

As he went on to explain, the existence of social instincts would lead animals to associate with others of a like nature, to take pleasure in their company, and through the familiarity and sympathy born of this association to perform various services for them "of a definite and evidently instinctive nature." Second, Darwin continued, "as soon as the mental faculties had become highly developed," each animal would become aware of its actions, of its past, and live in anticipation of its future. If not quite "conscious" of the instinctive drive to fulfill its needs, "images of all past actions and motives would be incessantly passing through the brain of each individual," which would prompt feelings of discontent should those instincts remain unfulfilled. Third, the acquisition and development of language would allow the expression of the wishes of the members of the community, and as a result, "the common opinion how each member ought to act for the public good, would naturally become to a large extent the guide to action." Finally, Darwin suggested, habit and the frequent repetition of such actions would strengthen such moral rules.[104]

Significantly, this account of the evolution of human morals from their sociable nature provided Darwin with a telling response to an old argument that the Duke of Argyll, George Douglas Campbell, had brought up against the possibility that natural selection could possibly account for human evolution. Argyll was in some measure sympathetic to the idea of evolution and had been described to Darwin by the Anglican theologian and naturalist Charles Kingsley, who knew him well, as "ready to hear all reason" on the matter.[105] However, despite this, by the late 1860s he was more widely known for airing his reservations about the efficacy of Darwin's ideas in the periodical press, and he finally brought these observations together in his 1868 book *The Reign of Law*. In it, Argyll had cited the French statesman-turned-historian and man of letters François Guizot in support of his own belief that unless man had first made his appearance upon the earth with fully developed social and ethical faculties, he could not possibly have survived the cutthroat struggle for existence that Darwin had portrayed in *Origin*. It was "a physical impossibility that Man—the human pair—can have been introduced into the world except in complete stature—in the full possession of all his faculties and powers," for in "no other condition could Man, on his first appearance, have been able to survive and to found the human family." Argyll continued, "There is undoubtedly much to be said in support of M. Guizot's position. . . . Man as a mere animal is the most helpless of animals. His whole frame has relation to his mind, and apart from that relation, it is feebler than the frame of any of the brutes."[106] Without intelligence and mind, Argyll proposed, it would be impossible for man to have survived in a world in which his life was threatened at every turn by savage beasts with tusks, sabered-teeth, and massive jaws. Man must have been placed on the earth with all his faculties. Argyll cited Wallace as being in agreement with him on this point, referring to Wallace's comments in a review of the first edition of *Reign of Law*, although it is clear that at the time he wrote the review, Wallace was in fact quite critical of Argyll's appeal to Divine oversight.[107] Further, Argyll pointed out that even if early man could have somehow eked out a tentative existence in such a hostile environment, natural selection certainly could not account for the full development of his mind and intellect from such primitive raw material.[108]

Darwin took the Duke's objection seriously. *Reign of Law* was, he wrote to Kingsley, "very well written, very interesting, honest & clever," although its author was also, in parts, "very arrogant" (notably because Argyll had the temerity to dismiss the views of John Stuart Mill on a particular issue!). "Clever as the book is," Darwin concluded, "I think some parts are weak."[109] Turning the Duke's argument to his advantage, he pointed out that Argyll's observation only shed further light on human origins and in fact undermined the very argument for human exceptionalism that he was using it to support. Embracing Wallace's move away from Malthusian individualism, Darwin sidestepped the force of the Duke's argument by suggesting that sociability among early humans would clearly have improved their chances of survival and reproduction.[110] Individuals did not survive or die alone but in the context of the communities of which they were a part.

In *Descent*, Darwin emphasized that cooperation was an alternative strategy to that of individual struggle, and, as Wallace had pointed out, it had been an eminently more successful one. The Duke was right to say that in an out-and-out physical contest with such a ferocious beast as a gorilla, a man would have little chance; however, Darwin took from this a decidedly different inference than had either the Duke or Wallace. Bridging the divide between man and beast, Darwin saw sociability and mutual aid among animals, where Wallace had seen none. Certainly, Darwin conceded, it was unlikely that man had evolved from a solitary and pugnacious primate like the gorilla; if this were the case, one would also have to account for the loss of otherwise useful weaponry such as the gorilla's powerful jaws and immense physical strength. Rather than this being evidence that man had not evolved at all, though, Darwin argued that it was much more likely that man shared his ancestry with a smaller, sociable, and clearly intelligent species such as the orangutan or the chimpanzee.[111] Arboreal in their habits, combining for mutual aid, and concerned to warn their fellows of approaching danger, these social primates had characteristics that Darwin thought likely to be found among the forbears of mankind. He pointed out that Wallace himself had noted the propensity for Orangs to engage in primitive tool use.[112] As Darwin later wrote to John Morley, the editor of the *Pall Mall Gazette* on this matter, "I do not think that there is any evidence that man ever existed as a non-social animal."[113]

Certainly, reasoned self-interest could account for a lot in terms of human morality. Adam Smith had pointed out as much in terms of the division of labor and his famous statement about the butcher and the baker. If everyone looked out only for their own best interests, they would quickly realize that helping others might occasionally work out to their benefit in terms of reward or reciprocity. Indeed, Smith had argued—and many of Darwin's contemporaries clearly agreed—that the very best of liberal society could be gained through exactly this process. Darwin fully allowed that "from this low motive" of self-interest, man's ancestor "might acquire the habit of aiding his fellows." Further, he argued, "the habit of performing benevolent actions certainly strengthens the feeling of sympathy which gives the first impulse to benevolent actions." And as a result, "each man would soon learn from experience that if he aided his fellow men, he would commonly receive aid in return." Moreover, if such actions were repeated often enough to become habitual, they might quite easily become heritable. Darwin, like many of his contemporaries, believed that "habits . . . followed during many generations tend to be inherited."[114] This much might be had from sympathy, but there remained several problems with an explanation of the development of human morals from reason alone. First, there was the problem of accounting for cases of instantaneous action—of a man rushing to save a drowning child, for instance, where there was no pause for the rational calculation of pleasures and pains, of costs and benefits. Second, what of the issue of discernment and discrimination? Smith's account of reasoned sympathy suggested that humans should recognize all men as beings who felt pleasure and pain like themselves, and thus extend moral consideration to them all equally. Yet Darwin had witnessed the extent to which the members of one tribe while acting with great consideration toward members of their own tribe might delight in the sufferings of a stranger. Third, what of those actions of self-sacrifice, of those who gave their lives in the defense of others? Where was the calculated self-interest in this? Finally, of course, if morals were the product of reason and reason was a purely human attribute, then the existence of morality did indeed open up an abyss between man and animals—a difference of kind rather than, as Darwin hoped to show, merely a difference of degree.[115]

Darwin's account of the evolution of other-regarding sentiments from the social instincts, and the "parental and filial affections" in particular, could quite account for each of these problems. It could account for the instantaneous and uncalculated act of the man who rushed to save a drowning child. It might also readily explain acts of discernment or discrimination. In light of the close association and familial relationship, an organism might understandably favor its closest relatives, and tribes might favor their own members over strangers. Certainly, Darwin believed that it was the case that only a truly other-regarding sentiment might account for actions of supreme self-sacrifice in which an individual gave up its life so that another might live. Finally, of course, the parental and filial affections, like all instincts, were evident across the animal kingdom, and thus provided the bridge that Darwin sought between man and his animal origins.

As I have suggested, though, grounding the origin of ethics in the social instincts of animals rather than in human reason was important for Darwin for another and much more explicitly political reason. The social instincts, the parental and filial affections among them, had existed in nature long before rationality had emerged, and therefore must clearly predate the development of the rational, calculated self-interest that the Manchester school of political economy claimed as the be-all and end-all of man's ethical nature. Thus, Darwin concluded, "the reproach of laying the foundation of the most noble part of our nature in the base principle of selfishness is removed."[116]

Taking Wallace's 1864 paper as a lead in refuting his 1869 recantation, Darwin pointed out that assuming that the instinct for parental and filial affection was just as variable as was any other natural character, they too would quickly become subject to natural selection, not through the competition of one individual with another, though, but between one community and another. The truly other-regarding instincts that derived from the parental and filial affections would initially be limited to immediate family members, but, and if it was indeed the case—as contemporary anthropologists suggested—that early humans were polygamous in their sexual relations, then the whole tribe or clan would come to share strong moral bonds too. Under such conditions one might quite easily imagine that it could become considered a high moral action to kill, maim, or murder members of other tribes.[117] Indeed, the combination of care for one's own kin and total lack of sympathy for strangers that Darwin had witnessed on his voyage might readily be explained as the outcome of long-continued competition between communities for scarce resources. Darwin cited Bagehot's "Physics and Politics" in recognition of the fact that in the context of warring tribes, "a tribe including many members who, from possessing in high degree the spirit of patriotism, fidelity, obedience, courage and sympathy, were always ready to aid one another and to sacrifice themselves for the common good, would be victorious over most other tribes; and this would be natural selection."[118] As a result, those societies that extended their moral communities not only to include their "in-laws," as it were, but also non-related individuals, might readily find themselves at an even greater advantage over less-cohesive societies.

However, and despite the evident strengths of this form of group selection, Darwin was aware that the idea also had inherent weaknesses—weaknesses that, if he had not thought of them before, had been raised by the political essayist William Rathbone Greg and his own cousin, Francis Galton. Both men argued that while such an account of the evolution of ethics might sound quite plausible at first glance, it could not withstand closer scrutiny. Further, both Galton and Greg saw that evolution had significant social implications, and were concerned that however desirable the evolution of ethical behaviors might be in the short term, in the long run, the results of human ethical behavior might actually undermine the future progressive development of society. Despite their shared concern, they proposed very different solutions.

Greg's 1868 article for *Fraser's Magazine* on the subject was not in fact written in opposition to Darwin, but rather was a critique of the very paper by Wallace that Darwin thought so important. The title of the article, "On the Failure of 'Natural Selection' in the Case of Man," is indicative of the argument of the essay. Greg was a formidable critic; he had long been an active campaigner for political reform, and as a Manchester mill-owner's son he was an ardent advocate of free trade. As Darwin was only too well aware, Greg was well connected and his opinion was both informed and influential.[119] Darwin had other and more personal reasons for taking Greg seriously, too. The two had been students together in Edinburgh and had both been active members of the Plinian Society, a society dedicated to the presentation and discussion of papers in natural history. Having been elected to membership on the same evening as Darwin in November 1826, Greg had immediately proposed a paper that aimed to demonstrate that "the lower animals posses every faculty & propensity of the human mind."[120] His interest in human evolution was thus of long standing. Greg thus acknowledged that Wallace had written an "admirable paper" and had made a significant contribution to the discussion of human evolution by noting the effects of competition between groups. However, he thought that Wallace had unfortunately not taken full account of all the relevant factors and that, as a result, his argument was "by no means the whole of the case."[121]

Where Wallace had suggested that the ethical nature of man would tend to maintain the evolution of society in the ascendant, Greg was concerned that in fact the opposite would be true. Not only would the evolution of ethics serve to maintain the unfit, whose continued fecundity would weaken the society that preserved them, but—and using the opportunity to attack aristocratic privilege and inherited wealth—Greg argued that the existing political conditions worked to ensure that the sons of rich men were rewarded regardless of talent. Further, the sort of charitable provision being urged by many of their contemporaries, and which seemed a logical outcome of the spread of the sort of ethical sentiments that Wallace had described, would not work to advance society either, but would undermine it instead.[122]

Greg was thus fearful that "existing society, which is the result of the operation of this law [of natural selection] in past ages, may be actually retarded and endangered by its tendency to neutralise that law."[123] As Wallace had acknowledged, as ethical sentiments grew, so "the action of natural selection is therefore checked; the weaker, the dwarfish, those of less active limbs, or less piercing eyesight, do not suffer the extreme penalty which falls upon animals so defective."[124] Greg pointed out that this preservation of the unfit would not only burden the productive members of society, but that the detrimental effects of their continued existence would only be exacerbated across the generations as they continued to reproduce a progeny who were as much a drain upon society as themselves.[125]

Greg was convinced that this tendency of ethics to thwart biological and social progress was hampered rather than helped by the prevailing politics of nineteenth-century society. Just as hereditary privilege would favor the parasitic aristocratic class, so charity would preserve the weakest and most sexually profligate. Indeed, rather than working to give an advantage to the most temperate and hardworking members of society—the entrepreneurial middle class—the arrangements of society were such as to favor "those emasculated by luxury and those damaged by want." "Thus the imprudent, the desperate, those whose standard is low, those who have no hope, no ambition, no self-denial,—on the one side, and the pampered favourites of fortune on the other, take precedence in the race of fatherhood, to the disadvantage or the exclusion of the prudent, the resolute, the striving and the self-restrained."[126] Without significant political change, Greg believed that degeneration and not progress loomed on the horizon of mankind.

Greg maintained that the solution to the problem was to be found in laissez-faire and the abandonment of hereditary privilege, and like Herbert Spencer, he remained hostile to any form of state welfare. In contrast, Francis Galton was one of a growing number of people who made the case that, given the circumstances that Greg had pointed out, state intervention was exactly what was needed. To Galton's mind, however, it was not state charity that was required, but the state regulation of reproduction. "No one, I think, can doubt . . . that if talented men were mated with talented women, of the same mental and physical characters as themselves, generation after generation, we might produce a highly-bred human race," he wrote.[127] Greg's commitment to laissez-faire left him horrified at such suggestions. Any such measures would be the worst form of despotism. He welcomed the gradual spread of liberal politics and ideas that had characterized the nineteenth century, and tried to remain optimistic for the future, but could only articulate the vague hope of relief in the form of "the slow influences of enlightenment and moral susceptibility, percolating downwards and in time, permeating all ranks."[128] This did little to alleviate his concern that the evolution of ethics would ultimately undermine the future development of society, however.

Darwin was aware that Galton was also deeply concerned about the effect that various forms of political organization might have on the future progress of human evolution. He had been deeply impressed with Galton's work on heredity—in particular his 1865 essays for *Macmillan's Magazine*, "Hereditary Talent and Character" and "Hereditary Genius: The Judges of England between 1660 and 1865," but also Galton's 1869 book, *Hereditary Genius*, which he thought "remarkable." Galton had emphasized Wallace's point that mental characters were just as heritable as those that were purely physiological, and that variations in intellect between one man and another could be significant. As Galton made clear in both his papers on the subject and in his book, he believed that while it was possible to hold land and title regardless of one's biological worth, the beneficial qualities of one's remote ancestors who might have legitimately won title through their merits might easily have been bred out of a family where the title still remained. In the middle-class professions, however, mediocrity was not so easily tolerated, in consequence of which, high attainment was indeed a fair measure of innate talent. These professions, perhaps unsurprisingly given Galton's own background and class allegiance, were those that typically occupied men of his own type. The law and science were particularly relevant examples, he argued. What was particularly problematic, however—as Greg too suggested—was that there was a differential birthrate between these high-quality members of society and those who were of less biological worth. Further to this, Galton firmly believed that the state of things with respect to these professions was indicative of marked differences that prevailed across society. The middle class was indeed made up of people who, by and large, had more biologically advantageous traits than members of the aristocracy or the working class. This explained why they gravitated to the professions they did, and why they were so successful at them. Those who lacked these traits and were unsuccessful would quickly fall down the social ranks to where they might be a better fit. Similarly, professions such as those opened up by the advance of science and industry might allow the rise of the most able men from the lower ranks of society. Men like Huxley had proven just such a path. Thus, Galton argued, a society that rewarded talent and penalized idleness would ensure future social progress. However, this was far from a description of Victorian society. In "Hereditary Talent and Character," Galton pointed out that not only were there political and cultural barriers that worked to preserve aristocratic privilege, but the most talented men deferred marriage through prudential considerations as they tried to establish themselves. Marrying late, they reproduced in far fewer numbers than their less-worthy countrymen.[129]

Importantly, significant for the later development of biological thinking on the effects of heredity in populations was the statistical framework that Galton employed to address this question. He suggested that given that society was divided into what were effectively castes, "any agency, however indirect, that would somewhat hasten the marriages in caste A, and retard those in caste B, would result in a larger proportion of children being born to A than to B." Over time, this process "would end by wholly eliminating B, and replacing it by A." At present, he suggested, social arrangements worked to ensure that the worst members of society were outbreeding and thus would ultimately eliminate the best members. This would result in the eventual eradication of the best men and the evolutionary degeneration of society as a result. This was a conclusion he shared with Greg, but unlike Greg, Galton believed that state intervention might be a legitimate recourse in order to prevent any such outcome. Galton argued that while this much was certainly true, it was by no means inevitable. Society might intervene to initiate agencies that promoted marriage among the best and discouraged marriage among the worst. Indeed, it was Galton's hope that this approach to the management of human reproduction would be adopted in the future. "If it was generally felt that intermarriages between A and B were as unadvisable as they are supposed to be between cousins, and that marriages in A ought to be hastened . . . while those in B ought to be discouraged and retarded, then, I believe, we should have agencies amply sufficient to eliminate B in a few generations."[130]

Thus, unlike Greg, Galton was quite willing to see the state intervene to prevent the sort of evolutionary degeneration that they both feared lay in store. Where Greg was skeptical of the possible success that might be achieved through laissez-faire, Galton denied it outright. Even voluntary charity might undermine social progress. He looked forward to a day in which society would take the matter of sexual selection in hand and ensure that the best men married the best women and that the reproduction of the unworthy would be curtailed. This might develop not only a physically healthy race, but also, if the selection were aimed at such an outcome, a morally healthy race as well. Galton concluded that just as "by selecting men and women of rare talent, and mating them together, generation after generation, an extraordinarily gifted race might be developed, so yet a more rigid selection, having regard to their moral nature, would, I believe, result in a no less marked improvement of their natural disposition."[131] This comment of Galton's about the significance of sexual selection for the evolution of morality as well as for the development of desirable physiological characters is telling, for although Darwin never made it explicit in his chapter on the evolution of morals, in a later chapter of *Descent* in which he dealt with sexual selection among humans, he seemingly suggested that sexual selection might also play a significant role in the evolution of the moral sentiments. However, by Darwin's reckoning, no eugenic intervention by the state would be needed to facilitate the best of outcomes.

In *Descent*, Darwin acknowledged and responded to both Greg's and Galton's concerns. He believed that Greg had identified a significant problem but had overlooked the solution. Darwin thus believed that Greg's fears were unfounded and Galton's eugenic solutions were unnecessary. Darwin admitted that, as Greg pointed out, the tendency to aid the poor and sickly worked to undermine the fitness of society, and in a passage that appears too to acknowledge Galton's critique of unmanaged reproduction, Darwin wrote: "We civilised men . . . do our utmost to check the process of elimination; we build asylums for the imbecile, the maimed, and the sick; we institute poor laws; and our medical men exert their utmost skill to save the life of every one to the last moment. . . . Thus the weak members of civilised societies propagate their kind. . . . No one who has attended to the breeding of domestic animals will doubt that this must be highly injurious to the race of man. It is surprising how soon a want of care, or care wrongly directed, leads to the degeneration of a domestic race; but excepting in the case of man himself, hardly anyone is so ignorant as to allow his worst animals to breed."[132]

However, Darwin did not believe that the aid that civilized men felt compelled to offer to the less fortunate among them was sufficient to undermine the general trend toward social progress. This was fortunate, for as Darwin was clear to point out, the alternative was not only morally unthinkable, but would undermine the ethical framework that made society possible. "Nor could we check our sympathy, even at the urging of hard reason, without deterioration in the noblest part of our nature," he wrote.[133] "We must therefore bear the undoubtedly bad effects of the weak surviving and propagating their kind," but with the consolation—and here, doubtless mindful of Galton's "two caste" illustrations—that "the weaker and inferior members of society do not marry so freely as the sound."[134] However, while Darwin hoped that the unfit might refrain from marriage as an increasingly general trend, he recognized that this "is more to be hoped for than expected." As Greg had indelicately put it, the poor man, like "the careless, squalid, unaspiring Irishman multiplies like rabbits," while the "frugal, foreseeing, self-respecting, ambitious Scott . . . marries late, and leaves few behind him." Even so, Darwin sought to make the case that the processes of natural selection in civilized society favored the numerically fewer offspring of the selfless and industrious over and above the more-abundant weaker and less-virtuous individuals. "The intemperate suffer from a high rate of mortality, and the extremely profligate leave few offspring," he argued. The worst types also tend to cluster together in urban slums in which "it has been proved by Dr. Stark from the statistics of ten years in Scotland, that at all ages the death-rate is higher in towns than in rural districts, 'and during the first five years of life the town death-rate is almost exactly double that of the rural districts.'" Given that these figures generalized across rich and poor, and across the fit and the unfit, Darwin believed it was a safe presumption that the death rate among the children of the worst types was much higher than the 50 percent indicated. Further, although the reckless tended to marry young (one reason why they had numerically greater offspring), they often married too young. As Darwin noted, "With women, marriage at too early an age is highly injurious," and, quoting the figures of Dr. Farr, an authority of the subject, he concluded that "'twice as many wives under twenty die in the year, as dies out of the same number of the unmarried,'" adding, "The mortality, also, of husbands under twenty is 'excessively high.'"[135]

The statistics suggested too that those who remained married, those, we can assume, who were fit and thus survived into middle age, tended to enjoy a greater longevity and thus had more years in which to reproduce. Thus, even though the industrious and moral males tended to marry late and have fewer offspring overall, they also tended to live longer and have children when they were older, and were in a sufficiently stable and prosperous position to give them the best start in life and the likelihood of inheriting some modest property in due course. Thus, where Darwin concurred with both Galton and Greg that the inheritance of vast sums of money, of land and of title, interfered with evolutionary progress, the "moderate accumulation of wealth" typical of middle-class families did not "interfere with the process of selection," for they and their children were likely to "enter trades or professions in which there is struggle enough, so that the able in body and mind succeed best."[136]

In addition to the factors that worked to promote those he believed to be the fit and the moral, Darwin also noted numerous others that took a toll upon the unfit and the immoral, and worked to limit their number: "Malefactors are executed, or imprisoned for long periods, so that they cannot freely transmit their bad qualities. Melancholic and insane persons are confined, or commit suicide. Violent and quarrelsome men often come to a bloody end. The restless who will not follow any steady occupation—and this relic of barbarism is a great check to civilisation—emigrate to newly-settled countries, where they prove useful pioneers."[137] Citing E. Ray Lankester's work "Comparative Longevity" (1870), he reported that "intemperance is so highly destructive that the expectation of life of the intemperate, at the age of thirty for instance, is only 13.8 years; whilst for the rural labourers of England at the same age it is 40.59 years. Profligate women bear few children, and profligate men rarely marry; both suffer from disease."[138]

With all these fates set to befall the unfit, there was little to fear in the direction of Greg's concerns. Nevertheless, Darwin conceded that if these checks, "and perhaps others as yet unknown, do not prevent the reckless, the vicious and otherwise inferior members of society from increasing at a quicker rate than the better class of men, the nation will retrograde, as has too often occurred in the history of the world. We must remember that progress is no invariable rule."[139] Although the historian Diane Paul has concluded that this sentiment was representative of Darwin's final thoughts on the matter, citing the fact that "Wallace noted that that in one of their last conversations, Darwin had expressed gloomy views about the future," it is not so clear that this was his view in the 1870s.[140] After all, his final words in his chapter on the development of the intellectual and moral faculties among primitive and civilized nations were much more optimistic: "It is apparently a truer and more cheerful view that progress has been much more general than retrogression; that man has risen, though by slow and interrupted steps, from a lowly condition to the highest standard as yet attained by him in knowledge, morals and religion."[141]

Reinforcing the class politics that ran throughout his work, Darwin also drew sharp lines of distinction between the relative merits of a liberal bourgeois lifestyle, not only over that of the lower classes, but also that of the aristocracy as well. With the Whig middle class in the ascendant, belief in continued progress seemed warranted. The middle class was largely responsible for the organization and distribution of the bulk of the charitable work that helped the deserving poor to raise themselves, and they were the champions of education and industry, too. Further, it was also the case that upon the death of the head of a middle-class household his estate was generally bequeathed equitably to his children—a much better arrangement than the primogeniture favored by the aristocracy. In terms of social evolution, "Primogeniture with entailed estates is a more direct evil," Darwin concluded. Nevertheless, the existence of the aristocracy showed that they had certainly been favored by natural selection at an earlier stage of social development, for (and here doubtless recalling the Fuegians) "any government is better than none," but their day in the sun was coming to an end. Even though the eldest sons of the aristocracy would often have their choice in the marriage market, "as Mr. Galton has shewn, noble families are continually cut off in the direct line."[142]

Even though this was as much as Darwin had to say on the evolution of ethics in his chapter on morality, as I have suggested above, this was not all that he had to say. Indeed, he had long been interested in the effects of sexual selection, and having read Galton, it seems that he recognized that sexual selection might do much to effect the evolution of morality as well as morphology. The question at the heart of the evolution of the moral sentiments was to explain how genuinely other-regarding sentiments could possibly evolve from the process of natural selection, which, in the face of it, appeared to only reward selfishness. At the very best—and this was certainly so in the case of organisms that were unrelated—it might account for what is today referred to as "reciprocal altruism," in which an apparently altruistic behavior might evolve if it does so in a context in which others reciprocate.[143] As I have shown, though, Darwin was unhappy with this conception of the origin and development of the moral sentiments, because it ultimately grounded the highest qualities of humanity, our morals and conscience, in the "low motive of self interest." However, while Wallace and now Darwin had suggested that truly other-regarding qualities might have evolved as being advantageous to the community or group to which the individual belonged, the problem remained that such sentiments would ultimately be self-defeating.

In light of Wallace's 1864 article, Darwin had appreciated that "a tribe including many members who, from possessing in high degree the spirit of patriotism, fidelity, obedience, courage and sympathy, were always ready to aid one another and to sacrifice themselves for the common good, would be victorious over most other tribes; and this would be natural selection."[144] What greater indication of the other-regarding sentiment was there than to risk one's life for the welfare of one's fellows? However, all things being equal, surely such sentiments would put the person who expressed them at a distinct disadvantage in the struggle for existence. Coming to the fore in times of battle might well serve to benefit the group, but surely how much better to be in the second or third rank of the defenders—or better still, to take the very back seat when it came to the fighting? However, and as Darwin sought to show in his own account of the natural history of mankind, as in the rest of nature, all things are not always equal.

If the brave and selfless individuals always came to the fore in circumstances requiring defense of the community or in the attempted conquest of another nation, they would surely suffer the heaviest loss of life. One might reasonably expect, therefore, that selection would act to quickly eradicate their kind, leaving only their less-courageous fellows to sire subsequent generations. Thus, and as Darwin freely acknowledged: "He who was ready to sacrifice his life, as many a savage has been, rather than betray his comrades, would often leave no offspring to inherit his noble nature. The bravest men, who were always willing to come to the front in war, and who risked their lives for others, would on an average perish in larger numbers than other men."[145]

Contrary to this conclusion, however, Darwin pointed to a number of prevailing circumstances that militated against the presumption that in this circumstance all things were in fact equal. The first echoed the account he had given in *Origin* of the evolution of sterile and distinct castes in social insects: the relatives of those who died would carry the same characteristics. However, he also appealed to two other mechanisms by which a willingness to risk one's life for others might have evolved in human societies. Both related to the significance that man universally attached to the praise or blame that he might receive from his fellows. One was the consequence of man's propensity for mimicry and imitation—by seeing courageous men praised, even the timid might be encouraged to acts of heroism; the other, which would only further enhance the effects of the first, was the influence of sexual selection.

In *Origin*, Darwin had tackled the problem of the existence of sterility among the several castes of workers in the social insects. In explanation of how sterility could possibly become a heritable trait, Darwin had turned once again to artificial selection as an analogy. Just as a stockman who prizes the meat of a slaughtered animal might reliably turn to the parental and filial stocks with confidence that they too would carry these prized characters, if beneficial traits were either coincident with sterility, of if, as was the case with the social insects, it benefited the community to have a division of labor between the workers and the reproducing queen, then Darwin could see no reason why this could not develop by natural selection. So too with the bee's sting: it might kill the bee, but served to perpetuate the hive. And so it was with the tendency for bravery and a willingness to risk self-sacrifice among men.[146]

To the extent that a willingness to risk one's own life for the welfare of others was an inherited character, such a tendency would be found in the relatives of those who had fought and died on the field. As Darwin had stated in the case of the social insects, "This difficulty, though appearing insuperable, is lessened, or, as I believe, disappears, when it is remembered that selection may be applied to the family, as well as to the individual, and may thus gain the desired end."[147] Thus, although those individuals that came to the fore in battle did so at the peril of their lives, and might die disproportionately in comparison to their more-timid fellows and thus leave fewer progeny, to the extent that bravery was a heritable character, the brothers, or perhaps, even, the sisters (although Darwin was not explicit on this last point), might perpetuate this character despite the ultimate sacrifice made by their sibling. As Darwin noted, "Even if they left no children, the tribe would still include their blood relations; and it has been ascertained by agriculturists that by preserving and breeding from the family of an animal, which when slaughtered was found to be valuable, the desired character had been attained."[148]

If this alone would not convince his readers, there were other factors to consider. In light of the presence, even in man's progenitors, of an incipient form of conscience, the propensity of individuals to keenly feel the praise or blame of their fellows would be significant. Darwin cited Bain as his source for the importance of public sanction, but it had been there in Smith's *Theory of Moral Sentiments* too.[149] Darwin was anxious to point out to his readers that he was by no means stretching his speculations beyond observed evidences, which by analogy would support his point. Everyone was aware that higher animals, such as the dog, recognized the difference between praise and blame, between pride and shame, and that in comparison with the slow mechanisms of heredity alone, such peer pressure was potentially a "much more powerful stimulus to the development of the social virtues." Darwin contended that it was obvious, given the tendency of the progenitors of man to approve or judge the actions of their fellows, that "members of the same tribe would approve of conduct which appeared to them to be for the general good, and would reprobate that which appeared evil." Thus, in light of the capacity for imitative behavior among both apes and men, not only would the brave be encouraged to risk their lives for the acclaim they would receive by doing so, but as Darwin was clear to point out, even those men who were not "impelled by any deep, instinctive feeling" to risk their lives might be induced to do so after seeing their braver comrades receive the acclamation of their tribe. Even if the brave man died in action, his heroism might inspire others, which would strengthen by exercise not only the "noble feeling of admiration," but the likelihood of those who were inspired to acts of bravery to be brave in similar future circumstances.[150] A third reason to expect that bravery and a willingness to risk one's own welfare for the good of the group would persist among the men of a community, and indeed, probably the most important, was sexual selection.

Darwin had broached the subject of sexual selection in *Origin*, but returned to it in detail in *Descent*, and not least because he believed that it played a most significant part in the evolution of morality.[151] Indeed, in the second edition of *Descent*, published in 1874, Darwin quoted the popular German philosopher Arthur Schopenhauer in support of the significance that he attached to sexual selection in this respect: "The final aim of all the love intrigues, be they comic or tragic, is really of more importance than all the other ends in human life. What it all turns upon is nothing less than the composition of the next generation." And underscoring the shift in his own views from his former focus upon the survival of the individual to that of the community, he included the philosopher's comment that "it is not the weal or woe of any one individual, but that of the race to come, which is here at stake."[152] Not only had Darwin appealed to sexual selection as the means by which sexual dimorphism had been affected across the animal kingdom, but he had also argued that the race-specific conceptions of beauty in the opposite sex explained the differences between the various races of mankind.[153] Now, following Galton, he also invoked sexual selection as playing a significant role in the origin and development of the moral sense. However, unlike his cousin's account of things, sexual selection might work to increase the morality of a community without the need for the state to intervene to manage reproduction.

Darwin conceived of there being two aspects to sexual selection in nature. One he called "the law of battle"; the other, and that which was to prove the more controversial, he called "female choice."[154] To distinguish sexual selection from the struggle for existence that faced each and every animal, Darwin defined sexual selection as the struggle not to survive per se, but to leave progeny. Darwin recognized that sexual selection could explain the evolution of any number of physiological traits that might not only add nothing to an organism's chances of survival, but which might actually undermine its longevity. A case in point was the showy but weighty plumage of the peacock's tail. This would certainly attract a mate, but at the same time would also advertise its presence to predators as well as slowing its escape if this was the case.[154] In such instances, the balance between natural selection and sexual selection acted as a cost-benefit exercise, and when it came to his analysis of sexual selection in man, Darwin presented the evolution of potentially self-sacrificial behaviors as analogous to such a development. Both the law of battle and female choice would play distinct but connected roles in the evolution and persistence of such altruistic behaviors.

Darwin's law of battle described the conflict and competition between males of the same species to mate with the female, whereas female choice, as the name suggests, described those cases in which the female of the species actually had a say in selecting or rejecting the advances of one male over another, either by rational deliberation or unconscious instinct. While Darwin had spent a great deal of time discussing the implications of competition between one society and another, he noted that one of the primary reasons for conflict, even more compelling than a struggle over food, was competition between the men for the possession of females, and that according to contemporary authorities on the conditions of early man—notably, John Lubbock, McLennan, Morgan, and Bachofan—this had always been the case. In terms of how Darwin had discussed his theory of sexual selection, this would be the "law of battle." In a tone that was clearly apologetic for bringing such an indelicate topic as the sexual behavior of primitives and savages to the attention of his Victorian readers, Darwin introduced the euphemism: "What Sir J. Lubbock by courtesy calls communal marriages; that is, all the men and women in the tribes are husbands and wives to each other." Darwin noted that "all those who have most closely studied the subject, and whose judgment is worth much more than mine, believe that communal marriage was the original and universal form throughout the world, including the intermarriage of brothers and sisters." Exclusive "marriage" to any one particular female, and by extension the modern form of marriage, he explained, had most likely had its origin in the capture of women from another tribe in battle. Referring to Lubbock's address to the British Association from earlier that year, he informed his reader that "Sir J. Lubbock ingeniously accounts for the strange and widely extended habit of exogamy, —that is, the men of one tribe always taking wives from a distinct tribe,—by communism having been the original form of marriage; so that a man never obtained a wife for himself unless he captured her from a neighbouring and hostile tribe, and then she would naturally have become his sole and valuable property. Thus the practice of capturing wives might have arisen; and from the honour so gained might ultimately have become universal."[156]

Once a measure of exclusivity became the norm—and Darwin suggested that it would quickly have become so—then competition between men, even of the same society, would similarly become the order of the day. "With barbarous nations," he noted—pointing to the aboriginal Australians as his example—"the women are the constant source of war both between the individuals of the same tribe and between distinct tribes."[157] Given continued competition for females, only those men who were the most self-assertive, the most pugnacious, and the most successful in battle would secure themselves a mate, and thus these characteristics would tend to prevail over timidity in subsequent generations. This much might account for the evolution and persistence of bravery and pugnacity, but what of the willingness of individuals to turn these qualities to the defense of their communities? After all, to risk their lives for no direct benefit to themselves would seem to undermine, rather than enhance their reproductive success.

In order to explain this, Darwin turned to the other aspect of sexual selection, "female choice." While perhaps controversial among his peers for casting women as the agents rather than the objects of sexual selection, female choice made perfect sense in light of what Darwin had already said about the importance of praise and blame in molding the social behavior of members of a community. In those societies in which women had a say in the choice of their mate, "the women would generally choose not merely the handsomer men, according to their standard of taste, but those who were at the same time best able to defend and support them." Further, it would follow in light of what Darwin had already said about the importance of praise and blame, that women's "standard of taste" would also be influenced by those who were most highly regarded in their society. If this was the case, as Darwin had convincingly argued, then those men who put themselves at risk in defense of their fellows would be the most favored, and thus it would follow that even though they would die in disproportionate numbers compared with the more timid men in that society, those brave men who did survive would assuredly sire a disproportionate number of offspring. Such success with the ladies would certainly prompt imitation, driving even the timid to attempt brave actions. Female choice might thus explain not only the spread of bravery in a population, but a willingness to defend the commonweal despite the inherent danger in doing so. The timid man who could not bring himself to risk his life in times of conflict would be hard-pressed to find a bride.[158]

While female choice may not have been overtly practiced among the English Victorian middle class (although anyone who has read a Jane Austen novel or two will recognize that much of female middle-class society revolved around the intrigue of ensuring that the right male gave suit to the right female), what was more important in accounting for the origins of present-day sentiments were the actions and intentions of our forebears. Darwin documented a welter of examples of more-primitive societies in which female choice was prevalent. Notably, in each of the cases female choice worked in conjunction with the law of battle. While a woman captured in combat would become the exclusive property of her captor, from the evidence of at least some native societies it seems that the men were anxious at the same time to make a good show of themselves in the process. Darwin noted tribes in which warriors deemed it their highest aspiration "to render themselves attractive to the ladies and conspicuous in war."[159] Perhaps closer to the publicly acknowledged norm though was the trend among the Kalmuck people.

Darwin here related that in order to win the bride of his choosing from her family, an intending husband must first catch her in a race—she being given a judicious start upon her suitor. As Darwin noted, "No instance occurs of a girl being caught, unless she has a partiality to her pursuer." This was also the way among tribes of the Malay archipelago, as Darwin quoted John Lubbock from his influential 1865 work *Prehistoric Times*: "The race 'is not to the swift, nor the battle to the strong,' but to the young man who has the good fortune to please his intended bride."[160]

While Darwin grounded the origin of other-regarding behaviors in the social instincts, he was careful to acknowledge that they would have also been very much developed by intellect and reason. Indeed, it was through the development of reason that humanity had become conscious not only of their actions, but of how others might view their actions. Thus, they might use reason to conquer their instincts. Those who were instinctively timid might, for fear of mockery, raise themselves to heroic action. Seeing the rewards in high praise—and the attention of the ladies—a timid man might conquer his fear and imitate the actions of the brave. Indeed, Darwin reflected, this was the very definition of bravery, was it not?—the ability to act despite one's fears. Such actions would lead at least to sympathetic feelings toward those the brave man defended, if not indeed to feelings of love, and these in turn would spread: "As the feelings of love and sympathy and the power of self-command become strengthened by habit, and as the power of reasoning becomes clearer so that man can appreciate the justice of the judgments of his fellow-men, he will feel himself impelled, independently of any pleasure or pain felt at the moment, to certain lines of conduct. He may then say, I am the supreme judge of my own conduct, and in the words of Kant, I will not in my own person violate the dignity of humanity."[161]

Darwin clearly saw this as the foundation of liberal humanist ethics, of a politics that would spread from the family to the tribe, to the nation and race, and eventually to include men and women of all races. The fight against slavery was perhaps not Darwin's motivation in theorizing the common ancestry of all life on earth, as Adrian Desmond and James Moore have recently contended, but it was certainly his ending point: "As man advances in civilisation, and small tribes are united into larger communities, the simplest reason would tell each individual that he ought to extend his social instincts and sympathies to all the members of the same nation, though personally unknown to him. This point being once reached, there is only an artificial barrier to prevent his sympathies extending to the men of all nations and races."[162]

Indeed, ever an animal lover, Darwin noted that "sympathy beyond the confines of man, that is humanity to the lower animals, seems to be one of the latest moral acquisitions. . . . This virtue, one of the noblest with which man is endowed, seems to arise incidentally from our sympathies becoming more tender and more widely diffused, until they are extended to all sentient beings."[163] For Darwin, this was not only an emotional attachment, but a logical outcome of his own thoughts on the evolution of morals. In Notebook B he had long since pressed his thoughts on this matter to their logical conclusions: "If we choose to let conjecture run wild then our animals our fellow brethren in pain, disease, death & suffering, & famine, our slaves in the most laborious works, our companions in our amusements. They may partake from our origin in these one common ancestor; we may be all netted together."[164]

Thus, far from being grounded in the low motive of self-interest or the relentless war of each against all, the view of life that Darwin found so full of grandeur was that which told the story of the evolutionary development of the liberal humanism he held dear, of the spread of a real and genuine altruism he hoped to see become universal among men, and which would ultimately be extended to all sentient beings.

If Darwin had hoped that the inclusion of female choice and Kantian imperatives in his hereditarian account of the origin and development of human morality would appeal to Cobbe, he was to be disappointed. He had suspected that she would not be satisfied with his account of the evolution of morals, but he had reason to hope at least for a sympathetic hearing. In their discussions in the small Welsh village of Caerdeon in 1868, Darwin and Cobbe had agreed that Mill had paid insufficient attention to heredity and instinct and that in consequence he had placed too much emphasis upon the role that education might play in the formation of human character. However, and despite this apparent agreement, Cobbe's reactions to Darwin's own view of things in *Descent* were deeply critical.

Despite her enthusiasm for hereditarian explanations, Cobbe was horrified to see where Darwin had taken the logical implications of her own opinions on the matter, and in consequence her review, "Darwinism in Morals," was a significant retraction of her earlier enthusiasm for Darwin's views. She was clear to state that while she felt that there were no theological grounds for rejecting the idea of evolution per se, and that there could be no faulting the "true philosophic spirit of its author," she was adamantly opposed to the conclusions that Darwin drew regarding the origin of the human moral sense. Where the doctrine of common ancestry and of descent with modification were "topics which properly concern the journals of physical science," in treating of morality as he had, Cobbe continued, "Mr. Darwin gives to a Theological Review the right to criticise the present volume." As she went on to explain, and as intimated above, what she found most deeply troubling about Darwin's argument was the relativism that was at the heart of his suggestion that human morality was nothing more than the accidental outgrowth of social instincts that had been molded solely by the contingencies of circumstance and selection. By such an account there could be no absolute standard of morality, no "supreme and necessary moral law common to all free agents in the universe," either by the lights of Kantian idealism or by the more conventional road of Christian piety. Cobbe lamented that Darwin's account of things would not only "crush the idea of Duty level with the least hallowed of natural instincts," it did not "involve any higher agency for its production than that of the play of common human life, nor indicate any higher nature for its seat than the further developed intelligence of any gregarious brute."[165] Indeed, "in extreme cases (such as that of the bees), the moral sense, under conditions of the hive, would . . . impress it as a duty on sisters to murder their brothers."[166] It was this view of life, in which the morality of an act was a function of no more than the historical "accident" of expediency in contingent circumstances that Cobbe found so unpalatable.[167] Shying away from what she saw to be a worse version of the utilitarian views that Spencer had outlined in his *Principles of Psychology*, and with nothing of Mill's good points to recommend it, she warned her readers of the perilous moral consequences of treating humanity as just another animal. Kantian idealism may not have all the answers, she acknowledged, but to follow Darwinism in morals was to pave a perilous road. In acknowledgment that her own Kantian predilections could not give a complete account of things, she confessed that "a philosophy founded solely on the consciousness of man, *may*, and, very likely, will, be imperfect; and certainly it will be incomplete. But a philosophy which begins with inorganic matter and the lower animals, and only includes the outward facts of anthropology, regardless of human consciousness—*must* be worse than imperfect and incomplete. It resembles a treatise on the solar system which should omit notice of the sun."[168]

Darwin had already read Cobbe's review in draft, which she had sent to him via Fanny Wedgwood. Darwin had not given it close attention at the time, feeling "too much volatilized in this dreadful London to read the article with care." Cobbe had most likely asked for Darwin's comment with a view to making sure she had not misrepresented his views before the review went to press.[169] Darwin, however, declined. Even from such a cursory reading he could see that they were poles apart. "My dear Fanny," he replied, "I hope you will thank Miss Cobbe for her kind & liberal offer. Our differences, however, are too fundamental ever to be reconciled." When the review appeared, Darwin left it for his wife Emma to respond. She wrote to thank Cobbe for the review on her husband's behalf; it had been waiting for them upon their return to Down after a few days in London. Ever the diplomat, she told Cobbe, "Mr. Darwin is reading the Rev. with the greatest interest & attention & feels so much the kind way you speak of him & the praise you give him that it will make him bear your severity, when he reaches that part of the review."[170] In an earlier letter, Emma had been more forthcoming, stating of her husband that "he knows so well how much you & many others will disapprove of the moral sense part that he will not be surprised at any degree of vigour in your attack." For her own part, however, she confessed, "Speaking in my own private capacity I quite agree with *you.* I think the course of all modern thought is 'desolating' as removing God further off. . . . So you see I am a traitor in the camp."[171]

Not all of the reviews of *Descent* were so harsh. There was one in particular that had been published in the *Pall Mall Gazette* in March of 1871 that Darwin thought offered a more positive and thoughtful critique.[172] The review appeared anonymously, and so Darwin wrote to the editor, Frederick Greenwood, to inquire after the identity of the reviewer.[173] He was surprised and pleased to receive a letter from John Morley revealing himself as the author of the review, and the two subsequently exchanged several letters. Morley, who was a staunch defender of Mill's views about both politics and morals, had argued that Darwin was wrong on two counts: first, he suggested that Darwin had misinterpreted Mill, confusing the standard of utility with its goal; and second, he argued that Darwin had been wrong to suggest that Mill's views on the origin and development of morals were incompatible with his own. In fact, in light of this, Morley was unsure that Darwin had actually contributed much of significance to the debate about the origin of moral sentiments that Mill had not already said.[174]

Morley's concerns related to what were clearly differences between how the two men interpreted some key elements of utilitarian philosophy, and as a result, differences in the manner and extent to which they perceived Darwin's theory of the origin and nature of the moral sense to differ from that offered by utilitarianism's most-famous proponent. In *Origin* Darwin had clearly felt that, despite the Malthusian slaughter, natural selection worked, if not in accordance with a preordained teleology, then certainly toward generally progressive ends. Despite the struggle, the war, and the death, he had argued that

FIGURE 3.1. John Morley, 1st Viscount Morley of Blackburn, 1838–1923. Photograph by Elliott & Fry, published by Bickers & Son, 1886. Darwin was deeply impressed by Morley's review of *Descent of Man*. (NPG Ax27808; © National Portrait Gallery, London)

"the vigorous, the healthy, and the happy survive and multiply." Ultimately, it was "from the war of nature, from famine and death, [that] the most exalted object which we are capable of conceiving . . . directly follows."[175] The consequentialism and apparent utility of these conclusions could hardly be denied.[176] However, in *Descent* Darwin had clearly sought to put some intellectual distance between his own views and those of the utilitarians, who emphasized pleasure and pain as the source of all moral judgment. As Morley pointed out, however, so too had Mill.

In a move to develop utilitarianism beyond what he perceived to be the limitations of a Benthamite reading of utility, Mill had sought to explain the political development of a truly other-regarding sentiment and had attempted to derive "virtue" from utilitarian association. Darwin had taken pains to distinguish his own views from Mill's on two points. As we have seen, first he sought to differentiate the utilitarian end of achieving the greatest amount of general happiness from the "general good," as he had defined it in terms of the well-being of the group that resulted from natural selection. Second, and more fundamentally, he had argued that the origin of the moral sense was not acquired solely in the lifetime of an individual, as Mill insisted, but rather that it lay in the social instincts that had long been selected for and inherited from our pre-human ancestors. Morley had taken Darwin to task over each of these points.

In relation to the first point, that Darwin had misread the standard of utility for its motivation, Morley had written, "Mr. Mill, to whom Mr. Darwin refers, has expressly shown the Greatest Happiness principle is a *standard*, and not a *foundation* [of utility]," and thus he concluded that Darwin's critique was misplaced. Continuing, he implied that the second point too was not of the import that Darwin attached to it, and that the standard of utilitarian morality, the Greatest Happiness principle, retained "its validity as a standard of right and wrong action[, in such a way as to be] just as tenable by one who believes the moral sense to be innate, as by one who holds that it is acquired."[177] Morley's words clearly spurred Darwin to revisit Mill's essays, even to the point of enlisting his son William to the task. Upon reflection, he quickly conceded that he had blundered on the first point, but after a detailed rereading of *Utilitarianism*, far from being a moot point, he reiterated that the second difference was fundamental.[178] Not only was it the very point at stake in his discussion of the evolutionary origin and nature of the moral sense, but whether morality was innate or acquired also shed significant light upon what could reasonably be expected of man as a moral being, not only in the present but in the future. William Darwin confirmed his father's view that Mill had been less than clear on the matter, and thought it "very extraordinary that he should recognise the social instincts to be natural to Animals, which he can hardly put down to intellect, and should consider them almost entirely the result of intellect & association in man." He concluded that, ultimately, Mill "must have been very close to allowing the moral faculty to be inheritable, but rather in a muddle on the whole subject."[179]

Darwin had stated that his intention in *Descent of Man* had been to demonstrate that "man, like every other species, is descended from some preexisting form."[180] In the process of making this case, he had addressed the manner of this development, and, as he had suggested he might in the last pages of *Origin*, his doing so had shed light "on the origin of man and his history." Through a focus on sexual selection he had also suggested a novel approach to the explanation of the differences between the human races. However, the most pressing question of the day, and that which elicited the most critical response, was the origin and nature of the human moral sentiments. Even Wallace had ultimately decided that they were beyond the explanatory power of natural selection. Darwin had taken it upon himself not only to show that they could quite easily be explained by his theory, but he took the opportunity to show that their origin lay not in the self-regarding and rational sentiment of "sympathy," but in the genuinely other-regarding social instincts—and in the parental and filial affections in particular. "Thus," he concluded, "the reproach of laying the foundation of the most noble part of our nature in the base principle of selfishness is removed."[181] He had demonstrated that the claim in the Manchester press that he had given an endorsement to every cheating tradesman was in error.

As I shall show in chapter 4, debate over the politics to be derived from different readings of evolution became increasingly significant in the second half of the nineteenth century. It played a role in the development of what Michael Freeden has termed the "new liberalism," in which many liberals moved away from laissez-faire economics and toward an embrace of legislation and regulation, and it was also instrumental in what quickly became known as the "socialist revival" of the 1880s.

4

Liberals and Socialists: The Politics of Evolution in Victorian England

> Everybody has read Mr. Darwin's book; . . . every philosophical thinker hails it as a veritable Whitworth gun in the armoury of liberalism.
>
> T. H. HUXLEY, *1860*

> I am a socialist because I believe in evolution.
>
> ANNIE BESANT, *1886*

By the end of the 1860s the vast majority of men of science accepted that evolution had occurred, even if there were few of them who were convinced that natural selection was either a sufficient or even a primary cause of the origin of new species. Nevertheless, Darwin had succeeded where the author of *The Vestiges of the Natural History of Creation* had failed. Where the anonymous "vestiginarian" had made transmutation a sensation of society conversation, Darwin ultimately succeeded in making it something that could be treated as respectable science. This had as much to do with context as content; a lot had changed in England between the 1840s and the middle 1860s, both in science and society. The "young and rising naturalists" to whom Darwin had appealed in *Origin* had come to the fore, and a few key individuals had not only done "good service by conscientiously expressing his conviction" on the immutability of species, but had taken up influential positions in the scientific societies and were reshaping science in the process.[1] Darwin's work had implications far beyond the scientific community, however. The press had made Darwin a nationally recognized figure. Caricatures and cartoons of apes and angels made good copy and sold papers, and among a broader public too, it was natural selection that became the focus of debate.[2] In contrast to the discussion among naturalists as to whether selection could account for new species, in the press and in society more broadly natural selection was accepted not only as the primary cause of speciation but in the process was also described very much in terms of being the driver of evolutionary progress. More specifically, the focus was upon what it was that was that was being selected, and as a result, upon what exactly the mechanism of natural selection said about humanity and our relationships with each other and with the rest of the natural world. These were significant questions, and the answers were controversial. They reveal that the place and meaning of evolution in English politics was more complex and pervasive than many commentators have recognized.

The Whiggish neo-Malthusian emphasis upon progress through competition that Darwin had employed in *Origin* was read by many as a clear endorsement of the political economy of capitalism. The fact that the question of whether Darwin had provided a defense for the actions of "every cheating Tradesman" could be raised in jest was a clear indication that transmutation was no longer necessarily associated with revolutionary politics, but could be articulated to be quite compatible with even the most extreme conception of laissez-faire political economy. However, exactly what evolution said about human nature or about politics and society was hotly disputed, and over time an increasing number of people argued that evolution endorsed a collectivist politics. As I shall show in the first half of this chapter, evolutionary arguments became central to the development of what Michael Freeden has described as the "new liberalism" of the 1860s as liberals moved away from laissez-faire political economy. In the second half of the chapter, I go on to show that some took these collectivist arguments to socialist conclusions and in the process sparked the English "socialist revival" of the 1880s and 1890s.

As Freeden has pointed out in his *New Liberalism*, the period between 1859 and the First World War was a significant one in the development of English liberal politics. With only a few notable exceptions, across this time period liberals became increasingly disillusioned with the social consequences of laissez-faire, which many struggled to reconcile with their progressive and humanitarian ideals. Instead, they turned to interventionist and collectivist solutions to the social problems that resulted from industrialization and urbanization. As Freeden puts it, the premise of this "new liberalism" was "an ideology of social reform."[3] Following the appointment of Lord Palmerston as prime minister in 1859, and the subsequent formation of the Liberal Party around William Gladstone, political as well as public opinion grappled with the "condition of England" question. Free trade and rapid industrialization had made England the workshop of the world, but this had come at a cost. Vast wealth stood in the face of poverty, vice, squalor, and disease; unregulated working practices such as the truck system of wages and the prevalence of child labor raised questions about the ethics of industrial production; and the inner-city slums were perceived to be breeding grounds of crime, unrest, and moral degeneracy.[4] As the Tory statesman Benjamin Disraeli had pointed out in the 1840s, England had effectively become "two nations,"[5] and Palmerston and Gladstone were bent on trying to make it whole again. Palmerston had overseen the introduction of the Factory Acts, the Truck Acts, and the regulation of child labor in the 1850s in an attempt to ameliorate the worst effects of the "let-alone" economy, and after he became prime minister, liberals abandoned laissez-faire in droves. Instead of free trade, liberals now looked to social legislation to secure the health as well as the wealth of the nation.

My concern here, however, is not with the social or legislative history per se—although there are instances where I shall engage with legislative efforts and their effects—but with the ways in which key liberal thinkers theorized their response to the social and economic developments that were reshaping the world they lived in. We have already seen that liberal theorists like Mill had made an effort to develop a more-collectivist liberal politics; the self-interest that had been at the heart of Adam Smith's liberal political and moral economy was no longer deemed a satisfactory account of human motivation. We have seen too that, like Mill, Darwin was also concerned to account for the origin and development of other-regarding sentiments, although in *Descent* he had diverged from Mill by grounding his own account in our evolutionary natural history. They were far from alone in attempting to theorize liberal collectivism; Darwin had cited a number of thinkers on this subject in *Descent*, reaching back to the eighteenth-century philosophers Smith and Hume in particular, but as Edward Manier notes, also referring to contemporaries. Perhaps the most notable of Darwin's contemporary influences was James Mackintosh, whose *Dissertation on the Progress of Ethical Philosophy* Darwin read and annotated in detail.[6] Others, including Darwin's college friend William Rathbone Greg, his cousin Francis Galton, and the liberal essayist Walter Bagehot had been important too, and each of these men were significant contributors to the debate over the direction that liberalism should take in the second half of the century. The relationship of the individual to society was the crux, both in terms of one's relationship to others, and to society as a whole. This was by no means a new debate, of course, but in the second half of the nineteenth century it was reframed in the context of a vigorous discussion of the implications of evolution for each of these questions. Darwin, Malthus, and Lamarck provided the backdrop for this debate as belief in economic and evolutionary progress were undermined by economic depression, industrial competition from abroad, and as the century went on, growing concerns about moral and evolutionary degeneration. For many, how one interpreted the Malthusian elements of Darwinism was decisive. In relation to my broader thesis regarding the development of Malthusian and anti-Malthusian political traditions, it is not my contention that liberals embraced Malthus and socialists did not, for this is far from the case.

Rather, both within the context of nineteenth-century debates about what it meant to call oneself liberal, and later what it meant to call oneself socialist, one's position on Malthus and evolution was often crucial.

In the context of such a study, the debate between Herbert Spencer and Thomas Huxley over the role of government in liberal politics would seem the obvious place to start; by the 1860s each had become thought of as a prophet of evolutionary ideas. Spencer had attacked the 1870 Education Bill and in doing so had stretched his friendship with Huxley to the breaking point, and Huxley wrote his 1871 essay "Administrative Nihilism" as a response to Spencer's insistent defense of laissez-faire. Huxley, deeply impressed by the importance of Malthus, was all for expanding the role and reach of government, especially in the realm of education and in support of scientific and technological research. Spencer's persistent suspicion of government, his Lamarckism and deep ambivalence about Malthus's conclusions, informed his defense of laissez-faire. While most liberals sought an increasing but limited role for the state, from the mid-1880s there were a number of former radicals who started to call themselves "socialists" and who argued that the state might usefully take a hand in the administration of all aspects of social life. Huxley set himself to carve out a middle ground.

As we have seen in chapter 2, Spencer argued that the progressive social, economic, and political development that had become the hallmarks of industrial England were only possible as a result of a freedom of contract. Since the 1840s he had made the case that government could not compel progress, and indeed he maintained that the problems that were becoming so apparent to those who were so concerned about the condition of England would only be exacerbated by state intervention. Spencer's politics were congruent with his understanding of biology. The agency and autonomy of the individual was a central part of Spencer's interpretation of the Lamarckian inheritance of acquired characters and motivation for his opposition to state interference; actions by the state would only diminish the need for men to forge their own independence.

Spencer's defense of laissez-faire was thus an important issue far beyond how it might serve the business interests of England's industrialists, and his motivation to facilitate individual agency and responsibility was one that resonated with many of his contemporaries. The widespread support for this kind of moral emphasis among dissenters upon respectability, self-reliance, thrift, and independence was reflected in the popularity of Samuel Smiles book *Self Help*, which John Murray had published in the same year as Darwin's *Origin*. Although Murray was less than confident that Smiles's book would be a commercial success, it sold over twenty thousand copies within the year, and had global sales of a quarter of a million by Smiles's death in 1904. Smiles's own intellectual development toward these views and the extent of their popularity is informative, but so too is the speed with which that popularity declined. By the 1890s, Smiles's emphasis upon individual character was no longer in vogue, and in 1896 Murray declined the offer to publish Smiles's book *Conduct*, which was written on the same theme.[7] Smiles was a Scottish dissenter, a surgeon, and a general practitioner who pursued a similarly varied career to Spencer. He worked on the radical *Leeds Times*, becoming its editor is 1838, and he involved himself in local radical efforts to advance household suffrage. In his position as editor he wrote over six hundred leading articles in which he variously opposed the Corn Laws, promoted free trade, and attacked the aristocracy. He also corresponded with the famous radical manufacturer Richard Cobden, and like many of his contemporaries in the radical movement Smiles believed that industry, thrift, and independence were the basis of a just polity—one based upon merit rather than title. As the cultural historian Patrick Joyce has recorded in his *Visions of the People*, although "Rochdale man" was a term originally coined by John Bright (who was himself from Rochdale) to describe someone who embodied the virtues he held dear, the term became a part of mainstream popular radicalism as a shorthand for a certain understanding of middle-class respectability that was based upon the self-educated, independent, and industrious cooperator.[8]

Like many of his contemporaries, Smiles was initially a thoroughgoing radical who hoped to see the day in which peace, justice, and social harmony prevailed across the nation as a result of a reform of the institutions of government. However, in the 1840s he became increasingly uncomfortable with the turn toward violent means advocated by the "physical force" Chartists under the leadership of Feargus O'Connor, a man whom Smiles thought "loud and mouthering."[9] O'Connor was the editor of the *Northern Star*, a paper that used increasingly confrontational rhetoric to press the Chartist claim. As sales of the *Northern Star* increased, so those of the *Leeds Times* fell. Disillusioned with the direction that radical politics was taking, Smiles left the *Leeds Times* in 1842. After several years of writing for a living, in 1845 he took work as the assistant secretary of the Leeds and Thirsk Railway.[10]

Over the following years Smiles became increasingly liberal in his outlook, encouraging each individual to aspire to high standards of personal morals just as the radical liberal statesman and theorist Richard Cobden had done. Placing less and less emphasis upon institutional reform, Smiles focused upon the importance of individual character and the right moral upbringing. In 1837 he wrote *Physical Education; or, The Nurture and Management of*

Children. Smiles continued his study of the development of moral character throughout the rest of his career, publishing works entitled *Character* (1871), *Thrift* (1875), and *Duty* (1880). None of these were as popular as *Self Help*, however, which began life as a lecture on the subject that he had given in 1845 and grew to embody his thoughts on the importance of self-reliance and the development of moral character and emphasized his growing belief in individual solutions to social problems.[11]

Smiles began *Self Help* by quoting such notable authorities as Mill and Disraeli. He quoted Mill in classical liberal mode to argue that "the worth of a State, in the long run, is the worth of the individuals composing it," and Disraeli to support his belief that "we put too much faith in systems, and look too little to men." Smiles went on to make clear that, like Spencer, he believed that anything that was done for people by external agencies acted to undermine their independence and character. It was only through having to rely upon themselves that men learned to persevere: "The spirit of self-help is the root of all genuine growth in the individual; and, exhibited in the lives of many, it constitutes the true source of national vigour and strength. Help from without is often enfeebling in its effects, but help from within invariably invigorates. Whatever is done FOR men or classes, to a certain extent takes away the stimulus and necessity of doing for themselves; and where men are subjected to over-guidance and over-government, the inevitable tendency is to render them comparatively helpless."[12]

The success of *Self Help* was indicative of prevailing liberal sentiment at midcentury. Classical liberal appeals clearly still found a broad audience among dissenters in 1859, although as mentioned, poor sales figures of his later works show that Smiles was riding the last wave that could carry such a limited conception of the role of government. In 1861, the journalist-turned-social-researcher, Henry Mayhew, published a new edition of his social survey of the living conditions of London's urban poor, *London Labour and the London Poor.* This edition was more statistical in its analysis than the first edition of 1851, but it was also more damning of the society that would tolerate such deprivations while claiming to be the most-civilized nation in the world. Freeden has highlighted the significance that liberals were coming to attach to statistics and empirical social science, and Mayhew's surveys and reporting had a lot to do with this development. While not the only reason for a shift away from laissez-faire, it was certainly the awareness of the scale of the social impact of unrestrained capitalism that led many liberals to seek government regulation of labor and state provision of municipal sanitation, education, and charitable relief.[13] Ironically, even as Spencer's works were reaching the height of their popularity, those who defended unfettered laissez-faire found themselves in a rapidly diminishing minority. Mayhew's account put a human face on the statistical evidence that all manner of vice and disease thrived in the midst of degrading poverty, and those who argued that the state should not interfere found themselves engaged in a rearguard action against reformers who were steadily chipping away at the fundamental principles of classical liberalism. Even Smiles came to endorse state education.[14]

Gladstone, who had accepted the invitation to serve as chancellor of the exchequer for the second time in his life in 1859, had done so as a member of Palmerston's cabinet. He too was a classical liberal at heart and had been a firm advocate of the 1832 reforms. His outspoken views on the extension of the franchise had earned him the nickname "the People's William," a reputation that was only enhanced by his fiscal policies, even though he opposed further electoral reform. In 1860 he had proven his mettle as a radical liberal by securing a free-trade agreement with France, and the following year had succeeded in forcing the abolition of tax duties on paper through the House of Lords. This had a significant impact on the radical press in particular, but it also facilitated a veritable explosion in the number of journals and papers published on all manner of subjects. Thus, even before he was elected prime minister in 1868, Gladstone was the people's champion as far as both many radicals and liberals were concerned; he had done away with the hated "tax on knowledge" and was lauded by even the radical atheist cooperator George Holyoake as "the only British Minister who ever gave the English People a right because it was just they should have it."[15]

In his long career in politics Gladstone would serve as prime minister four times. In his first premiership, he reluctantly oversaw the extension of the franchise under the Second Reform Act, and despite his own preferences in political economy, he also administered the subsequent expansion of government regulation of education, welfare, and industry. In this much, he followed Palmerston's lead. Despite Gladstone's reluctance, Spencer royally abused him for betraying the fundamental tenets of classical liberalism. Following Lord Palmerstone's death in 1865, Gladstone shepherded in the Representation of the People Act—or the Second Reform Act, as it was more popularly known. The Reform Act was more far-reaching than Gladstone had intended, more than doubling the electorate when it became law in 1868.[16] Although the nation did not descend into anarchy as a result of giving workingmen the vote as many opponents of the 1868 act had feared, the nation's economy did falter, although for different reasons. The cotton famine brought on by the American Civil War had thrown many in the textile districts out of work.

The rise in unemployment, especially in the radical heartland of the provincial cotton districts, prompted a dramatic increase in the number of charitable organizations across the North and Midlands as well as in the capital. However, concern that charity was undermining the motivation to self-help, a number of prominent old-school liberals formed the Charity Organisation Society in 1869. Its founding is indicative that not all liberals had given up on classical theory and the moral and social significance of self-help. It included among its members a number of prominent liberal politicians and commentators as well as some notable conservatives, who despite their differences were united in their belief that the well-meaning social concern that was finding its expression in private charity was also getting out of hand. Gladstone was among the founding members, perhaps repenting of his role in reform, along with Helen Bosanquet and Octavia Hill as well as John Ruskin and Cardinal Manning. Between them, they sought to coordinate and limit the charitable relief that was being distributed to London's poor and the needy. There were so many different charitable organizations at work, in the capital in particular and especially in the East End and on the Embankment—a notorious district inhabited by homeless and destitute families—that they felt that relief was being distributed indiscriminately. This only encouraged a culture of dependence, and as a result they drew a distinction between the "deserving" and the "undeserving" poor. Employing "scientific principles to root out scroungers," they sought to help the needy to help themselves but resisted giving simple handouts to those who lacked the moral characteristics of thrift and sobriety.[17] "Rochdale man" stood in stark contrast to the image of the scroungers who were presumed to make up the undeserving "residuum."[18] By the late 1860s, "the residuum" was a term that was loaded with Malthusian and evolutionary assumptions. At the time of the census of 1851 London was a city of over two million people, over twice the size of Paris, its closest rival in terms of numbers. Throughout the second half of the century, these number grew exponentially, and were projected to be five million by the turn of the century.[19] The numbers seemingly supported the Malthusians, and social surveys such as Mayhew's only underlined the problems that resulted from such rapid and unregulated expansion. The crowded slums appeared to breed men and women who were beyond self-help and who would only continue to propagate their kind if nourished by the charity of their betters. The kindness of indiscriminate giving was short-sighted, the members of the Charity Organisation Society argued; it was feeding the problem, not providing a solution, and threatened to undermine society as a result.

Although the early years of the 1870s saw unprecedented economic expansion, the hope that this exponential increase in trade, investments, and profits would continue indefinitely proved short-lived. By 1873 the economic boom years that had helped to make liberals of former Chartists were over. The "Long Depression" that lasted well into the 1890s undermined economies across the globe and further weakened faith in laissez-faire economics. Emigration and free trade—the primary weapons in the armory of classical liberal economics—had been found wanting in the face of a poverty that threatened militancy. Even Henry Fawcett, who like Spencer was a diehard free-trader, reluctantly concluded that the prosperity that free trade had brought to the country had been dissipated among the multitude. Writing in 1871, Fawcett lamented the situation in explicitly Malthusian terms, while also invoking liberal middle-class standards of moral respectability. "Unhappily in this prosperity there were the germs of future poverty," he wrote. "The people did not become more prudent; the additional wealth which was then obtained did not generally lead to more saving; a greater amount was spent on drink and the number of marriages rapidly increased."[20]

It was not only Fawcett who feared the consequences of the dissolute and indigent "residuum." The work-shy "social problem group," as they would later be termed, seemed to revel in their lawlessness and amorality.[21] As Gareth Stedman Jones has pointed out, within three years of the publication of *Descent of Man* the liberal economist J. E. Cairnes was effectively writing the obituary of classical economics. The weight of the "undeserving poor"—the growing "submerged sixth"—had little hope of rising, he argued, and unless something was done they would surely drag the rest of society down to their level. "The problem of their elevation is hopeless," he concluded. "As a body, they will not rise at all."[22] In contrast to the optimism that Darwin and Spencer shared, to Fawcett and Cairnes the prospect of social degeneration loomed on the horizon.

By the time that Fawcett was prepared to recognize the limits of laissez-faire, across the nation, even reluctant expansionists perceived a need to institutionalize at least a minimal social safety net for the poor in order to ameliorate what they now recognized as the effects of a free-market economy. The motivations of these expansionist liberals—or the "new Tories," as Spencer derisively called them—were multifarious. Some were genuinely prompted to charity by humanitarian concerns; others were moved by deep political and ideological convictions about liberal ethics; many were merely being pragmatic. Growing unrest and fear of the "mob," particularly in London's notorious East End, led many in the middle class to see the future of British politics as a Hobson's choice between reform or revolution—there was hardly a city in the nation that had not seen sizable demonstrations in favor of reform, and many had endured riot and unrest.[23] Further, socialism loomed from across the channel; the red flag of the Commune flew over Paris only months after *Descent of Man* was published.[24] Even as the economic crisis sparked new waves of emigration from Europe toward the New World, the days of unfettered faith in free trade were over. The Malthusian struggle for existence was not something that the majority of liberals could stand by and watch with a clear conscience. Much to Spencer's distress, even his erstwhile friend and colleague, Thomas Huxley, was among them.

It was though his friendship with Spencer that Thomas Huxley had first fallen in with the radical crowd at the *Westminster Review*. In the 1850s, recently returned from his own voyage as assistant surgeon under Captain Owen Stanley on HMS *Rattlesnake*, Huxley relied upon any work that Chapman could give him to supplement his meager income—Huxley had a small grant from the Royal Society to finance his work, but it didn't go far. Reviewing and translating for twelve guineas a sheet was hack work and a waste of his talent, and he was bitter about it.[25] He had been elected a Fellow of the Royal Society in 1851 for his *Rattlesnake* memoirs, and was awarded the Royal Medal in 1852, but he still lacked a professorship or professional post. Indeed, Huxley's biographer, Adrian Desmond, is doubtless correct that the vitriol in Huxley's attack upon the still formally anonymous *Vestiges*, which was then going through its tenth edition, was as much about Huxley's psychology as about the relative merits of the book. Huxley suspected that Chambers was the author, and he was clearly galled that such a rank amateur and mere popularizer should enjoy such success while he, for all his expertise, "FRS," and Royal Medal, could only mark time. There was still no paying position for him in science. "Mr. Vestiges" would serve as whipping boy to his own lack of advancement, and he damned the book as "so much waste paper." In his frustration, not even his friends in the *Westminster* circle were safe from Huxley's vituperative insistence upon expertise; he even slated G. H. Lewes as an amateur, reserving his praise for the carefully demonstrated expertise of Darwin's barnacle monographs.[26]

Huxley's fortunes finally changed in July 1854 when a position opened up for him at the Government School of Mines in Jermyn Street; Robert Jameson, Darwin's old Edinburgh geology tutor had died, and Edward Forbes, who had previously held a meager lectureship at Jermyn Street, stepped in to fill the gap, leaving open a course of lectures in natural history and palaeontology.[27] Huxley jumped at the chance. "I find it very hard work," he wrote, "but I like it." He picked up more work from the Department of Science and Art at Marlborough House and the London Institution as well, teaching science teachers, as well as taking on the post of naturalist to the Geological Survey.[28] Finally he received a permanent appointment at Jermyn Street as Lecturer on General Natural History, and although the pay for each of these posts was abysmal, combined, Huxley netted 700 pounds for the year. Huxley could finally afford to marry and to turn to his research, but he did not leave off promoting what science could do for the nation—and this at a time when the state was finally taking notice.[29]

The developments in Huxley's career made explicit in practical terms the theoretical differences between himself and Spencer. Where Spencer had thought that Huxley's aim, like his own, had been to free society from the state, in fact Huxley's efforts had been to wrest the power of the state from the hands of Anglican privilege—a difference that was now made clear. Huxley had done most of his early scientific work while in the pay of Her Majesty's Royal Navy, and he now saw his growing connections and employment in various state agencies as an opportunity to weave science and the need for scientific literacy into the very fabric of government. This was fitting vengeance for years of being passed over for less-talented men who had happened to enjoy the connections of faith and title. Huxley was teaching science teachers, had been appointed examiner at the War Office as well as at London University, and was also working on the government Fisheries Commission. With growing Whitehall contacts, Huxley was also successful in his campaign to develop a London University science faculty and to introduce the Bachelor of Science degree.[30]

It was shortly before his appointment at Jermyn Street that Huxley met Frederick Dyster, and through him that he came to rub shoulders with the Christian socialists Frederick Maurice and Charles Kingsley.[31] Huxley had first met Dyster, a Christian Socialist and Church of England alderman, on a well-earned holiday occasioned by his honeymoon at Tenby. A chance meeting on the beach revealed a shared interest in the inhabitants of seashore rock pools.[32] Both men were keen on the importance of education too, not only to advance science, but to improve society. Huxley subsequently introduced Dyster to the director general of the Geological Survey, Henry de la Beche, who was also the director of Huxley's Government School of Mines. Like Huxley and Dyster, de la Beche was a dedicated "merit and education" man who along with Maurice and Kingsley endeavored to make some sort of education available to workingmen in the capital. Chartists would need education if they were to have the franchise—at least that was Kingsley's view.[33] It was no coincidence then that 1868—the year that reform was enacted—also finally saw the establishment of the Working Men's College in London's Red Lion Square. Organized by Dyster but initially funded largely through Maurice's efforts, its success owed as much to Huxley as anyone. As Desmond has noted, the working-class audience for public science lectures was significant and could fill even a large auditorium at sixpence a head with ease; the crowds flocked to hear Darwin's Bulldog. Tapping into the long working-class tradition of auto-didacticism, the Working Men's College was a huge success. Huxley proudly confessed his "plebian" roots and addressed his audiences with a frank honesty and an uncondescending openness they clearly found attractive. With John Tyndall joining him in writing for the *Westminster Review*, science was finding its place in the liberal politics of statecraft, just as it has in the radical politics of reform.[34]

By the 1870s it had become impossible for Huxley and Spencer to hide their differences, and it was Huxley who carried the majority of nonconformists with him on this matter, including powerful industrialists who had the money and position to support his attempts to institutionalize science. Also by 1870 Huxley was president of the British Association for the Advancement of Science and could rely on wealthy men like the engineer and baronet, Sir Joseph Whitworth. Whitworth supported Huxley's attempts to establish government support for science and science teaching; British industry needed trained men if the nation was to compete with Germany, France, and the United States.[35]

Despite the fact that to the public Huxley and Spencer were both still "Darwinians," to Huxley, the social and political consequences of laissez-faire were unconscionable, and adamant that political economy was as suitable a subject for anthropological study as the origins of primeval man, he could not help but see this as a topic upon which he was not only qualified to speak, but upon which he felt morally obliged to speak. From his earliest publications Spencer had made it clear that he saw laissez-faire as the scientific route to individual and therefore social progress, and the fact that in the fifth edition of *Origin*, which had been published in 1869, Darwin had adopted Spencer's "survival of the fittest" to describe his central idea, only furthered the already popular impression that Darwin and Spencer were in accord. This much was too much for Huxley, and regardless of the fact that he knew his friend would be offended, when Spencer pressed the issue over the 1870 Education Act, Huxley could no longer remain silent. The expression of their differences was to cast a shadow over the friendship they had built, despite Spencer's insistence that "this passage of arms was carried on in perfectly amicable spirit, and left the relations between us undisturbed."[36]

Pressure toward state intervention had been mounting, and in the aftermath of reform it was in education that the first real moves were made. The Liberal Party member of Parliament William Forster had sponsored a bill that would place an obligation upon local government to provide schooling for children between the ages of five and twelve. It became law in 1870 and as a result, across England and Wales pupils of that age were to be tested upon approved standards in reading, writing, and arithmetic.[37] While there were several lines of argument against state involvement in education, Spencer's objection flowed from his belief that children should be free to explore their own education and be led by the experience.[38] This was perhaps unsurprising given both Spencer's own childhood education and his Lamarckism, but Huxley would have none of it.

In 1871 Huxley put pen to paper to write "Administrative Nihilism," an essay that was as much a stinging attack on Spencer as it was a spirited defense of Forster. "There is a minority, in whose judgement all this legislation is a step in the wrong direction, false in principle, and consequently sure to produce evil in practice," Huxley thundered. Their presumptions, he wrote, were wrong. Rather, he maintained that "the great attempt to educate the people of England which has just been set afoot, is one of the most satisfactory and hopeful events in our modern history."[39]

Before tackling Spencer head on, however, Huxley dispensed with the stock arguments against a national education. There were those who argued, for instance, that it was courting disaster to educate the lower classes, for in doing so they might become dissatisfied with their station. What would become of England if all aspired to become gentlemen and ladies? To Huxley, the hypocrisy of such a view was rank. Merit, not birth, should determine who should rise and who should fall. "A new-born infant does not come into the world labelled scavenger, shopkeeper, bishop or duke," he railed. Further, it was the very people who made this kind of argument, those from the upper reaches of the middle class, who set so much stock in education as a means to secure the social advancement of their own offspring. While Huxley was under no illusion that a national scheme of education would level the playing field, he did believe it would go some way toward removing the artificial barriers that seemed put in place purely to maintain title over talent. Playing upon the fears of those who saw state action as a safety valve for social discontent rather than its cause, and in full knowledge that his readers would know that socialism was even then engulfing Paris, he pointed out that England's choice was stark: either allow social mobility for those with talent through education, or invite the social revolution of the masses. "What gives force to the socialistic movement which is now stirring European society to its depths, but a determination on the part of the naturally able men among the proletariat to put an end, somehow or other, to the misery and degradation in which a large proportion of their fellows are steeped?"[40]

Although Huxley had felt the sting of being left out in the cold by those who had connections for most of his life, his intentions were clearly reformist rather than revolutionary. Like Kingsley and Maurice, he sincerely felt that the workingman had unjustly been excluded from the franchise, but he too thought that without an education there was little hope of their wielding their newfound political power with any acumen. The workingmen of England were full of honesty and integrity, but this did not mean that they were immune to being misled by demagogues who would be only too happy to make martyrs of them to fulfill their own ambitions. To be sure, Huxley had no intention of teaching men what they should think, but he did have a deep and abiding belief that society had a moral obligation to teach its citizens how to think. Indeed, looking to Paris, he was fairly certain that an uneducated civilization would not survive. Ever the meritocrat, education might make gentlemen and ladies of all who were willing and able, regardless of whether one said "How awfully Jolly!" or "What a lark!"; and while his middle-class readers might have been chilled by his warning of revolution, Huxley's workingmen must have cheered at his confession that "some inborn plebeian blindness of my own" prevented him from seeing the presumed superiority derived from birth. To Huxley's mind, it was not title or wealth that marked out a gentleman or a lady from a scoundrel, but "thoughtfulness for others, generosity, modesty, and self-respect."[41]

Adopting a more serious tone, Huxley turned at last to the misplaced views of the man he had once rated "one of the profoundest living English philosophers," Herbert Spencer.[42] Spencer was wrong, Huxley argued, not in his ardent opposition to government overreach—certainly any state action needed full justification. Rather, Spencer fell into error because he opposed all state action on principle rather than a careful consideration of the merits of each case. His blanket refusal to consider the matter led him to oppose without question all kinds of legislation that Huxley believed was beneficial, such as the legislation aimed at the maintenance of public health, including the Vaccination Act, the Contagious Diseases Acts, the Sanitary Acts, and all attempts on the part of the state to prevent the adulteration of foodstuffs. Further, if Spencer was consistent, then he must not only be opposed to Forster's Education Act but also to "all attempts to promote the spread of knowledge by the establishment of teaching bodies, examining bodies, libraries, or museums, or by sending out of scientific expeditions."[43] This was the crux of their dispute.

Huxley also cared deeply about liberty, of course; but this was no grounds for denying any positive role for government. "Is the fact that a wise physician will give as little medicine as possible any argument for his abstaining from giving any at all?" he asked.[44] Surely the answer was no and thus neither did the fact that laissez-faire was good for business warrant its universal and indiscriminate application. The problem, Huxley lamented, was that "men

FIGURE 4.1. Thomas Huxley, 1825–1895. (Courtesy History of Science Collections, University of Oklahoma Libraries)

have become largely absorbed in mere accumulation of wealth," to the exclusion of all moral considerations.[45] Indeed, while the classical liberal ideal that Spencer clung to had been established to preserve the liberties of each against all, Huxley argued that in light of the increasingly complex web of interrelationships that characterized the development of modern society, it was no longer sufficient. Rather, the state was not only a legitimate authority in arbitrating the inevitable conflicts that ensued when one man's liberty clashed with another's, it was also quite within its remit in establishing certain positive requirements of its citizens whenever their inaction might have an adverse effect upon the liberties of others.[46] Huxley sought to show that even Thomas Hobbes and John Locke, each of whom were frequently invoked in defense of only the negative role of government, had in certain instances recognized that the state might legitimately expect certain positive actions from its citizens to best serve the interests of all.[47] If, as Huxley argued, "in a properly organised State" the government was no more than the "corporate reason of the community," then there was nothing to be feared. The very existence of society was testament to the benefits that individuals gained by giving up some of their personal liberties in the interests of the commonweal, "for no rational creature can be supposed to change his condition to be worse," he said, quoting Locke.[48]

Spencer, of course, had famously described society as a social organism, and it was clear that this was no idle metaphor on his part. However, this was the very point at which Huxley thought he might undermine his friend's attempt to give laissez-faire a biological grounding. Taking Spencer's argument that society was an organism seriously, Huxley countered: "Suppose that, in accordance with this view, each muscle were to maintain that the nervous system had no right to interfere with its contraction, except to prevent it from hindering the contraction of another muscle; or each gland, that it had a right to secrete, so long as its secretion interfered with no other; suppose every separate cell left free to follow its own 'interest' and *laissez-faire* lord of all, what would become of the body physiological?"[49]

To Huxley's mind, if Spencer's own analogy was properly applied, then it refuted rather than endorsed the conclusions he hoped to draw from it. Rather, society was more akin to "the synthesis of the chemist." Society was made up of "a number of primitively independent existences" brought together "into a complex whole" in order to achieve a mutual advantage; "independent elements are gradually built up into complex aggregations—in which each element retains an independent individuality, though held in subordination to the whole." Huxley thought the natural condition of man akin to that of isolated atoms, attracted to mutual action by the promise of mutual gain. Unlike in Spencer's model whereby men thereby adapted to conditions of mutualism (or Darwin's account of human evolution in which men had always been social creatures), instead men would always be self-interested rational actors—simply ones who would realize that they might serve themselves by contracting not to abuse or rob their fellows. Thus, unlike Hobbes's model of government, government was not some external imposing force but rather was the embodiment of the collective will of the people. There was no socialism here—merely liberal individuals contracting to restrain their natural self-assertiveness to reap their own best reward. "Hence," Huxley concluded, "if the analogy of the body politic with the body physiological counts for anything, it seems to me to be in favour of a much larger amount of governmental interference than exists at present."[50]

It was to Kant that Huxley turned in drawing his conclusions. The German idealist philosopher had argued that "the means by which Nature has availed herself in order to bring about the development of all the capacities of man, is the antagonism of those capacities to social organisation." This was the inherent contradiction of the human condition, "the unsocial sociability of mankind."[51] This view of man as subject to his own essential internal conflict would remain the central theme of Huxley's understanding of man and his evolved condition. In chapter 5 we shall have cause to return to Huxley's later expression of these ideas, both in his 1888 essay "The Struggle for Existence in Human Society" and in his final statement on the matter, his 1893 Romanes Lecture, "Evolution and Ethics," which was published in 1894 with an introductory "prolegomena," for clearly he was at odds not only with Spencer in this reading of human evolution, but with Darwin too. In 1871, though, Huxley gave fair notice of his position. While laissez-faire might be an apt description of the ethics of the natural world, the existence of unrestrained competition in nature could not be taken as a guide to action simply because humans had evolved the capacity for ethics. Science, technology, and education, as well as compassion, all worked to raise us above the brute struggle for existence.

Huxley's was certainly an important response to the social and political problems posed by human evolution, and arguably one of the most influential, but there were others.[52] Some of the more prominent were those offered by the mathematician and philosopher William Kingdon Clifford and the author and critic Leslie Stephen. Clifford died tragically from overwork and tuberculosis in 1879, and Stephen and the philosopher Frederick Pollock had a hand in editing and publishing a posthumous collection of his essays.[53] In a number of these essays Clifford acknowledged his debt to both Darwin and Spencer, and in "On the Scientific Basis of Morals" in particular, he reflected upon the implications of Darwin's evolutionary account of morals in *Descent of Man* for our conceptions of the right and the good. It was in this essay too that he articulated a conception of what he termed the "tribal self"—the ethical subject of his theory. According to Clifford, humans had evolved in society, and in a similar vein to both Spencer and Darwin, he too described an evolution of an ethics of social utility in which the individual would come to identify ethical action with those actions that aided the community. As

Clifford concluded, "Although the moral sense is intuitive, it must for the future be directed by our conscious discovery of the tribal purpose which it serves."[54] In his own work, *The Science of Ethics*, published in 1882, Stephen similarly argued that it was the evolution of a cooperative social morality that held society together and which found expression in many of the social institutions that were in evidence in Victorian society. Clearly more influenced by Spencer than by Huxley, he argued that the family, the friendly society, and the political club were all social institutions that nurtured the moral sentiments of respectability, independence, and a belief in collective advancement through individual effort.[55] On the eve of the 1890s, Patrick Geddes and J. Arthur Thompson published *The Evolution of Sex*, in which they sought to establish a collectivist-liberal ethics based on cooperation rather than on competition, in which sex was central. Sex was other-regarding as much as it was self-regarding, they argued. Sexually reproducing organisms could therefore not be entirely self-interested. "In the hunger and reproductive actions of the lower organisms, the self-regarding and other-regarding activities of the higher find their starting point," they wrote. Love, self-sacrifice, and cooperation were the signs of true Darwinian fitness.[56]

In reaction to this determined effort on the part of liberal naturalists, philosophers, and theorists to provide a scientific and thus naturalistic account of human social ethics, there were notable and popular accounts that sought to include faith and religious belief in human evolution. The most significant of these were Henry Drummond's *Ascent of Man* and Benjamin Kidd's *Social Evolution*, both published in the same year of 1894. I shall have cause to consider both of these authors and their works in chapter 7. While the new liberals sought to increase the role of the state, there was debate as to where the limits of state actions should lie. It would be wrong to think that there was a simple continuum from laissez-faire to a centralized state, with classical liberalism at one end and socialism at the other, however, for as I have shown, Walter Bagehot was not alone in making at least a nominally liberal argument for the supremacy of the state over the rights of individuals, and similarly there were plenty of socialists who appealed to the same Godwinian radical tradition that had inspired Spencer.

Before moving on to discuss the various individuals and organizations that made up the British socialist movement and the various ways in which they appropriated Darwinian ideas, it is pertinent to say a little about terminology. From what I have already written in this chapter, it must be apparent that what passed as the new liberalism at the end of the century was remarkably similar to what would have readily been identified as socialism in the mid-1880s. It was only through the politics of the period that what it meant to call oneself a socialist coalesced. The very breadth of what historians have only labeled after the event—"the socialist revival"—raises questions of definition. Indeed, it could be argued that the socialist movement included the cooperative movement, the Labour Church, and the Salvation Army, as well as the Social Democratic Federation, the Fabian Society, the Socialist League, the various Clarion Clubs, and more. On the other hand, there are still arguments among present-day socialists over the extent to which one or another of these organizations was or was not "really" socialist. These are arguments that are measured according to the particular political rubric of the disputants, however, and thus are rarely either resolved or productive. Here, I purposefully make no attempt to define socialism, but instead count as socialist those who called themselves socialist.

In paying attention to "actor's categories" in this way, my intention is to take seriously the contention that what it means to be a socialist has its own history—one that has also been inherently tied up with the history of what it means to call oneself a Darwinist. It is for this reason that I also accept as a "Darwinist" anyone who identified themselves as such. Throughout the 1880s and 1890s, at least in the political sphere, those who did identify themselves as "Darwinists," or "Darwinian," included many who adhered to the belief in the inheritance of acquired characters that historians have come to identify as "Lamarckian," as well as those who emphasized one conception of natural selection or another.

What came to be known as the "socialist revival" thus encompassed an eclectic mix of very different individuals, organizations, and ideals. What initially united them though was the shared conviction that there was something deeply immoral about industrialism and laissez-faire, as well as the belief that another world was possible. Among socialists the view that men were very much a product of the society in which they lived was a commonplace that they had inherited from the English radical tradition. It was an equally common view, therefore, that society seemed bent on degrading the people rather than raising them up to their full potential. The blatant disregard of the impact that capitalist social organization had upon those who lived and worked under it was felt by many to be a deeply moral offense, and thus, whatever else it later became, in the first instance English socialism was a movement inspired by moral outrage.

The explosive growth of socialism in Britain can only be gauged in light of the fact that in the early 1880s there were no openly declared socialist organizations in the country. As Fiona MacCarthy has noted, at that time, "the politics of protest were a confused amalgam of the London working-men's and Radical clubs, the remnants of the Chartists and the more recent influx of foreign refugees from Austria and France after the Commune, from Bismarck's Germany and from the repressive Russian regime."[57] The mid-1880s were a period of agricultural depression, and as a result urban populations swelled as those who could no longer make a living from the land moved into the cities in search of work. The growth of urban poverty and of unemployment, and the slum housing conditions, were a fertile ground for the spread of socialist politics.

"Socialism," wrote one commentator in the pages of the *Contemporary Review*, "is identified with any enlargement and Individualism with any contraction of Government."[58] While this polarization of the debate was an oversimplification, at the time it was written in the early 1880s it was at least indicative of the lay of the land. The moral questions that liberals were asking themselves found their answers in the extension of government in an attempt to ensure the good of the nation as a whole. Some decried this as socialism, while others embraced it as a new liberalism. They were moral questions, however, and just as Freeden has suggested that the new liberalism of the second half of the nineteenth century was an attempt to ground political action upon a scientific conception of ethics, Willard Wolfe has argued that what historians have identified as the English socialist movement—to the extent that it *was* a defined set of ideas—was largely defined in terms of values rather than theories. Socialists appealed to "cooperation" and "association" or, more simply, to "moralization," rather than to any broader principles of economic theory. The moralization of men, Wolfe suggests, was offered as the means and mechanism by which socialists articulated their aims and intentions, although in practice these were often vaguely stated and were little more than their radical forebears had demanded. High on the list of priorities was an ill-defined aspiration to social equality and an end to corruption in government. Utilizing the rhetoric of the "rights of freeborn Englishmen" in justification of land nationalization and "united property"—which might mean anything from the most minimal reform of rents to out-and-out communism—what counted as socialism covered a multiplicity of opinion.[59] Wolfe's drawing of a distinction between values and theories in this way is characteristic of scholarship on this early period of English socialism, and it is certainly the case that an emphasis upon morals was a fundamental part of socialist identity in the late nineteenth century, and not least because it infused the movement with such vitality and millenarian hope.[60] Quoting Ignazio Silone, Wolfe makes the significant point that "on a group of theories one can found a school; but on a group of values one can found a culture, a civilization, [and] a new way of living together among men."[61] Indeed, adherents of all stripes often referred to "the religion of socialism," which relied upon the creation of a thriving socialist culture as an alternative to the new commercial leisure-culture of gin-palace and the music hall, and which provided a basis for morals that was not necessarily derived from religious belief.[62] Wolfe's point is well made. The socialist revival of the 1880s was indeed a broad church that encompassed radicals and communists as well as all shades of opinion in between, and as the historian Stephen Yeo has made clear, it also relied heavily upon the rhetorical strategy of "making socialists"—an attempt to change the world by moralizing one person at a time.[63]

But this is not the whole of the story, for as we have already seen, natural history was just as concerned with moral development as was political economy, and this was especially so insofar as natural history spoke to both human nature and to human social development. Therefore, socialists as well as liberals sought to ground their conception of scientific ethics upon the Darwinian science of evolution.[64] Thus, the very fact that Annie Besant could utter "I am a socialist because I believe in evolution," is indicative of the fact that it was not that socialism was without a theoretical underpinning, but that Wolfe was not looking in the right place to find it. Socialists were looking to the new science of biology rather than exclusively to theories of economics to inform their broader strategies for change. Of course, to contemporaries the two were not separate disciplines at all.[65]

Thus, the strategy of "making socialists" that dominated the early years of the socialist movement was grounded in a distinctly Lamarckian biology. It was not simply a matter of converting individuals to a particular economic outlook, or even a particular lifestyle, although this was certainly a part of what it meant, but rather the aim and intention behind the change in lifestyle was to bring about a biological and moral adaptation to a different way of living, one that was believed to be heritable.[66] The prevalence of Lamarckian ideas in English radicalism has been well established, and thus we should not be surprised that they carried over into the socialist movement. After all, the movement was largely born out of the radical clubs in the nation's capital.[67] However, the same pressures that had pushed liberals to embrace state solutions to social problems also influenced socialism. This was one reason for the popularity of the strategy of "permeation" among the leaders of the Fabian Society. Under Sidney Webb, and with the ardent support of the philosopher David George Ritchie, who joined the Society in 1889, Fabians sought to infiltrate the ruling party in order to pressure them into passing ever more socialistic legislation. Socialism was not something that would require a revolution at some point in the future to bring it about, they argued, but rather had begun with the first piece of social legislation that had long-since passed into law. This was how evolution worked, they claimed. It was a slow, gradual, and incremental change, just as Darwin had made clear.[68]

At the same time, and in contrast to the Fabian strategy, Hyndman was pressing for a historical-materialist analysis of human history. Based upon his interpretation of Marx's *Capital*, he argued that the social and economic structure of society developed along an inevitable path. Throughout human history mankind had evolved as a result of tool use and the application of labor to nature. At periods in that history, one form of labor won out in a violent struggle with its predecessor and through revolution a new era was born.[69] As these two aspects of the socialist movement came to dominate discussion, just as Spencer's Lamarckian politics of laissez-faire was marginalized among liberals, so too the strategy of making socialists was marginalized among socialists. By the end of the century, non-state socialism was only to be found among the nation's few anarchist organizations.

As the historian Stephen Yeo long ago pointed out in his article on the socialist revival, the movement was based upon much more than a critique of the political economy of laissez-faire. Rather, it was an attempt to establish an alternative way of living, to build a uniquely socialist culture, and to bring about a "new life" of the socialist millennium as a means to provoking social change. As William Morris, one of the most influential socialists of the day made clear, socialism was "emphatically not merely a 'system of property holding,' but a complete theory of human life . . . including a distinct system of religion, ethics and conduct."[70] Upon conversion to socialism—and joining the movement was usually described in such explicitly religious terms—a new comrade would enter into the "new life" of the socialist community. As Chris Waters has shown, this would typically include participation in a socialist culture of meetings, activities, and friendships that were intended not only to undermine the appeal of the growing capitalist leisure culture, but also to foster an environment that would nurture social—and socialist—behaviors and provide a radically different space within which nascent socialists could practice socialist living and enjoy each other's comradeship.[71]

Across the country, socialists sought to live differently and encouraged others to join them in doing so. The aim was to show people what life might be like under socialism—to demonstrate that there was an alternative to the incessant work, the dire standard of living, and the degrading struggle for life that was the very essence of the life of the laborer under capitalism. Morris wrote, "The great mass of the oppressed classes are burdened with the misery of their lives, and too much overwhelmed by the selfishness of misery, to be able to form a conception of any escape from it."[72] This was what the strategy of making socialists set out to change. Converts were encouraged to take up "rational recreations," often physically demanding activities that were intended to foster intellectual, physical, and moral improvement.[73] This was a tactic that do-gooding liberals had attempted to press upon the working class as a means to accommodate them into their own moral culture, but in this instance such activities were deliberately adopted as a means of ushering in the new moral world of socialism. Participation in these socialist groups increased dramatically in the 1890s following the founding of the *Clarion* newspaper in Manchester and the subsequent establishment of "Clarion Clubs" across the Midlands and the North of England in particular. The *Clarion* was immensely popular; it was the first newspaper to forge a new and jocular style of journalism. Readers and local groups of socialists joined Clarion Clubs dedicated to cycling, rambling, singing, swimming, amateur botany, and more. Between them, these clubs took hundreds of members out of the crowded and unhealthy slums and tenements in which they otherwise spent their lives, as a means to aid their escape from "the sordid ugliness of modern competitive, commercial life."[74] In the 1890s, the countryside could be reached by only a few hours' walk, even from London and Manchester, and workingmen and women went in droves. Alongside these activities, other aspects of the socialist culture that gained popularity included reforms in diet and dress. Rational dress and Jaegerism, vegetarianism, and temperance all found their place in the movement, as did discussion of the possibilities for new domestic living arrangements—of reorganizing cooking, laundry, and other domestic chores on a collective basis.[75]

Importantly, and what has, I believe, previously gone unnoticed, is the fact that this strategy of "making socialists" had distinctly evolutionary aspirations.[76] The immersion of new converts into a cultural environment of socialist living was understood to have a Lamarckian transformative effect. Just as its practitioners believed that life under capitalism worked to destroy the bodies, minds, and spirits of those who lived under its yoke, so by living the new life of socialism they believed that those bodies, minds, and spirits might be reclaimed, revitalized, and restored.[77] The truly revolutionary promise of Lamarckian biology though was that these acquired effects might be transmitted and built upon across the generations. In this way, the common argument that socialism could not work because people were naturally selfish and individualistic held no water, for under new conditions of life humanity could be expected to rapidly adapt to their environment. Through the exercise of their own free agency, individual converts might form habitual socialistic behaviors that through repetition might become instinctive and therefore heritable. Thus, not only might they become transformed into socialists in and of themselves, but their newly acquired socialist traits might thus be passed on to their offspring and increased across the generations.[78] It was the belief that not only was another life possible than the one demanded under capitalism, but that the life one lived mattered in terms of the future evolution of society, that one's life informed the development of the many and diverse aspects of socialist culture.[79] The fact that the clarion call for a different way of living was sounded from Manchester, the spiritual home of free trade but also the site of slum tenements, destitution, and starvation, is not insignificant.[80]

In the early years of the socialist revival there was little disagreement. As Yeo has noted, "Organisational affiliation was an accident of time, place or convenience." The experience of Ben Turner, who was active in the movement, was typical. He held a number of memberships in the various socialist organizations of his area, including the Social Democratic Federation, the Fabian Society, and after 1893 in the Independent Labour Party, as well as other organizations, and like him, "Many in the movement could have said, 'I belonged to all these bodies neither caring much for dissentions.'"[81] While there was agreement upon the general moral aim of socialism, however, "dissentions" grew as the issue of strategy became more pressing. As I have suggested, at the one extreme was the Social Democratic Federation, which advocated Marxist revolution, while at the other, the Fabian Society argued for the gradual "permeation" of the governing party in order to influence future legislation to more collective ends. As these two alternatives took center stage, the emphasis on making socialists became marginalized. Indeed, as I shall show in chapter 6, this was in large part because it was deemed to be out of step with the latest evolutionary science, as well as being at odds with the broader turn to state action in English politics. The strategy of "making socialists" had rested not only on the inheritance of acquired characters but—and, in this much, it drifted from what Lamarck had actually argued—on the active will of the individual to change. In contrast, the Marxist Social Democratic Federation ultimately came to argue that individual agency was all but irrelevant. Evolutionary forces, revolutionary forces, could not be resisted. Key members of the Fabian Society, citing the gradualism that Darwin had emphasized throughout *Origin*, argued for the management of a gradual legislation of ever more socialistic legislation. Socialism was not something to be trusted to the multitudinous free will of a democracy—not an uneducated one, at least—but was best overseen and administered by committee.

The Democratic Federation was founded in June of 1881, an amalgamation of several London workingmen's and radical clubs that met in Westminster Palace chambers. Its founder and subsequent chairman was Henry Meyers Hyndman. Hyndman was a Cambridge-educated man of independent means who appreciated both Henry George's *Progress and Poverty* (1879) and Karl Marx's *Capital* (1867). He had picked up a French edition of *Capital* on a business trip to Salt Lake City, Utah, in 1881, and had returned to London with ambitions of reviving the Chartist movement as a means to bring about a redistribution of land and capital. He hoped that the Federation would become a national organization that would attain to a mass membership on that basis.[82] Like Marx and Engels, Hyndman was also concerned to present his own brand of socialism as fully in line with Darwinian science.

Hyndman had written a small volume called *England for All*, which he published in 1881 as the founding statement and manifesto of the Federation. As the first and foremost of the many books that were subsequently written on socialism in England, it was highly influential; published in a "cheap" edition later that year, under the longer title of *The Textbook of Democracy: England For All*, it quickly attained a wide distribution.[83] The book is a predictably hybrid work; both George and Marx speak loudly throughout, with Hyndman struggling to reconcile these two very different sets of politics with the German socialist Ferdinand Lassalle's emphasis on the role of the state in transforming society as well as Lassalle's version of the "iron law of wages."[84] Hyndman thus perpetuated the hybrid nature of nineteenth-century British socialism: the uneasy fit between the long-established English radical tradition that emphasized an agrarian politics of the people's rights invested in their claim to the land (a restatement of which occupies the first chapter of the book), and the Marxist analysis of labor and capital (which are the subject of the two subsequent chapters).

English socialism inherited a lot from the radical agrarian politics that centered the connections between the land and the people—a conception of the human–nature relationship that was deeply at odds with the picture of nature as "red in tooth and claw" that was associated with Malthusian interpretations of Darwin. It was the prevalence of the radical agrarian tradition that had prepared the ground for the enthusiastic reception that Henry George's *Progress and Poverty* received in England, a book which did as much as any other to shape the development of British socialism. (Alfred Russel Wallace, for instance, thought the book excellent and recommended it to Darwin enthusiastically; Huxley, on the other hand, hated it.[85]) George was a Californian land reformer who argued that the social and economic ills of society could be rectified not only by the recognition that people had a right to the things they created from their own labor, but also through the acknowledgement that the people had a rightful claim to the land. George argued that this could be achieved by the implementation of a single tax on land ownership. George thus appealed to both radicals and socialists, and when he did visit Britain and Ireland on a lecture tour in 1882 he was championed across the board. Notable, and also a part of Huxley's rejection of George's views, was that in addition to assuming that the relationship between man and nature was ultimately a harmonious one, George dedicated a large part of *Progress and Poverty* to an explicit rejection of Malthus's conclusions. Asserting that the very reverse of Malthusian presumptions were true, George claimed that "in any given state of civilization a greater number of people can collectively be better provided for than a smaller. I assert that the injustice of society, not the niggardliness of nature, is the cause of the want and misery which the current theory attributes to over-population."[86]

George asserted that the poor were made wretched not by the workings of any of nature's laws, but by the injustice of a system that would hoard up wealth among the few while the many were left destitute. Hyndman was impressed; their views were so similar in so many respects that he opened *England for All* with a chapter on "The Land"—after all, the radical tradition had been his own political starting point.[87] "Possession of the land is a matter of such supreme importance to the liberty and well-being of Englishmen," he stated, "that the only marvel is not that there should be a growing agitation on the subject today, but that the nation should ever have been content to bear patiently the monopoly which has been created during the past 300 years." By invoking the rhetoric of "Merrie England"—a vision of the Golden Age of the freeborn Englishman that was generally presumed to have existed prior to the imposition of the "Norman Yoke" of William the Conqueror—he aimed to ignite a sense of outrage at the clear injustices that had been perpetrated upon the workingmen of England ever since. Even under the medieval feudalism of the fifteenth century, he argued, the workingmen of England had "owned the soil and lived out of it." They were better off by far under feudal times than they ever had been, or ever would be, under the tyranny of capitalism.[88]

Like George, Hyndman was also ardently opposed to the Malthusian assumptions that were offered as an explanation and justification for poverty. Even radicals like Charles Bradlaugh and the Fabian Annie Besant had argued that Malthus's basic premise was true, but he was certain that there was nothing about nature that demanded that people starve. This was little more than blind idiocy, and he said as much in *England for All*: "'Oh Yes,' say the followers of Malthus, by no means confined to Mr. Bradlaugh and Mrs. Besant, 'but this over-population is at root of the whole mischief. If only the working class would keep itself under restraint, and not breed at such a terrible pace, they would at once raise their wages by the eternal law of supply and demand. They have to thank their own early marriages and excessive birth-rate for much of their present misery.'" This was a fable of misinformation that sought the solutions of the "Labor Question" in "abstention from marriage and Malthusian devices," he argued.[89] With a just distribution of goods and the equal participation of all in labor, nature would provide an abundance for everybody.

Such a romantic conception of the relationship between the English people and English soil was not totally fantastic, for even in the 1880s many of those who made up the urban work force were recent migrants from countryside to town, and a life on the land, even if not in their own direct experience, had often been within that of their parents or grandparents. Certainly, the recognition of an earlier time when people had experienced the world differently and enjoyed a more intense relationship with nature and the seasons was well within working-class cultural memory throughout the second half of the nineteenth century.[90] Indeed, the popularity of authors like Thomas Hardy, who explored exactly these tensions, reflected and promoted the idea that life had been very different before the industrial age, and that in many ways it had been better. If life had been so different so recently, Hyndman suggested, it was not unreasonable to hope that it might once again be different in the future.

"The Land Question," and all it entailed, was a starting point in English radical politics, and its resolution remained the goal of many of those who laid the foundations for what was to become the British socialist movement. Indeed, indicative that this was a fundamental aspect of British socialism and not merely a withering vestige from an earlier age is the fact that William Morris's *News from Nowhere* and Robert Baltchford's *Merrie England*, both books immensely influential in the movement for generations, were grounded very much in this appeal to a radical agrarian past.[91]

Potentially at odds with this agrarian ideal though, Hyndman had also reproduced Marx's chapters on "Labor" and "Capital" from *Capital* almost verbatim in the chapters immediately following his early emphasis upon the land. Neither Marx nor Engels found such plagiarism flattering: in the *Communist Manifesto* (1848), which went through a number of English editions throughout the 1870s, 1880s, and 1890s, Marx and Engels had famously rejected the "idiocy of rural life" in favor of the intellectual as well as the economic merits of industrial and urban society, and they also thought Hyndman's focus on the land a red herring.[92] However, it was with the publication of *England for All* that Hyndman declared the Democratic Federation an avowedly socialist organization, even though it would be some years before he would openly identify Marx as the "great thinker and original writer" to whom he was "indebted . . . for the ideas and much of the matter contained in Chapters II and III."[93]

One thing that was significant about Hyndman's adoption of Marx was that in the process he embraced an analysis of capitalism that had little room for the very personal and lifestyle politics that were central to the idea of "making socialists." Consequently, Hyndman came to view such a focus on the individual as merely a distraction from the real aim of what he, like Marx, was now referring to as "scientific" socialism—in part to distinguish it from the "utopian" schemes that did not address underlying economic concerns, of course, but also to ally it with Darwin's name and ideas. Just as Marx had recognized the significance of Darwin's views for his own, so too did Hyndman. Indeed, a survey of the socialist press of this period shows that in the course of the 1880s and 1890s there were more articles published that sought to explain Darwin's views and their importance for socialism than there were on Marx and his relevance.[94] Thus, the battle to define what it meant to be a socialist throughout these formative decades of the movement was fought out across a field in which the meaning, and thus the mantle, of Darwin was all important.

In 1876, with the publication of *The Part Played by Labour in the Transition from Ape To Man*, Engels had set the stage for a specifically socialist reading of Darwin. There he had argued that labor rather than the development of the brain had been the prerequisite that had enabled human evolution from their simian origins; indeed, it was the intellectual demands of work, of creating tools and solving tasks, that had driven the development of the human brain. However, Hyndman recognized that the debate about human evolution had long since moved away from monkeys and was now being argued out in the context of anthropology. The evolutionary development of social relations, of how one tribe won the necessary resources from nature and from other tribes, was the ground on which discussion of human evolution now took place. In this much Hyndman was aware that both Marx and Engels had made a significant contribution to this debate, one that not only focused on the social and economic relations as far as they spoke to the conquest of nature and other nations, but that had as its driving dynamic the tensions that this very process elicited between members of the community. Historical materialism not only updated Darwin's account of imitation, tool use, and intertribal conflict, making industry and commerce central to his understanding of what allowed one society to win out over others, but it also explained how, through the process of dog-eat-dog individualism and the ruthless division of labor, ultimately real other-regarding social bonds would be forged that would lead to the overthrow of capitalism and the establishment of the next epoch in human evolutionary history. Hyndman clearly believed that selection, to the extent that it operated on humans, now operated not on physiology but on humans' abilities to create and use technology, and thus ultimately on the mode of production of one society over another.

Following Marx, Hyndman argued that labor was necessary to win liberty from the constraints of nature, and was the force that would forge socialism out of a world that was dominated by capitalist social relations. Hyndman was adamant, though, that this was no admission of Malthusian claims. Although Hyndman made no explicit reference to Darwin in *England for All*, he did freely utilize the terminology of social evolution, and he ensured that readers of the Social Democratic Federation's weekly journal, *Justice*, were well versed in the implications of Darwin's work for socialism.

In his chapter on "Labour" Hyndman recapitulated the value theory of labor that was common to Marx as well as Ricardo, Smith, and a whole host of other political economists. "There is, of course, nothing new in all this," he wrote. "That natural objects are of no value unless human labor is expended on them is a truth as old as the world."[95] Under capitalism and the combined division of labor that had been so lauded by Adam Smith, and the subsequent mechanization of production that was everywhere becoming symbolic if not entirely characteristic of the industrial revolution, the ability of men to make anything of worth was removed from their own independent abilities and placed in the hands of that class of men who owned the means of production—the means of creating value and thus of making a living.[96] Consequently, skilled men who previously had been of independent means found themselves forced to contract themselves to work for others, and to make matters worse, they were in competition with others in a similar situation to their own in order to do so. So much for Spencer's freedom of contract and laissez-faire! It was in the face of such circumstances, of the desperate race that saw men vying with each other to sell their labor-power more cheaply and to submit themselves to longer hours than their competitors or face starvation, Hyndman noted, that the owners of industry shook their heads resignedly and in unity "invoke[d] the sacred laws of supply and demand and freedom of contract, to sanction an amount of daily toil which leaves a man or a woman utterly exhausted at its close, which weakens health, reduces vitality, and hands on a broken constitution to the progeny."[97]

For all the moral evils of such a system, and for all the ill-effects that it inflicted upon generations of working people and would continue to do so into the future, as Marx had pointed out, capitalism was bound to fail. Through its own inherent contradictions, the capitalist mode of production sowed the seeds of its own destruction. In the very process of stripping men of their skills and their independence and of subordinating them to its inexorable machinations, it forged a class of men who depended upon one another and who could depend upon one another—men who ceased to see their interests as mutually antagonistic, but rather had by the very conditions of their employment and as a direct result of their relations to the means of production, come to talk not of "I" but of "we." "When once laborers are collected together in one building," Hyndman wrote, "to do separate tasks at the bidding of an employer, they cease to be separate individuals, and become an organism"—he was almost transcribing Marx; they were "bound to exercise their collective capacity in accordance with the rules of capital," but the trick had been effected.[98] Now their labor was no longer that of isolated individuals, exercised for their own personal gain, but was social, coherent toward a collective end. Of course, and as Marx had made clear, under capitalism the collective labor of the workers was bent upon enriching their employer rather than themselves, but this too had its consequences: not only did they come to see their interests as being in common with their fellows, but they also saw them as opposed to the interests of their employers. Without anyone lifting a finger to consciously bring it into being, class society had been born.

True to the alienating and individuating nature of capitalism, while capital grew increasingly opposed to labor, and labor to capital, capitalists were also forced into an ever more bitter rivalry with one another. Each would be forced to increase the efficiency of his operation; he might invest in machinery and other technological innovations—further divide his workers' labor—but ultimately, in order to undercut his competitors while maintaining their profit, he would lower wages. This much was merely a restatement of Ricardo's "iron law of wages," but as Marx and now Hyndman recognized, wages would not only be driven to the lowest level of subsistence for the worker and his family in this manner, but would be forced even below this meager threshold. Workers would be squeezed, forced to endure their degradation or be put out to seek a better deal elsewhere if they could. Even if they were successful, though, their place would surely be taken by someone more desperate than themselves, often by women or children, whose dexterous fingers, long-established lower price, and easily bossed and nonunionized labor were all weapons in the inexorable campaign to drive costs down and preserve profits.[99]

Even with all these measures, this would only suffice for as long as one innovation replaced another in a bitter arms race to the death. In an unremitting struggle for existence between manufactory and manufactory, men would be thrown out of work, usurped in their enslavement by women and children only to then be driven into competition with them to accept an ever lower wage or starve. The very last thing to fall would be the profits of the industrialist, but ultimately even this last bastion would fall, and one capitalist would capitulate to another as the most efficient and grinding operation won its own brief stay of execution in the bankruptcy and failure of its competitor.[100]

No wonder that shortly after having read *Origin*, Marx had written to his fellow socialist and countryman Ferdinand Lassalle, "Darwin's book is most important and suits my purpose in that it provides a basis in natural science for the historical class struggle," even if "one does, of course, have to put up with the clumsy English style of argument."[101] Competition, profit, and survival were the natural laws of bourgeois political economy, and men, women, and children might all be sacrificed to the ever blind wheels of industrial production. Labor—the workers, the process, and the product—were all insignificant as they became mere means to an end in pursuit of profit. Any altruist in such a world of devil-take-the-hindmost would as surely fall by the wayside as surely as if he were to cut his own throat. The effects were a moral outrage, but the causes—and thus their remedy—were economic, to be found in the amoral and inexorable mechanism of the mode and the means of production. Marx realized this, and so too did Hyndman—or at least in the early days he caught glimpses of it despite his prevailing radical concern with the land. At times, though, he grasped the determinism of the system clearly: beginning with the expropriation of the land from the people, capitalism had subsequently effected the socialization of their labor. Currently, that labor was turned to the production of individual profit for the owners of capital, but the day would come, and come soon, Hyndman believed, when the injustice could be tolerated no longer. On that day, the workers would rise up and assert their rights to the fruits of their own labor and emancipate themselves and their labor power to social ends. Meeting the needs of the many rather than filing the pockets of the few, then too they might reclaim their heritage in the land, finally reclaiming England for all.[102]

This much seemed to be the message that Hyndman offered in *England for All*, but the extent to which he really comprehended the full implications of Marx's critique of labor and capital are questionable in light of the policies he adopted. Of course, it may well have been that in light of the historical-materialist basis of Marx's theory, in which one could neither hurry nor deter a revolution once the conditions for its fulfillment had arisen, that Hyndman was simply anxious to at least do something while he bided his time. He continued to endorse the radical calls for reform: for the eight-hour day; for free and compulsory education; for the provision of clean and healthy municipal housing; and for cheap transportation to enable workingmen to live at a distance from their place of employment.[103] He further supplemented these commonplace demands with the call for manhood, or adult, suffrage; for triennial or annual parliaments; for equal electoral districts; and for payment of members of Parliament. But, and despite the fact that many middle-class liberals decried such measures as rank socialism, at most this was to resuscitate the demands of the People's Charter. "They are but means to an end," he wrote of these measures, "yet it is humiliating to remember that they were demanded in 1848 by a powerful organization, and now here we are in 1881 still without them."[104]

Hyndman distributed his book to members of the Federation at the end of the first week of June in 1881, the same week that a cheap edition rolled off of E. W. Allen's press in Ave Maria Lane. The first pressing having sold well among his middle-class socialist colleagues, Hyndman was anxious to put it into the hands of workingmen, "at a price which will bring it within the reach of all to induce them to combine for their own cause."[105] At this point, even Engels, despite his intense personal dislike of Hyndman, thought that there was hope for the Federation's aim of reviving Chartism. Further, by 1883 the radical outlook of the Federation was rapidly being replaced by a more overtly socialist stance—Hyndman had, after all, given a coherent call for the nationalization of key industries and services under the auspices of a socialist state, and had given a fair account of Marx's materialist theory of history.[106] Indeed, the Marxist elements of *England for All* and the tenor of the Federation's meetings attracted men and women who had become openly enamored of revolutionary socialism and, in equal numbers, had alienated those radicals who were unwilling to embrace the increasingly revolutionary direction in which Hyndman was leading the Federation.

Among those who joined were several who later became influential in the socialist movement, such as Harry Quelch and John Burns. Indeed, the latter was to become the first workingman to attain a cabinet post in Parliament, but in these early years he was out-and-out for revolution. The Federation also attracted disaffected members of the middle class, such as H. H. Champion, an ex-army officer who had resigned his commission in the Royal Artillery, and James Joynes, a radical who had been forced to leave his post as a Master at Eton after having been arrested in Ireland in the company of Henry George. Other significant members included Eleanor Marx, Karl Marx's daughter, and her partner Edward Aveling, as well as Andreas Scheu, an Austrian-born anarchist, and the journalist Ernst Belfort Bax. Of them all, it was Bax who best understood the economic side of *Capital*, and he helped to steer the Federation down this new path accordingly, despite his firm belief that psychology and ethics were thoroughly independent from economics.[107] This was a view he discussed at length with the artist, craftsman, and recent convert to socialism, William Morris.

William Morris joined the Federation in 1883. Formerly involved in liberal radical politics, Morris had grown increasingly disillusioned with a system that was so blatantly careless of its citizens. "I am going in for socialism: I have given up these Radicals," he declared to a friend.[108] At the time Morris joined the Federation, Hyndman had not yet declared it an openly socialist organization, but Morris hoped that it would shortly do so. On this score at least, he was not to be disappointed. Hyndman declared the change in August 1884, and changed the organization's name to the Social Democratic Federation to indicate the transformation. Soon afterward the Federation launched a weekly propaganda journal named *Justice.*[109] This development was hardly unexpected. In Hyndman's *Historical Basis for Socialism in England*, published late in 1883, he had restated his views, but this time around it was clear that the full import and significance of Marx's work had sunk in. He gave an account of England's economic development from what he considered to be the glory days of England's Golden Age in the fifteenth century through to the present, and in doing so he said enough to bring Morris on board.[110]

In many ways *Historical Basis of Socialism* was a much more compelling book than *England for All*, which had been too much caught up in the radical politics of the day—of land and Ireland and the colonies. Next to these issues, Marx's theory of surplus value and historical materialism were as liable to confuse as to convert the earlier book's readers. Even if these issues were both symptomatic of the growing tensions between radical and socialist politics, the earlier book was only a first attempt to synthesize a distinctly British Marxism. *Historical Basis,* on the other hand, spoke much more in terms of the recognition that a bridge had been crossed. In it, Hyndman recognized the moral strength of radical grievances but offered socialist analysis and answers. Here he did not hold back from acknowledging his debts, mindful that it had cost him dear in terms of influence to have alienated Engels by his earlier failure to acknowledge Marx as his source. "My indebtedness to the famous German historical school of political economy headed by Karl Marx with Friedrich Engels and [Johann Karl] Rodbertus immediately following I have fully acknowledged throughout," he advertised clearly in the preface.[111] All the same, he was also clear to attempt to allay the fears that socialism might be seen as a foreign doctrine, the concern that he claimed had led him to omit Marx's name up to this point.

He painted a genealogy of true English lineage. "Socialism is no foreign importation into England. Tyler, Cade, Ball, Kett, More, Bellers, Spence, Owen read to me like sound English names: not a foreigner in the whole bunch," he declared.[112] Again he appealed to the Golden Age of England's past, but in doing so cited the facts and figures of the historian, economist, and free-trade liberal Thorald Rogers's *History of Agriculture and Prices in England* (1866–1902). Even from such a hostile witness it was clear that the standard of living of the workingmen of England had fallen dramatically. The enclosure of lands, the turning over to pasture of tillage that had formerly kept hundreds of men and their families fed and in comfort, followed by the enactment of harsh penalties including the branding and hanging of those made destitute by these changes, made, and still make, grim reading. However, as Hyndman made clear, while this might be a moral outrage, moral outrage alone could affect nothing if the economic conditions were not ripe for change. Social institutions, custom, and religion might retard the development of socialism, but they must eventually give way to the economic conditions of society—just as in "the evolution of a species the hereditary tendency struggles with the growing adaptation to altered conditions of life."[113] But in any society, unless the economic conditions of production changed, Hyndman argued, "it is never possible to make a revolution; it is only possible when a revolution has already begun in the existing conditions of society, to give it outward legal expression and consequent accomplishment." He continued: "To wish to make a revolution is the madness of foolish men who have no idea of the laws of history. Equally foolish and childish is it to attempt to stave off a revolution which has once developed in the bowels of a society, and to withstand its legal expression."[114]

Morris grew increasingly concerned with the direction in which Hyndman was taking the Federation. Not only was he disconcerted by Hyndman's autocratic style of leadership and his tendency to attempt to provoke violent confrontations at the demonstrations that their group sent speakers to, but more than anything he was concerned at the turn away from the strategy of making socialists and the focus upon effecting change in the quality of individuals lives as part and parcel of a socialist strategy for change. Morris left the Federation at the end of 1884 to found the Socialist League, taking the larger part of the Federation's executive committee with him. Historians of English socialism have quite rightly speculated on how the future of socialism in England might have been very different had Morris stayed and taken control of the Federation, and while this is indeed a very interesting question, it is not my immediate concern here. One consequence of Morris's cessation from the Federation and the founding of the Socialist League was that his views were all the more easily marginalized by the larger and already established organizations—not only the Social Democratic Federation, but also the Fabian Society.

The Fabian Society had its origins in the Fellowship of the New Life, an organization set up by the radical J. Morrison Davidson in 1882. This was a group of middle-class men and women who, as their name suggests, aspired to a new way of living that was based upon an ethics of mutual help and simple living. As Norman and Jeanne MacKenzie suggest in their book *The Fabians*, this quest for a new basis for moral living was in many ways a response to the Fellowship's belief that Darwin had made conventional religion impossible.[115] The group's original membership included men and women who were later to become prominent in the socialist movement. Davidson, however, had no sympathy for the socialistic sentiments or political aspirations that motivated many within the Fellowship, a difference in outlook that ultimately led to the departure of those members with socialist convictions in 1883. For Davidson and those that remained, personal spiritual renewal was what was important. It was those who left—among them Edward Pease, Frank Podmore, and Hubert and Enid Bland—who subsequently founded the Fabian Society in January of 1884.

The Society attracted all comers who sought to make a difference. Vegetarians, teetotalers, dress reformers, and the like came together with those who sought variously to legislate the reform of factories and child labor, to nationalize land and capital, or to provide clean water and improve education. In the early days, all were welcome, although membership was always exclusively middle class and intellectual. Despite their intellectualism, the Fabian Society was always dominated by a bias for the practical and had little time for grand theoretical statements. As a result, and despite their differences, the Fabians could all have shared in the sentiments of Sidney Webb, who having joined the Fabians shortly after its foundation, declared to H. G. Wells that he simply "wanted to get things done."[116]

Thus, in spite of the fact that the Fabian Society later became an intellectual and exclusive middleclass organization that pursued an administrative policy at the municipal level of government that was derided by more-revolutionary socialists as "gas and water socialism," the Fabians had originally been as much a part of the ethical socialism of the "new life" as any other socialist organization of the time.[117] When Morris had described socialism as more than a system of property holding, as a new religion and ethics, he had done so in a review of the *Fabian Essays* (1889).[118] These essays, written by some of the most prominent of the first generation of Fabian socialists, including Sidney Webb, George Bernard Shaw, and Annie Besant, quickly became one of the more significant texts that defined the early years of English socialism. Looking back upon the early days of the Society from the position of 1892, Shaw recalled that there had indeed been little to choose between the many and diverse socialist organizations of the mid-1880s.[119] It was only after the 1885 general election and what came to be known as the "Tory Gold scandal" that the movement began to splinter as the different aims and intentions of the various factions became apparent.

Hyndman's parliamentary ambitions had culminated in the Federation standing candidates in the November general election that year. However, it later became apparent that the money to finance the candidates had been provided by the Tories in an attempt to weaken the liberal vote. The scandal seriously damaged the credibility of the Federation within the wider socialist movement, and the poor showing of their candidates was a humiliating testament to the true measure of their following. The vast majority of those Fabians who also held membership in the Federation were led to resign by the fiasco, and the Fabian Society was moved to denounce the political intrigues of their former friends and allies. In response, Hyndman emphasized the one area where the Federation was having success—in its appeals to the unemployed workers movement.[120]

The Federation's hyperbolic revolutionary propaganda, coupled with Hyndman's advocacy of violent revolution at every opportunity, gained him the ear of many of London's unemployed, even if they did not subsequently flock to join the Federation. The press, of course, was quick to blame any disorder upon socialism—Hyndman was living up nicely to their expectations of the socialist bogey. Riots followed a protest of the unemployed in February 1886, and this was quickly followed by the events of "Bloody Sunday," 13 November 1887. What had started out as a protest in favor of free speech and against unemployment had ended in the assault of the crowd by armed police, the reading of the Riot Act, and the clearing of Trafalgar Square by a regiment of guardsmen with fixed bayonets. As a result of the ensuing melee, three men were killed and two hundred were treated in hospital for injuries. All of those injured were civilians.[121]

In the aftermath, the Fabians disassociated themselves from the Federation and the advocacy of such revolutionary strategies. In doing so, they sought to emphasize not only that they held to a different strategy to bring about socialism, but that they held to the only scientifically correct strategy to do so. Certainly, Hyndman followed Marx in claiming that *they* were the scientific socialists, but this much, the Fabians argued, was simply not true. The science in question, of course, was evolution. From January 1887, the Fabians had appended a subtitle to their journal, *The Practical Socialist*, which read "*A Monthly Review of Evolutionary or Non-Revolutionary Socialism.*"[122] Hyndman responded to this barbed criticism by claiming that "those who try to draw a distinction between evolution and revolution or speak of evolutionary and revolutionary Socialism and Socialists, misunderstand the entire theory of sociological development as formulated by the whole Scientific School."[123] But for all his protests, the Fabians became increasingly influential.

Many among the Fabians took their conception of social evolution from Spencer as much as from Darwin. From Spencer they took the idea of society as an organism and an emphasis upon efficiency as an adaptation to the conditions of life. However, and what distinguished many of the Fabians from the vast majority of the rest of the socialist movement, like Darwin, they accepted the doctrine of Malthus as an accurate description of the conditions of existence in nature. Thus, to the majority of Fabians, socialism was an attempt to organize society in such a way as to be best adapted to a world in which Malthus prevailed.

Further though, and what again set them apart from socialists in other branches of the movement, was that they accepted Malthus root and branch. Annie Besant, of course, was also a leading neo-Malthusian influence among the Fabians. She had made her name (or her reputation, one might perhaps better say) in 1877 as a result of her collaboration with the notorious atheist and founder of the National Secular Society, Charles Bradlaugh. The two had put their names to a republication of Charles Knowlton's birth-control manual under the title *Fruits of Philosophy*. Knowlton, an American physician, had written this small pamphlet as a practical guide on how a married couple might still enjoy intimacy without the burden of children. It was direct, and to the Victorian public it was a scandal. Indicted for "obscene libel," both Besant and Bradlaugh were found guilty, given heavy fines, and sentenced to six months imprisonment. The book was deemed more likely to corrupt the reader than to limit fecundity.[124]

Notably, whereas Knowlton had subtitled his work *The Private Companion of Young Married People*, Besant and Bradlaugh's subtitle aimed for a more direct application to the pressing issues of their own times. The subtitle they chose, *A Treatise on the Population Question*, spoke volumes. As they wrote in the introduction to the work, "We believe with the Rev. Mr. Malthus, that population has a tendency to increase faster than the means of existence, and that *some* checks must therefore exercise control over population." They continued, whereas the checks that kept population down under present conditions were starvation and an enormous rate of infant mortality among the poor, "the checks that ought to control population are scientific. . . . We think it is more moral to prevent the conception of children than, after they are born, to murder them by want of food, air and clothing." In a conclusion that was later to cause consternation among many of Besant's socialist colleagues in other organizations (Bradlaugh, a radical individualist, had no truck with socialism), they wrote: "We advocate scientific checks to population, because, so long as poor men have large families, pauperism is a necessity, and from pauperism grow crime and disease."[125] This stance on Malthus and population was controversial to say the least. As Morris and Hyndman retorted, "This foolish Malthusian craze is itself bred out of our anarchical competitive system," and was no law of nature.[126] This much had been evident to the radicals! While it might have been acceptable or even expected in 1877, the fact that the Fabians could apparently endorse such weak analysis was incredible. "What is it produces value?" Morris and Hyndman demanded in a joint publication, and then answered, "—labour applied to natural objects. . . . Yet to provide more wealth we are to cut off the supply of labour by breeding no labourers."[127] It just didn't make sense. It was capitalism and the expropriation of labor that drove men into such grinding poverty, and here were men and women who called themselves socialists blaming the victims! Neither Hyndman nor Morris could let this pass, although it did nothing to stem the enthusiasm for population control in the Fabian ranks.

Among those Fabians who embraced Malthus, H. G. Wells and George Bernard Shaw were perhaps the most notable. Wells had imbibed his Malthusianism from Huxley, whom he had studied under briefly at the Normal School of Science at South Kensington in 1884, whereas Shaw took his Malthus from Besant. In his contribution to the 1889 *Fabian Essays*, Shaw observed that capitalism had blinded the working class to the consequences of their profligate rate of reproduction. The more they were degraded by the hardships of unremitting competition, the more they were "throw[n] back, reckless, on the one pleasure and the one human tie left to them—the gratification of their instinct for producing fresh supplies of man." "They breed like rabbits," he continued, "and their poverty breeds filth, ugliness, dishonesty, obscenity, drunkenness, and murder."[128] Inimitable Shaw. Chastising those who made their profit from the lives of others, he blamed them for causing the social degeneracy they complained of. "They poison your life as remorselessly as you have sacrificed theirs heartlessly."[129]

Beatrice Potter, who married Sidney Webb in 1892 and became an influential figure in the Society, had known Spencer since she was a child. A friend of her father's, he became her mentor and she became his lifelong friend as a result of his longtime acquaintance with her family.[130] As she grew up and eventually became a convert to socialism after seeing the dire social and economic circumstances in which many Londoners lived, she retained a Spencerian conception of evolutionary processes—of society as an organism and of the need for efficiency as a means to the better adaptation of society to the conditions of life—despite developing a very different understanding of the

FIGURE 4.2. "Socialism is presented as the answer to the Sphinx's riddle." Illustration by Walter Crane. (Frontispiece to Shaw et al., *Fabian Essays in Socialism*; courtesy History of Science Collections, University of Oklahoma Libraries)

role of the state in the evolution of society than her erstwhile tutor. In fact, by the time she published the first volume of her autobiography she could quite comfortably confess that she had long disagreed with the "deep-rooted fallacy" that pervaded Spencer's defense of laissez-faire, that "the system of profit making . . . belonged to the natural order of things," while the state did not.[131] Potter was by no means exceptional in being influenced by Spencer in the early years of the socialist movement, however. Indeed, by the 1890s there were few social theorists of any political stripe who did not think of society in organic and evolutionary terms, and this was certainly the case among the Fabians. Beatrice Potter, Sidney Webb, David G. Ritchie, George Bernard Shaw, and H. G. Wells all articulated conceptions of society as an evolved organism. There were differences between them, of course, but more telling were the similarities.

The overarching trajectory of Fabian thought matched the move from individualism toward collectivism that was a characteristic of liberal as well as socialist politics in the second half of the nineteenth century. The utopianism of the early years of the socialist movement—the anarchist influenza that Shaw had talked about, which was focused upon the development of the individual as a means to social change—was replaced with a focus first upon municipalism, and later, after the turn of the twentieth century, with a vision of society run by an efficient, technocratic, and centralized state. As the historian of Fabianism Peter Beilharz has noted, this was a trend that was present in embryonic form even in Sidney Webb's contribution to the *Fabian Essays.* "The perfect and fitting development of each individual is not necessarily the utmost and highest cultivation of his own personality, but the filling in, in the best possible way, of his humble function in the great social machine," he wrote.[132] Later on, he invoked more explicitly biological terminology, arguing, "We must have regard not only to the development of the individual, but also to that of the Social Organism."[133]

Fabian politics was not only functionalist, it was a populist ideology and appealed to the people as a whole rather than to one class or another. In that, Fabian philosophy was unlike that of the other Marxists in the Social Democratic Federation or that of Morris and the Socialist League. It was erroneous and therefore detrimental to social progress, the Fabians argued, to make "socialist" appeals to the interests of one class against another. James Ramsay MacDonald, another prominent Fabian and the man who would later go on to become England's first Labor prime minister, agreed. "The existence of class struggle is of no importance to Socialism. . . . It is the anti-socialist who makes class appeals; the Socialist makes social appeals," he wrote.[134] True socialism should aim to recognize and realize the proper functions of each class to best facilitate the common interests of society as a whole. Indeed, it was in light of this broader focus upon the social whole that the Fabians sought to deny the common socialist charge that the middle class was parasitic upon the productive members of society—those who produced material good through their labor. Instead, they argued that although the middle class was a relatively recent arrival on the evolutionary stage, its members were an essential administrative class-in-the-making and thus were a vital part of any truly socialist agenda. This was a point that Potter made repeatedly, and in her 1891 history *The Cooperative Movement in Great Britain* she cited Alfred Marshall, the founder of the Cambridge school of economics, to do so again: "It is sometimes said that traders do not produce; that while the cabinet-maker produces furniture, the furniture-dealer merely sells what is already produced. There is no scientific foundation for this distinction. They both produce utilities. The dealer in fish helps to move fish from where it was of comparatively little use, and the fisherman does no more."[135]

For the Fabians, producers of ideas as well as social administrators—like themselves—were similarly a vital and productive force in the coordination and development of society. It was true that the middle class had not yet recognized that this was their historic role in the progressive evolution of society, but they were coming to do so. Indeed, it was the Fabians who were realizing their own social utility by pressing them to do so. Again in unspoken acknowledgment of Spencer's Lamarckian influence, Webb argued that this much was a rule of sociology just as it was of biology, that a difference of function precedes a difference in structure. Clearly, the middle class had a function; they had just not yet become fully adapted into the structure of the social organism. This reflected the Fabians' overall conception of the evolutionary process as moving away from the pursuit of individual happiness and toward the pursuit of the collective interest. The development of reason meant that humans were in a position to consciously adapt to these ends. The evolution of mind and will, therefore, were not in and of themselves the end of human evolution, but the means by which individuals could adapt themselves to serving the social whole and becoming a part of the collective will of the social organism. To this end, Fabian politics—as well as the personal accounts of the lives of individual Fabians—were littered with appeals to service, to duty, and to self-sacrifice for the benefit of the good of mankind.[136]

This explains the Fabian ambivalence toward the trade-union organization that dominated labor politics of the 1890s. Trade unions were all well and good insofar as they ensured the health and safety of their members, but they could not be the be-all and end-all of a truly socialist politics, for there was more to society than the workers and more to socialism than production. It also explains the social makeup of the Fabian Society and the strategy they pursued. Throughout its existence, the Fabian Society remained an exclusively middle-class organization, and in contrast to both the Federation and the Socialist League, the Fabians did not seek to enlist the working class to their cause. As Potter none-too-delicately put it in her diary, their aim was to "make the *thinking* persons socialistic," not to "organise the *un*-thinking persons into socialist societies."[137] The Fabian conception of socialism, then, was not the leveling down, nor even the leveling up, of everyone to a position of equality, but instead was grounded in the assumption that the different classes in society fulfilled different necessary functions within the social organism. Socialism was thus a matter of organizing society in such a manner as to achieve social health both efficiently and effectively. It was in order to achieve this aim that, once the Society had come under the sway of Sidney Webb, the predominant Fabian strategy became one of "permeation," the attempt to influence the ruling political party to effect further social legislation. Informing this strategy was the underlying belief that the ills of society could be transformed by incremental and gradual reform from within the present system of government. Webb made it clear that he believed that the coming of socialism was not something that would break over the horizon like a new dawn at some point in the distant future, but rather it was something that was already in existence, even if only in incipient form. Indeed, the "death knell" of capitalism had long ago been sounded with the passing of the first Factory Act in the middle of the century, and socialism had been creeping in and spreading across the political landscape ever since. "Slice after slice have, in the public interest, been cut off the profits of land and capital, and therefore off their value, by Mines Regulation Acts, Truck Acts, Adulteration Acts, Land Acts," he wrote. As far as Webb was concerned, socialism was already here.[138]

One of the most ardent voices within the Fabian Society in support of Webb's permeation strategy was that of the Scottish philosopher and Oxford tutor David G. Ritchie. Indeed, Ritchie resigned from the Society in 1893 in protest at what he believed to be the first moves to abandon permeation as a strategy and to establish an independent labor party.[139] He also provided the most clearly articulated account of Fabianism as an evolutionary social program in his *Darwinism and Politics* (1889), published in the same year as the *Fabian Essays.*

For Ritchie, Malthus was the starting point. He thought the insights that Darwin had derived from Malthus's work of much greater importance than anything Spencer had had to say on the subject. Referring to Malthus's *Essay on the Principle of Population,* he wrote, "An economic treatise suggested the answer to the great biological problem [of evolution]; and it is therefore fitting that the biological formulae should in their turn, be applied to the explanation of social conditions." In light of this, he argued, "evolution has become not merely a theory, but a creed, not merely a conception by which to understand the universe, but a guide to direct us how to order our lives."[140] Ritchie was as aware as anyone that Darwin was being utilized by all and sundry for their own political purposes, but he sought to make the case that the only politics that Darwinism lent support to were the Fabian socialist views he went on to outline in *Darwinism and Politics.* Certainly, Darwin had never intended to give credence to the laissez-faire politics advocated by either the Manchester economists or Herbert Spencer. Ritchie pointed out that Darwin had been clear about his own political beliefs in correspondence with his cousin, Francis Galton. "In answering Mr. Galton's questions, Darwin describes his own politics as 'Liberal or Radical.'" And as Ritchie was clear to point out, "This was in 1873, by which time Radicalism was no longer bound to out and out laissez-faire."[141]

Darwin's account of human evolution in *Descent of Man* was a far cry from Spencer's "old-fashioned individualistic radicalism," Ritchie argued.[142] Darwin had been clear that "we must emphasise that the struggle [for existence] goes on not merely between individual and individual, but between race and race." However, Ritchie was keen to point out that this was no endorsement of interracial warfare or of war between nations. "War is 'natural' only in the sense of being the primitive form of struggle between races or nations, not in the sense of something which ought to be," he wrote. "It is easy for the historian to show how much service has been rendered to mankind by fierce struggles, by war, civil dissension, economic competition. But does it therefore follow that equally good ends can never be attained at less cost?" Ritchie sought to show his readers that rather than endorsing laissez-faire, in *Descent of Man* Darwin had argued that it was through the Malthusian struggle between groups that moral consideration for one's fellows had arisen. Those tribes that were the most socially cohesive and in which individuals were willing to risk their own lives for the good of the group would have been the most successful in the struggle for existence—"and so these virtues have come to receive special respect." This, Ritchie recognized, would lead to morals evolving in such a way that it would be laudable to defend one's own community or to be brave on the field of battle without it necessarily being commendable to extend moral consideration or bravery beyond these limits. Indeed, such behaviors would likely become institutionalized as customary, "and customs are laws in their primitive form—are habits regarded as right, because, having been adopted, they have proved conducive to the welfare and success of the tribe or nation." However, Ritchie pointed out, Darwin had thought that these characteristics would have been extended to others and in different situations as soon as the ability to reflect upon one's actions and to imagine how one's actions might be judged by others had developed, and in this circumstance, "the sphere of courage comes to be extended at least in the minds of some of the more reflective and sympathetic individuals." Thus, natural selection would continue to operate on these communities, selecting those customs that best served the community. In this manner, one would expect to see the spread of moral and social cohesion throughout any given society relatively quickly, just as Darwin had suggested. Moreover, this process would be hastened by the fact that, being conscious and rational animals with language, those who perceived that one course of action might benefit a community over another could argue for a change of action. Indeed, seeing a special evolutionary role for socialist agitation and education, Ritchie concluded that "this is in all ages the function of the political, religious, or social reformer—to save his people from destruction or decay by inducing them to change a custom which, however beneficial once, and in some respects, has now become mischievous." In the past, customs might have been altered by the dictates of a strong-willed leader who was able to enforce changes upon his tribe; in modern civilized societies, however, custom was established by law. Indeed, in the most civilized of nations laws were made and repealed through the institutions of representative government and as a result were amenable to change through reasoned debate rather than through either the arbitrary dictates of a tribal leader or the unwavering rule of natural selection. The role of the political activist was to influence the polity as a means to fostering a new and beneficial morality.[143] Thus, from Ritchie's perspective, the Fabian strategy of permeation was not just pragmatic, it was fundamental to ensuring the continued progressive evolution of social morals toward a more inclusive and socialist future.

Ritchie was moved not only to advance his own evolutionary argument in favor of Fabianism, but to undermine the claims of those who sought to wield Darwin in defense of their own particular politics. He not only attacked those who claimed that the aristocracy were the "fit," in Darwinian terms, by virtue of their position at the top of the social hierarchy, but he also took pains to attempt to rule out Spencer's understanding of evolution, Huxley's liberal arguments in favor of educational reform, and the evolutionary ideas put forward by his socialist colleagues who advocated the strategy of "making socialists."

In discussing Darwin's own political views and the exchange that Darwin had had with his cousin Francis Galton about hereditary genius, Ritchie pointed out that between them they had made it clear that the aristocracy was a dying breed. Not only had Galton confessed that he looked upon the peerage "as a disastrous institution owing to its destructive effects on our valuable races," but Darwin too had joked that primogeniture was a particularly bad way to select for quality![144] More urgently, however, Ritchie was concerned to tackle the "let alone" politics that Spencer had done so much to advance and which until very recently had dominated English politics and policy. For one thing, Ritchie argued, and in stark contrast to Spencer's assumptions, "we cannot be sure that Evolution will always lead to what we should regard as the greatest perfection of any species"; Darwin was not alone in pointing out that "degeneration enters in as well as progress."[145] Spencer had argued that "constitutions are not made, but grow," but as Ritchie pointed out, government was itself an outcome of evolving social forms. The actions of mankind had always determined his fate, either as an individual or as a member of a community. Now, however, in light of Darwin's work, they had the opportunity to affect their future consciously and in light of science. Thus, counter to Spencer's assumptions, he argued that "human societies do not merely grow but are consciously altered by human effort. . . . The teaching of evolutionary science, rightly understood, gives us no excuse for putting aside all schemes of social re-organisation as mere foolishness," for while the state of evolutionary science may not be such as to deduce exact legislative measures from nature, "a fair study of social evolution will at least indicate the direction in which we have to move."[146] As we shall see in the next chapter, Thomas Huxley had taken a similar line in his 1888 essay "The Struggle for Existence," which he had published in *Nineteenth Century*. In it, he had suggested that it was necessary for mankind to intervene in the natural processes of selection, but he had, for what to Ritchie's mind was an unfathomable reason, stopped short when it came to tackling the problem that Malthus had posed. Darwin had remarked that by establishing civilization and developing their capacities to subdue nature to meet their immediate needs, humanity had conquered one aspect of the struggle for existence. However, as Ritchie went on point out, "Professor Huxley then goes on to show how the struggle for existence appears in a new form through the zealous fulfilment of what we are told was the first commandment given to man—'Be fruitful and multiply.' But instead of arguing, as before that the further history of civilisation must consist in putting a limit to this new economic struggle, he avoids drawing any such inference, and very lamely concludes that we must establish technical schools."[147]

Ritchie agreed that Malthus had certainly pointed out a serious and significant problem for mankind, one that socialists both had to recognize and to deal with—and not by simply screwing as much labor out of a man for as small a wage as possible, as Huxley now seemed to suggest.[148] "We began by referring to Malthus, and with Malthus we must end," Ritchie wrote. His position on Malthus, while placing him at home among Fabians, brought him into conflict with the vast majority of the socialist movement. Like their radical forebears, by far the majority of English socialists thought that Malthus was mere capitalist apologetics and offered no great insight into the state of nature. Ritchie understated the case when he acknowledged that "socialists have usually brushed aside the Malthusian precepts and somewhat too lightly neglected the Malthusian arguments." Malthus's observations about resources and reproduction were true, he argued, but they were not insurmountable. "Here as elsewhere, human beings must raise themselves above unthinking animals and not trust to a kind Providence in which they take no part."[149]

It was to this end that Ritchie emphasized working conditions, the position of women, and the population question as the three central elements that a socialist program would have to address. While many who supported even the most radical politicians might oppose child labor and in some cases even might raise concerns about the condition of female labor, it was not sufficient to merely ameliorate these injustices, awful as they were. Rather, Ritchie argued, society needed to address the gross inefficiency of the labor system—or want of system—entirely. Some men were dreadfully overworked, spending fully eighteen hours or more a day slaving for a meager wage, while others could find no work at all. In addition to the unemployed at the bottom of the scale, society also supported a whole class of idlers at the top of the scale—the aristocracy, who consumed far beyond their needs. The unemployed at both ends of the scale were "the moral refuse produced by our economic system. This system is exactly what we find in nature, generally, but one would think that human beings would use their reason to discover some less wasteful scheme." A balance of leisure as well as labor was necessary if society was to function efficiently and healthily.[150] This was not only an argument against Spencerian laissez-faire, but against those who argued that socialism might be achieved by focusing upon the moral conversion of individual men and women rather than attempting to reform the system under which they lived. "It is little use preaching kindliness and consideration for others and hoping that sympathetic feelings will gradually become innate, if the society into which individuals are born be openly and confessedly a ceaseless struggle and competition," Ritchie argued.[151]

The position of women in society was also something that required state action. Ritchie was adamant that women deserved an equal place in society to that held by men. It may well be true that through the long, slow processes of human evolution women had evolved to be more retiring and less intellectually adept than men, but this did not make it either a situation that was right or without remedy. "Because a certain method has led us up to a certain point, it does not follow that the same method continued will carry us further." And in any case, he argued, "it is hypocritical to deny the political capacity of women simply because their political *in*capacity has through long centuries been diligently cultivated."[152] Ritchie was aware that the Lamarckian belief in the inheritance of acquired characters was under attack in biology, but while he admitted he thought that "one who is not a biologist has no right to a private opinion in a biological controversy," he confessed that he thought that environment did have a hereditary effect. "The same education and the same responsibilities will, in the course of time, put the average woman on the same level as the average man."[153] Even if this was eventually ruled out, however, he concluded that both natural selection and sexual selection would work to remedy the difference. Since societies had been selected one over another on the basis of the ingenuity of its men, then how much more would a society prosper if all its members, regardless of sex, applied their intellect and imagination to the problems they faced? Further, with a new ideal of womanhood—"the clever would be preferred to the stupid, and the mother of clever daughters to the mother of stupid daughters"—across a relatively few generations of conscious choice on the part of men, the average intellectual standard of women would be raised. Ritchie did remark, however, that he was skeptical of the claims of any great biological difference in the intellectual capacities of men and women. After all, "Little girls are certainly not on the average stupider than little boys."[154]

Ritchie was not blind to the fact that the evolution of female equality would have a marked effect on the population question. While he thought that a great part of what was passed off as a problem of overpopulation was in fact a result of the poor distribution of both labor and its produce, he did believe in the basic premise of Malthus's essay on population. Society needed to tackle the problem of reproduction as a part of any attempt to manage its production. However, the endeavor to raise women to equal participation in society as men would be a significant part of the solution. "When women have other interests in the world than those of maternity, things will not go on so blindly as before," he argued. "Fewer children will be born, but fewer will die, fewer will be sickly. Those who are born will be better and more intelligently cared for. Two healthy well reared children will be more useful to the community than a dozen neglected waifs and strays. . . . Rational selection will take the place of the cruel process of natural selection."[155]

Ritchie was thus determined that Malthus was something that socialists needed to address and that they could do so successfully, and he believed that in *Descent of Man* Darwin had provided society with a guide to how this might be done. Mankind had evolved to be an ethical species, and while this presented its own problems, they were not ones that could not be conquered by the application of reason to effect the necessary political changes. It was only through making institutional changes that human ideals and human ethics could be changed accordingly. Through the application of right government under the guidance of evolutionary science Ritchie hoped "that by degrees this mutual conflict" of capitalism "will be turned into mutual help."[156]

The second half of the nineteenth century was a tumultuous time in English politics. The effects of rapid industrialization and urbanization called for a rearticulation of the relationship between the individual and society, but it also challenged people to ask deep questions about human nature, human relationships, and the relationship between humans and nature. It was in this context that evolutionary theories of progress and degeneration became pertinent to political debate and that the moral and political significance of Malthus became central. Radical politics, which had been a very broad church in the early nineteenth century, had been split along class lines by the 1832 Reform Act. The middle-class Whiggish radicals went on to articulate a politics that by 1860 had taken on the name of liberalism; however, it was less than a decade later that liberalism erupted in a crisis of identity. Many liberals felt that the politics of laissez-faire could no longer adequately address the social, ethical, or economic problems that faced the nation. As I have shown here, how individual actors conceived of evolutionary processes determined where they stood in this debate.

Michael Freeden's contention is that a number of notable historians of English liberalism have misrepresented this period in the development of English liberal politics by maintaining that different interpretations of Darwinism informed each of these understandings of nineteenth-century liberalism. The war of each against all that Darwin had described in *Origin* seemingly supported the competitive politics of laissez-faire, while the group selection that Darwin, Spencer, and others had discussed seemingly lent credence to imperialist war between nations. What Freeden has pointed out, however, is that there was a middle ground between these two positions—and a vast middle ground at that—in which a political Darwinism based on a collective understanding of social evolution was marshaled to stress cooperation not only *within* groups but *between* groups as well. In the account of Darwin's views on human social evolution that I have offered in chapter 3, I have argued that this latter interpretation was Darwin's position.

As Michael Freeden has pointed out in his *New Liberalism*, past historians have suggested that this debate in the evolutionary politics of liberalism can be adequately described in terms of two camps each pressing their case. Bernard Semmel and J. D. Y Peel, both of whom have written extensively on Herbert Spencer, Social Darwinism, and the later development of liberal imperialism, for instance, have each argued that the history of English liberalism can be understood in terms of a move away from the politics of laissez-faire and toward a liberal imperialism in a quest for "national efficiency."[157] On the one hand were those who sought to defend free trade and industry and who pressed their case in terms of an individualist and competitive Social Darwinism. This is where Spencer has most often been placed in the literature. On the other hand, men like Bagehot have been characterized as social imperialists who grounded their politics on a group-selectionist reading of evolution in which evolutionary and social progress was maintained through international competition. However, this is by no means an adequate account, and for a number of reasons. I have shown in chapter 2 that such a presumption grossly mischaracterizes Spencer, whose Lamarckism led him to emphasize laissez-faire as furthering the economic conditions that would encourage the most-favorable adaptations among mankind. The very qualities that Samuel Smiles had advocated in *Self Help* were those which Spencer sought to see developed, and like Smiles, he believed that government intervention would undermine that possibility. Certainly, there were those who sought to utilize evolutionary arguments in order to mount a cynical justification of the actions of "every cheating Tradesman," but this was from what either Spencer or Smiles were talking about. Rather, Spencer believed that the most healthy and progressive society could only come about by ensuring the healthy and progressive development of the individuals who made up society. To this end, he maintained his opposition to government intervention regardless of whether government was run by corrupt Anglican aristocrats or "new Liberals" like his erstwhile friend and colleague Thomas Huxley. Smiles spoke for the majority of middle-class Victorians in voicing concern for the development of individual character and independence as the primary function of society. Further, and far from being adequately described purely in terms of social imperialism, the new-liberal turn to collectivism was a much more deeply self-reflective analysis of the state of mid-Victorian society. Men like Huxley, Fawcett, Cairnes, and Bagehot were deeply concerned to understand how the insights to be derived from evolution might be applied to ensure the welfare of the English people as well as the English nation as a whole.

To Huxley the world was very much dominated by the reality of Malthus's observations. In light of this, any insistence on total laissez-faire made no sense at all, especially when it came to preparing society and its members to ameliorate the struggle for existence where they could and to win out in that struggle where circumstances demanded conflict. At a minimum, education and scientific and technological researches were exactly the areas in which the state should be involved. Liberal politics had become a collective enterprise, Huxley argued, although where others took their collectivism to further extremes, Huxley sought to defend a middle ground.

By the 1880s, those who were unsatisfied with the collectivism of the new liberalism pressed on to socialist conclusions. However, the debates that had occupied liberals were repeated among the socialist movement. In the early years of the "socialist revival" a diversity of views coexisted without significant difficulties; all were in agreement that capitalism had no adequate response to the ethical wrongs it was causing. However, it was not long before disagreements over strategy erupted among the various leading lights of the movement, and once more at issue was how those who involved themselves in this debate understood evolution to apply to society, and again, where one stood on Malthus proved important. The majority of English socialists were deeply skeptical of Malthus. English socialism had its origins in the radical movement, and Malthus was still regarded as an apologist who attempted to ground social inequalities in nature. Without Malthusian competition as the driving force of evolution, therefore, socialists like William Morris emphasized Lamarckian adaptation to environment and circumstance. As I shall go on to show in chapter 6, this was a view that was quickly marginalized as other strategies came to the fore. Henry Hyndman was typical of many former radicals who turned to socialism. Impressed by Henry George and by Karl Marx, he agreed that Malthus's work was a fiction, insofar as it was an attempt to describe nature. There was competition in life, certainly, but it was the result of capitalist economic forces, and no natural and immutable condition. Under such circumstances natural selection selected those societies that had the most-sophisticated division of labor and the most-advanced technologies. As Hyndman became better versed in Marxism, he stressed that the struggle for existence among workers would ultimately be resolved into a struggle for existence between the workers and their employers. This being the case, the socialist revolution was all but inevitable.

The Fabian Society was more varied and complex in its politics, but its members were united in their belief that the evolution of society was marked by a gradual diminishment of class struggle rather than toward an ever greater exacerbation of class distinction and class conflict. While there were exceptions, the mainstream of Fabian socialism embraced Malthus as the starting point of their politics. In this they had much in common with many of the "new liberals" who sought to organize society so as to best surmount this struggle. Webb, Potter, and perhaps most clearly, D. G. Ritchie, argued this position. Social progress required the marshaling of scientific and productive forces to overcome the Malthusian challenge, and as we shall see in chapter 6, this also prompted some members of the Society to look to control reproductive forces as well.

Deeply influenced by Darwin's *Descent of Man*, Ritchie was not the only person in the socialism movement to argue that mutual help was what Darwin had pointed toward as the outcome of human social evolution, nor was he the only one to tackle Huxley on this basis. In fact, in the year after Ritchie's *Darwinism and Politics* appeared, the Russian émigré anarchist-communist Peter Kropotkin would also respond to Huxley's essay "The Struggle for Existence," in the first of a number of articles he would write for the periodical *Nineteenth Century*.

5

Malthus or Mutualism? Huxley, Kropotkin, and the Moral Meaning of Darwinism

I do not think that there is any evidence that man ever existed as a non-social animal.
CHARLES DARWIN TO JOHN MORLEY, *14 April 1871*

The political landscape of England had changed significantly between the publication and promotion of *Origin* and the last decades of the century, and Huxley as ever was concerned to have a say in its future development. In 1865 he had overseen the formation of the "X Club," a group of nine of the most progressive and influential men in science. In addition to Huxley, the group included George Busk, Edward Frankland, Thomas Hirst, Joseph Dalton Hooker, John Lubbock, Herbert Spencer, William Spottiswoode, and John Tyndall. Ostensibly a social group, they met for dinner once a month, although rumor had it that nothing happened in science without their say so.[1] By the 1870s Huxley and his colleagues in the "X" had successfully harnessed the "Whitworth Gun" of evolutionary naturalism to the liberal radicalism that had come into its own. They not only secured for themselves and their allies council positions in the most-influential scientific societies—the British Association, the Geological, the Linnean, and others—but memberships in the most-established London clubs as well, most notably the Atheneum. These were the real corridors of power.[2] As Ruth Barton has pointed out, the quiet capture of gentlemanly culture in this way was worth a hundred Oxford debates, even if Huxley would later wave off suggestions that such guerrilla tactics had ever been employed as fantastic. "The club has never had any purpose except the purely personal object of bringing together a few friends," he wrote.[3]

Huxley had been keen to push science as a means to improve the state of the nation's industry and economy, and for all his outspoken attacks on the church he had been able to count on the support of liberal Anglican churchmen as well as nonconformist industrialists. Huxley had long championed the importance of education, and of science education in particular; in the

FIGURE 5.1. "The New Science Schools, South Kensington, London." Artist unknown. (From *The Illustrated London News*, 1 July 1872; © Look and Learn / Peter Jackson Galleries)

early days of the 1860s he had received the support and encouragement of the liberal Anglicans Frederick Maurice and Charles Kingsley, but by the 1870s he could also rely upon the support—financial as well as political—of prominent liberal industrialists who could press his case in government. By 1870, Huxley was president of the British Association for the Advancement of Science, and as Adrian Desmond has pointed out, his celebrity was such that his picture graced the pages of both the *London Illustrated News* and *Vanity Fair*; he was indeed among the "Men of the Day." Sir Joseph Whitworth, of Whitworth gun fame, was not alone in lending Huxley his support. Whitworth hosted Huxley and his wife Nettie after the 1870 British Association meeting; he had connections in the War Office and thought nothing of donating a hundred thousand pounds to science for scholarships. At the same time, the liberal *Spectator* reminded Gladstone that the state had a duty to support science as a means to advancing the national interest. Gladstone responded by establishing a Commission on Scientific Instruction, naming Huxley as one of the commissioners. By 1871, Huxley was overseeing the move to the new Science Schools building at South Kensington; it was symbolic of a new era for science at the heart of government.[4]

Huxley had welcomed Darwin's *Origin of Species* as having grounded liberal politics in nature: Darwin had utilized natural selection to validate merit over privilege; it was a competition in which the best man won, not the man with the best connections. Times had changed, however, and liberalism was now at the center of government and Huxley was forging his own connections. As I have shown in the previous chapter, new liberals had followed this broader trend toward the embrace of government action for the good of society, and in *Descent of Man* Darwin had seemingly provided the biological justification for doing so—that it was a collectivist ethic that had facilitated social evolution. Some socialists had even made the case that Darwin had given grounds for believing that social evolution lent support to their own political agendas. As I shall show in this chapter, however, Huxley did not follow Darwin's lead very far, stopping far short of the conclusions that Darwin had drawn in *Descent of Man*. Huxley had formed his opinion of man's place in nature in the 1860s in light of his reading of *Origin*, and his reading of Malthus prevented him from ever thinking that mankind might evolve to a future of peace and harmony. Humankind were simply too much the product of bitter competition and struggle for any more recently evolved ethical feelings they may have to be anything but superficial. In *Origin* Darwin had described the conditions of the animal world as dominated by competition between individuals; why should mankind be subject to a different law? This had been Huxley's own experience of life, after all. However, unlike Spencer or other advocates of laissez-faire, who had argued that humanity should let this natural process take its course, Huxley argued that science, technology, education, and other cultural phenomenon should be put to use to mitigate such struggle. Huxley remained much more deeply impressed by Malthus than Darwin, however. In *Descent*, Darwin had suggested that the forces of natural and sexual selection, combined with moral restraint, judicious marriage, the imprisonment of felons, and the high mortality and low fertility of those of low morals, would maintain a progressive social evolution, but Huxley remained unconvinced. The truths that Malthus has pointed out might certainly be delayed, but they could not be denied.[5]

In his 1888 essay "The Struggle for Existence," Huxley stated his position clearly in a thesis that he would later develop in his famous Romanes Lecture of 1893, "Evolution and Ethics." Certainly, mankind had evolved to be ethical creatures, but only at a comparatively recent point in their history. As a result, Huxley believed that morality and the whole structure of civilization that mankind had erected over the centuries to secure themselves from nature were but a thin veneer over man's deeper natural propensities. Nature, the enemy, not only threatened society from without, but also from within. Society thrived by setting itself in opposition to the wild forces of nature;

men won the struggle for existence by conquering the natural elements and securing nature's resources. Man's very success in this endeavor had facilitated the erection of great civilizations, but, Huxley now contended, all the while this had gone on in denial of the fact that the polite conventions of civilization masked the deep-seated self-interest that Huxley believed was the essence of humanity. By Huxley's account, humans had first banded together not from some innate social instinct, but from a reasoned self-interested calculation; those who did so had been better placed to survive and reproduce than those who did not. This was a subtle but significant divergence from Darwin's position. Further, Huxley shared none of the optimism that was evident in Darwin's work, at least not when he wrote these essays. Huxley had struggled with depression all of his life, and in 1887 his daughter Mady died of pneumonia, exhausted from the trials of her own precarious mental health. Huxley was thus in no mood for optimistic conclusions. He admitted that ethical regard for others would certainly follow from the formation of society, but mankind should not thereby anticipate a millennium of peace and plenty. The very success in the struggle for existence that society secured would also be its undoing. Ultimately, and despite the kinds of social ameliorations that Huxley looked to government, science, and industry to provide, Malthus's population principle would see to it that human morals would in the end be self-defeating.

It was in response to Huxley's essay that the Russian anarchist-geographer and naturalist Peter Kropotkin wrote a series of articles that would eventually be brought together and published as the book *Mutual Aid* (1902). The first essay, "Mutual Aid among Animals," was originally published in the periodical *Nineteenth Century* in 1890. Indeed, Kropotkin continued his efforts to demonstrate what he thought was wrong with Huxley's Malthusian view of life for the next thirty years, publishing numerous articles and several books. His final project, in which he sought to synthesize his lifetime's work on the evolution of ethics, was never completed; he died in 1921 before he could finish it. The first volume of what was clearly intended to be a multivolume work was published posthumously in 1924 as *Ethics: Origin and Development.* Kropotkin argued that Huxley had seriously misrepresented Darwin's views on man and society, and that in fact his own account was much more true to what Darwin had written on the subject. Darwin had not been entirely correct, either, Kropotkin contended; even he had been misled by his political biases into taking Malthusian polemics at face value. As the historian Dan Todes has pointed out, Kropotkin offered his readers a view of Darwin without Malthus, an account of evolution that emphasized the evolution of cooperation that led toward an anarchist-socialist society in which adaptation to environmental circumstance and the inheritance of acquired characters played important roles.[6]

If Huxley had been unaware of the proliferation of socialist ideas and organizations that had erupted from the London radical clubs in 1883-84, the truth of the matter was brought home to him with a jolt in February of 1886. The omnibus in which he was traveling through the West End was caught up in the chaos of the "Black Monday" demonstration-turned-riot that had been organized by the Social Democratic Federation.[7] This was Hyndman at his most dangerous. Addressing a demonstration of the unemployed in Hyde Park, he had incited the crowd, stoking their already deeply felt sense of injustice. Taking their grievances to the streets, the demonstrators rioted. Smashing the West End shop fronts, they vented their anger that such wealth should stand in the face of their own privation.[8] However much Huxley might sympathize with the plight of the workingman, and he clearly did so, this was not the means to social change that he had given the best years of his life to realize. Education, thrift, and hard work had presaged his own achievements, and these were the qualities that he had sought to instill through his efforts as a teacher, lecturer, and now as a statesman of science. Huxley had argued vehemently against both laissez-faire and the institutional barriers to social advancement that had prevented those with talent from rising. This had been his message in "Administrative Nihilism" and he had been prepared to sacrifice a friendship of long standing over his convictions.[9]

With some justification, Huxley considered himself the workingman's friend; he cared about his "cloth caps" and did not want to see them led astray by demagogues. He had established classes for workingmen at Jermyn Street and had been on hand to help realize the aspirations of the Christian socialists. Maurice, Kingsley, and Frederick Dyster had sought to moralize the working class through education, and like Huxley, they saw science as a tool. The science of sanitation and health had been central to the Christian-socialist agenda, but so too had Darwinism. In his own way, Kingsley had done as much as Huxley to popularize Darwin's views.[10] Natural and sexual selection were among God's immutable laws, Kingsley had argued, and as with any other of God's laws, it was for man to discern their workings and effects and then place himself in a position to benefit from them accordingly, for by doing so he was doing God's will. Huxley may not have shared the half of Kingsley's convictions, but he certainly appreciated his efforts to get some modern science into the heads of working people—and doubly so if in the process he could alleviate their fears about the orthodoxy of evolution.[11]

Indeed, while the lectures that Huxley later gave to his aspiring science teachers at the Normal School of Science in South Kensington were designed to give a firm grounding in anatomy and physiology, and thus spoke little to evolution, his lectures to workingmen made it their focus. His lectures were well attended and his audience attentive. "My working men stick by me wonderfully, the house being fuller than ever last night," he had written to his wife Nettie. "By next Friday evening they will all be convinced that they are monkeys."[12] This had been in 1861. Huxley's lectures were published in the *Natural History Review* and were so popular that he reluctantly agreed that the following year's lectures might be taken down in shorthand and published verbatim; *On the Origin of Species, or the Causes of the Phenomena of Organic Nature* (1862) appeared before the year was out.[13] Darwin had loved it. "It cannot fail to do good the wider it is circulated," he wrote. The lectures were "simply perfect." "They will do good and spread a taste for the Natural Sciences."[14] Envious of Huxley's direct and easy lecturer's style, he had concluded, "What is the good of my writing a thundering big book, when everything is in this little green book so despicable for its size? In the name of all that is good & bad I may as well shut up shop altogether."[15] Not all who heard or read Huxley took the moral message as intended, however. Huxley used evolution to wage war upon Anglican privilege. Might not Darwin be taken up against capitalism as well as against "Old Corruption"? Many thought so. Karl Marx sat through Huxley's lectures attentively. He had read *Origin* twice over the years and had pondered this very thought.[16]

After all the promise of the 1870s, by the 1880s Huxley was dismayed at the direction that English politics was taking and irritated that Darwin's name and ideas were being dragged through the mud, not only by the "let-alone" school, but now by foreign agitators. Evolution did support a politics, but it was not that of revolutionary socialism—of that he was sure. Ever since he had taken the presidency of the Royal Society in 1883, though, he had tried to keep himself out of the political limelight. It did not fit a man well, he thought, to try and serve science and yet be seen to be a party man. Indeed, he had waved off requests to write on politics at the time. "I have other fish to fry," he wrote to his son when pressed for a piece for a political journal. This fit with his X Club agenda. "Such influence as I possess may be most usefully employed in promoting various educational movements now afoot," he wrote, "and I do not want to bar myself from working with men of all political parties."[17] Huxley was thus later greatly exercised when it transpired that the man who succeeded him to the presidency in 1885, G. G. Stokes, evidently felt no such compunction. Stokes had accepted an invitation to stand as member of Parliament for the University of Cambridge, a seat he won handsomely, much to Huxley's annoyance.[18] The educational movement then afoot was one he had had a hand in throughout that spring, and which was due to culminate with the delivery of an address to the Technical Education Association at Manchester at the end of November.[19] The city of Manchester was raising taxes to fund local technical education, an act which was not uncontroversial in itself. It was 1887.

Yet, just as he was preparing to travel to Manchester, tragedy struck Huxley's family for the second time. Back in 1860 his firstborn son, Noel, had been struck down by scarlet fever; Huxley's anguish as well as his resolve was evident in every line of the deeply personal letters he wrote to Kingsley. Kingsley had urged him to seek solace in the possibility of a reunion with his son after death, but Huxley would not take consolation in something he could not believe in. "My convictions, positive and negative on all the matters of which you speak, are of long and slow growth and are firmly rooted. But the great blow which fell upon me seemed to stir them to their foundation, and had I lived a couple of centuries earlier I could have fancied a devil scoffing at me and them—and asking me what profit it was to have stripped myself of the hopes and consolations of the mass of mankind? To which my only reply was and is—Oh devil! truth is better than much profit. I have searched over the grounds of my belief, and if wife and child and name and fame were all to be lost to me one after the other as penalty, still I will not lie."[20]

Principled science had been cold comfort, as despite his years of medical training Huxley could only watch his son fade away. Now his twenty-eight-year-old daughter Mady, who had always been emotional, was hysterical and sinking fast toward what he feared was dementia. Huxley's name brought the famed neurologist Jean-Martin Charcot to her bedside, and he agreed to take her into his care at the Salpêtrière Hospital in Paris. Mady made the journey, but a severe bout of pneumonia released her from her mental torment before Charcot even had a chance to see her.[21] Reeling, Huxley reached out to Spencer. Death had "carried her off without warning," he wrote.[22] It was one of the very few letters he wrote on the subject, and Mady's death brought them close again. Relief and grief came together and he poured it out.

The long and drawn-out nature of Mady's decline had pained the whole family. "I cannot convey to you a sense of the terrible sufferings of the last three years better than by saying that I, her father, who loved her well, am glad that the end has come," he wrote. "My wife is well nigh crushed by the blow. For though I had lost hope, it was not in the nature of things that she should."[23] Mady, whom Huxley had always called "my bright girl" and in whom he had seen "such bright prospects half a dozen years ago," was gone. "Rationally

FIGURE 5.2. Marion Collier (née Huxley), 1859–1887. Painted by John Collier, 1882–83. (NPG 6032; © National Portrait Gallery, London)

we must admit that it is best so," he wrote to Dyster, but he did not feel it to be true, adding, "but then, whatever Linnaeus may say, man is not a rational animal—especially in his parental capacity."[24] "Don't answer this," he wrote to Spencer. "I have half a mind to tear it up."[25]

Even in grief Huxley was unwilling to run from his obligations, and grateful for the occupation, he scribbled lecture notes and caught the train to Manchester, refusing to break his appointment for the 29th despite it all. It all came out in his lecture though—in floods. A loose cannon, he fired off shot in all directions. Any lasting reconciliation with Spencer was crushed as he lashed out again against the overwhelming carelessness of nature and those who would emulate its amorality by applying laissez-faire to society. He fired one more broadside at natural theologians and another at the romantic notions of those radicals and socialists who saw beneficence in nature. All who thought nature a benevolent guide to life were in his sights, and he was determined that they could not be more wrong.

His lecture was revised for publication in *Nineteenth Century* early the following year, appearing under the title "The Struggle for Existence: A Programme."[26] It might have been revised, but it was still raw. Huxley brought the full force of a merciless natural selection down upon humanity. Nature cared nothing for our trials. The only refuge from our heartless and isolated existence was to be found in family—and Huxley knew just how vulnerable that could be. Beyond that intimate circle, we might reach out to others, but in the full knowledge that any society we might forge from such connections was at best a fragile affair. In time, all our ambitions are destined to come to naught, he argued; the relentless laws of nature would not be denied. The whole essay was politics from start to finish—the very thing he had been anxious to avoid as a statesman of science. But in his pain he could not but wear his heart on his sleeve. It was bitter irony. Despite his years of ambivalence about natural selection in the formation of new species, now he could see the world in no other terms. Life was a ruthless and relentless struggle and even man's best efforts to sue for peace were ultimately futile. "The effort of ethical man to work towards a moral end by no means abolished, perhaps has hardly modified, the deep-seated organic impulses which impel the natural man to follow his non-moral course," Huxley wrote. "One of the most essential conditions, if not the chief cause, of the struggle for existence, is the tendency to multiply without limit, which man shares with all living things." But a consequence of this Malthusian imperative was the inevitable "re-establishment, in all its intensity, of that struggle for existence—the war of each against all—the mitigation or abolition of which was the chief end of social organization."[27] Nature was blind to the sufferings of her creation, and those who argued otherwise were pedaling delusion.

He had been here before, of course, arguing against those who saw nature as evidence of a benevolent and loving God. Like Darwin, he had pointed out that there was just as much evidence of suffering in the workings of the natural world as there was of anything good and pleasurable; "the optimistic dogma, that this is the best of all possible worlds" was "little better than a libel upon possibility." Even Kingsley, who had accepted the harsh reality of a Darwinian world had had no reasonable grounds for his belief that God kept a ledger and oversaw that it all worked out for the best. The idea "that the sentient world is, on the whole, regulated by principles of benevolence, does but ill stand the test of impartial confrontation with the facts of the case," he wrote. The application to the natural world of such anthropomorphic terms as were expressive of human morals was wholly inappropriate. The deer suffered that the wolf might eat, that was all. "Viewed under the dry light of science nature will appear to be neither moral nor immoral, but non-moral."[28] The tendency to view nature as a moral and beneficent entity was no longer restricted to the natural theologians—Georgite land nationalists and any number of socialists were also now pedaling some kind of reconciliation with nature as the reward for pursuing their various schemes.[29] This was as much idiocy as Spencer's belief that everything left to nature would bring on some wonderful utopia: "From the point of view of the moralist the animal world is on about the same level as the gladiator's show. The creatures are fairly well treated, and set to fight—whereby the strongest, the swiftest, and the cunningest live to fight another day. The spectator has no need to turn his thumbs down, as no quarter is given. He must admit that the skill and training displayed are wonderful. But he must shut his eyes if he would not see that more or less enduring suffering is the meed of both vanquished and victor." There was no need to go to the depths of hell to hear the wailings and gnashing of teeth that Dante had imagined in the *Inferno*, he wrote.[30] They could be heard throughout the natural world, and Huxley heard them loud and clear as he buried his dear Mady.

To Huxley's mind it now seemed that Darwin had been unwarranted in drawing the optimistic conclusion that he had done in *Origin*. There, Darwin had suggested that in spite of the suffering, the war, and the death that so proliferated throughout the natural order, "we may console ourselves with the full belief, that the war of nature is not incessant, that no fear is felt, that death is generally prompt, and that the vigorous, the healthy, and the happy survive and multiply."[31] Such optimism was unfounded. Mady's death, when it came, had been mercifully swift, but what difference had that made? It was a mercy for her, certainly, but had done little to assuage the fears or alleviate the long and anguished pain that he and his wife had endured. Any notion that the evolutionary ends justified the means was as untenable as it was illogical. "It is not clear what compensation the Eohippus gets for his sorrows in the fact that, some millions of years afterwards, one of his descendants wins the Derby," Huxley mused.[32] The jest was without humor. Indeed, circumstance had brought Huxley face to face with the realization that the whole progressive tenor of *Origin* was unwarranted. Darwin had recognized that there was no necessary tendency for organisms to develop in one direction over another—a barnacle was as well adapted as "the most exalted object of which we are capable of conceiving"[33]—but he had still convinced himself that progress was the likely overall outcome. It all depended upon the contingencies of time and place, of course. Both men realized that the very factors that made an organism fit in one environment might be the cause of its demise in another. To Huxley, it was clear that there was no more likelihood of progress than degeneracy. "Upward or downward," he wrote, "retrogressive is as practicable as progressive metamorphism." Indeed, in light of William Thompson's second law of thermodynamics, physicists were now convinced that life on earth would find its end in the entropy of sun-death and global cooling. "The time must come," Huxley concluded, "when evolution will mean adaptation to a universal winter, and all forms of life will die out, except such low and simple organisms as the Diatom of the arctic and antarctic ice and the Protococcus of the red snow." On Earth there was no inevitable progression, no gradual tendency toward perfection: "The course of life upon its surface must describe a trajectory like that of a ball fired from a mortar; and the sinking half of that course is as much a part of the general process of evolution as the rising."[34] Spencer had been wrong about this much as well. It was within this broad sweep of dispassionate events that the human drama of love and death played itself out. However, and contrary to those of the Manchester school who had attempted to co-opt Darwin to endorse their own selfish creed, Huxley argued that the fact that nature knew no morals did not give man leave to embrace the same dispassion as a virtue. Mankind's interests lay in opposing, not in mimicking the ways of external nature. He clarified this point in his 1893 Romanes Lecture. "There are two very different questions which people fail to discriminate," he wrote. "One is whether evolution accounts for morality, the other whether the principle of evolution in general can be adopted as an ethical principle. The first, of course, I advocate, and have constantly insisted upon. The second I deny, and reject all so-called evolutional ethics based upon it."[35]

While man was certainly a natural animal, he had evolved to oppose the hand that had reared him—albeit only in his most recent history. Ever since man had become a social animal he had learned to defy the rule of this primordial natural law. "Society differs from art in having a definite moral object," Huxley wrote, "whence it comes about that the course shaped by the ethical man—the member of society or citizen—necessarily runs counter to that which the non-ethical man—the primitive savage, or man as a mere member of the animal kingdom—tends to adopt." Rather, "the latter fights out the struggle for existence to the bitter end, like any other animal, the former devotes his best energies to the object of setting limits to the struggle."[36] But, and Huxley reiterated the point, such moral sentiments were only a recent development. "The first men who substituted the state of mutual peace for that of mutual war, whatever the motive which impelled them to take that step, created society," he wrote. In so doing, "they obviously put a limit upon the struggle for existence," at least between members of that society.[37] Prior to that time, man had lived and died with no more moral comprehension than an animal. "However imperfect the relics of prehistoric man may be," he went on, "the evidence which they afford clearly tends to the conclusion that, for thousands of years, before the origins of the oldest known civilisation, men were savages of a very low type. They strove with their enemies and their competitors; they preyed upon things weaker or less cunning than themselves; they were born, multiplied without stint, and died. . . . They were no more to be praised or blamed on moral grounds, that their less erect and more hairy compatriots."[38] Indeed, prior to the first formation of society, humans had set themselves against one another: "Life was a continual free fight, and beyond the limited and temporary relations of the family, the Hobbesian war of each against all was the normal state of existence."[39]

Mercifully though, this was not the be-all and end-all of human existence.[40] Having established society, mankind could find another aim. Although, like Darwin, Huxley did not openly speculate upon the reason for that first social act, it is clear that he saw it as the foundation of the ethics of traditional liberal political economy. Even though he had only recently said to Dyster that he believed that man was "not a rational animal" in some respects, he clearly conceived of man as a liberal rational actor. The ideal of the ethical man was to limit his own freedoms "to a sphere in which he does not interfere with the freedom of others" to seek the benefit of the commonweal in as far as his own benefit depended upon it; indeed, it had become "an essential part of his own welfare." The establishment of mutual peace was thus both a means to an end as well as an end in itself. The self-restraint that made society possible was the embodiment of "the negation of the unlimited struggle for existence." In short, he wrote, "social life, is embodied morality."[41]

However high man might have risen along this mortared trajectory, though, progress could not continue indefinitely. Indeed, no matter how sincere man's moral sentiments, being of such recent origin they were infinitely fragile compared to the long-inured self-assertion out of which long eons of struggle had forged human nature. As far as society had come along the upward path, it had "by no means abolished, perhaps has hardly modified, the deep-seated organic impulses which impel the natural man to follow his non-moral course."[42] The Malthusian urge to reproduce was the serpent that would upset any hope of an earthly paradise.

Although in the immediate future Huxley urged the provision of technical education and government sponsorship of science and industry as a means to give England a competitive edge over her rivals, in the longer term, ultimately sex and the progeny that would be its issue would swamp society and break the fragile bonds that made social life possible: Malthus would ultimately once again have his day. Echoing Darwin, who had made the Malthusian elements of natural selection abundantly clear, Huxley wrote: "One of the most essential conditions, if not the chief cause, of the struggle for existence, is the tendency to multiply without limit, which man shares with all living things," adding that "'increase and multiply' is a commandment traditionally much older than the ten," and perhaps the only commandment that man had ever followed with anything like consistency. The inevitable result, he concluded, would ultimately be a resurgence of the struggle for existence within human society, "the re-establishment, in all its intensity, of that struggle for existence—the war of each against all."[43]

Huxley had thus far laid out his position quite clearly: both the radicals and the socialists had entirely misconceived the state of nature, and as a consequence the state of man. As a result, not only was their critique of the present wrong, but Huxley believed that their hopes for the future were impossible as well. "No fiddle-faddling with the distribution of wealth" would help, he wrote, "and however shocking to the moral sense this eternal competition of man against man and of nation against nation may be; . . . this state of things must abide, and grow continually worse." But what then might be done? "It is the true riddle of the Sphinx," he wrote, "and every nation which does not solve it will sooner or later be devoured by the monster itself has generated."[44]

The Sphinx's riddle was an old legend. With a lion's body and the wings of a great bird, Sphinx had the face and breasts of a beautiful woman. Merciless and treacherous, she guarded the entrance to Thebes against all comers, and the failure to answer her question meant certain death. Thomas Carlyle, one of Huxley's favorite authors, had recalled the riddle of the Sphinx in his 1843 book *Past and Present*: "Nature, Universe, Destiny, Existence, howsoever we name this grand unnamable Fact in the midst of which we live and struggle, is as a heavenly bride and conquest to the wise and the brave, to them who can discern her behests and do them; a destroying fiend to them who cannot." In the myth, Oedipus answers the riddle and conquers the Sphinx, but who, Carlyle asked, would be the Oedipus of "the Labour Question," as it was termed? Carlyle had argued that "justice" was the answer, but this was really just to restate the question: "Answer her riddle, it is well with thee. Answer it not, pass on regarding it not, it will answer itself; the solution for thee is a thing of teeth and claws." "Thou art not her victorious bridegroom;

thou art her mangled victim, scattered on the precipices."[45] George and Marx were soon to be dead by the wayside, Huxley thought, fancying himself as the bridegroom.

Not incidentally, and as we shall see in chapter 6, Huxley's framing of what Malthus had to say to the nature and condition of man in this way resonated with one reader in particular—the young and aspiring science-fiction writer H. G. Wells. Wells, who had studied under Huxley briefly in 1884, would write *Time Machine* a decade later. Dominating the degenerate civilization that Wells's "Time Traveller" encounters in the dim-and-distant future is a colossal statue: "It was very large, for a silver birch-tree touched its shoulder. It was of white marble, in shape something like a winged sphinx, but the wings, instead of being carried vertically at the sides, were spread so that it seemed to hover."[46] The spread of the Sphinx's wings indicated that society had found no answer to the labor question; it had now become an issue of "teeth and claws" as she hovered over her victims.

In the short term, and in line with the main thrust of his speech, Huxley offered a two-pronged palliative to hold the Sphinx at bay. Well-directed labor would fend off the Malthusian monster, but a prerequisite too was social stability. The former demanded the recognition that the country was in a veritable struggle for its life with other nations and that to have a hope of winning out, labor must be efficient—and to be competitive, it must be cheap. The latter spoke to Carlyle's concern with justice: the remuneration for a fair day's labor must be a fair day's wage. Otherwise the cities would succumb to the misery and vice that was the breeding ground for Red revolution. Britain must produce more and better and more cheaply than any other nation—that was the bottom line—and must do so "without a proportional increase in the cost of production." Huxley argued that "as the price of labour constitutes a large element in that cost, the rate of wages must be restricted within certain limits." In short, "A moderate price of labour, is essential to our success as competitors in the markets of the world."[47]

While this was certainly much less than any socialist might want to hear, Huxley remained clear in his judgment that laissez-faire was also untenable. Indeed, it was the inveterate injustice that had prevailed when laissez-faire had had free rein that had fostered socialist extremism. Left to itself, society would swing from free trade to rabid revolution; under such conditions, the natural man, the animal, "preaches anarchy; which is, substantially, a proposal to reduce the social cosmos to chaos, and begin the brute struggle for existence once again." This much was evident from the merest acquaintance with inner-city life and the facts of the matter; and asserting his right to speak on such matters "as a naturalist," Huxley stated that it was "mere plain truth that . . . with every addition to the population the multitude already sunk in the pit and the number of the host sliding towards it continually increase."[48]

However, and moving to engage his Manchester audience, he suggested that there were yet weapons in England's armory that might satisfy the Sphinx. The means was the expansion of the state; the mode was sanitary reform and technical education. Just as sanitary reform would do its part to prevent the degradation of the prime of England's manhood and womanhood in urban squalor, so a technical education would fit the coming generation for the industrial war between nations. While the opponents of an education tax argued that it was some sort of grand imposition upon the liberties of free-born Englishmen, Huxley countered that under the circumstances, "an education rate is, in fact, a war tax, levied for the purposes of defense." The state, insofar as it was representative of the common interests of all, might quite legitimately ask, and even demand, that its citizens contribute to such an effort. After all, Huxley noted, "there is a manifest unfairness in letting all the burden be borne by the willing horse."[49]

Addressing his audience, Huxley outlined his continuing campaign to reform the nation's education. Despite the advances that had been made since the passage of the Forster Act, he argued the necessity of introducing science education at even the most preliminary stage, of the teaching of drawing and art rather than the exclusive focus on book-learning, and of providing training for more and skilled teachers to implement these proposals. He pointed out that with the replacement of crafts-guilds by industrial manufacture there was no longer anyone teaching the intricacies of a trade as the master craftsman had to his apprentices in bygone days. This could hardly be expected of the mill or factory owner, he argued, and thus such a duty must fall to the local municipality, to be provided in just the form of "special training which is commonly called 'technical education'" that his hosts in Manchester were advocating.[50]

After having raged against Spencer, against Kingsley, against George, and against socialism—and even against Darwin's optimistic conclusions—Huxley went numb. Mady's death brought on one of his debilitating depressions combined with pleurisy. Under doctor's orders, he fled the smog of the capital; at sixty-three, he was feeling old.[51] The harsh reviews of his essay, though, fired him up as much as the fresh air, and breaking free of his malaise and secure in the knowledge that the X Club's own permeation strategy had done its job, he thought he might make a go of things in political writing after all. Someone had to counter the outrageous claims of these revolutionary cum quack naturalists and his blood was up. He suggested a series of essays to James Knowles, the editor of the *Nineteenth Century*—they would

FIGURE 5.3. The Russian anarchist and naturalist Peter Kropotkin, 1842–1921. Kropotkin argued that cooperation and mutual aid were the most significant factors in the evolution of social species. (Courtesy History of Science Collections, University of Oklahoma Libraries)

make a fine companion to his biological lectures for workingmen. "I think I will make six of them after the fashion of my 'working men's lectures,'" he wrote, intending to take as his further subjects liberty and equality; the rights of man; property; Malthus; government; and the making and breaking of the law.[52]

Even though Karl Marx disagreed with Huxley over Malthus—Marx saw scarcity as purely a function of the market rather than any law of nature—it is easy to see why he found both *Origin* and Huxley's lectures so compelling. Nevertheless, whereas Huxley's repeated attacks on laissez-faire, natural theology, and teleology were welcomed by workingmen across the country, his rejection of Georgite and socialist politics left some of them disappointed.[53] This was nothing compared to the opinion of the Russian anarchist Peter Kropotkin, however. Kropotkin had been in exile in England since 1886, having recently been released from Clairvaux Prison in France where he had been incarcerated for his revolutionary activities. Kropotkin thought "The Struggle for Existence" an "atrocious article," objecting not only to the politics that Huxley had advocated, but also to the conception of nature that underpinned it.[54]

With only a few exceptions, where historians of science have considered Kropotkin they have tended to marginalize his views as inherently biased by his political commitments.[55] There is doubtless some truth in this evaluation, but there are also problems with such a dismissive treatment of Kropotkin. First, if his memoirs are accurate on this point, it was Kropotkin's study of nature and of native peoples that led him to frame his broader political conclusions, and not the converse; and second, to the extent that he did see his own politics reflected in the natural world, he was by no means exceptional in doing so.[56] Further, as Daniel Todes and, more recently, Mark Borrello have pointed out, the view that Kropotkin was an anarchist with only an amateur interest in natural history, which has been prevalent until recently, is no more an adequate description of him than it would be to say that Huxley was a liberal who dabbled in comparative anatomy.[57] Ultimately, Kropotkin would become a valued member of the British Association for the Advancement of Science, report to the Royal Geographical Society in London, and in 1896 was offered a chair in geology at Cambridge University. This last honor was one he declined, however, as it was conditional upon his abstention from further political activity.[58]

Kropotkin's interests and achievements in natural history were lifelong. He had grown up with an interest in the physical sciences, and as a result of his position in the Russian Imperial Court—his family had dynastic lineage—he studied all branches of science with some rigor. His interest in transmutation began in 1858 after having read three papers by Professor Roulier at the Moscow University, which he had then discussed with his brother, Alexander, who shared many of his interests. Indeed, Alexander recommended that he also read the public lectures of K. F. Rul'e, another Russian evolutionist. Todes concludes that in light of their studies, "both brothers were almost certainly evolutionists before reading *Origin*."[59] Intrigued by the species question, when they did read *Origin* the following year, they kept up a discussion and correspondence that "lasted for many years" on "various questions relative to the origin of variations, their chances of being transmitted and being accentuated across generations."[60] Kropotkin's interest in science though was not confined to personal discussion and debate with his brother. His position at Court allowed him to take a military post that took him to Siberia and Manchuria, from whence he pursued his interests in geology, geography, and natural history; while there he also sought to apply Darwin's insights to the ecology of the Siberian steppes.[61]

In 1863 uprisings among the serfs in Poland made him think deeply about politics, and he recalled that in Siberia, in addition to natural history, "I also thought a great deal during this journey about social matters," which "had a decisive influence upon my subsequent development."[62] Closer to home, too, although his father was regarded as a comparatively good master, even by his serfs, Kropotkin had seen men flogged for the slightest transgression, and marriage forced upon the young men and women who were considered his father's property, regardless of their personal sentiments.[63] Further, his life in the military had shown him just how much could be achieved through discipline and compulsion, as well as where such methods must ultimately fall short. His experience of the Siberian and Manchurian native populations showed him another side of life that was to radically shape his future, both in terms of politics and of natural history. He wrote of the Mongols, "To live with natives, to see at work all the complex forms of social organisation which they have elaborated far away from the influence of any civilisation, was, as it were, to store up floods of light which illuminated my subsequent reading."[64] The experience certainly shone a critical light on his reading of Huxley's work.

Without the imposition of arbitrary authority, men lived and worked with real enthusiasm and could achieve feats that would be impossible for pressed men. He had seen enough of both sides of the question to know that this was true, and in light of such compelling evidence he recalled: "I lost in Siberia whatever faith in state discipline I had cherished before. I was prepared to become an anarchist."[65] Viewing these natives as indicative of what mankind might achieve in their natural state, he saw their daily lives carried on in a spirit of cooperation and mutual aid, characteristics that he saw as predominant features throughout the natural world. He recalled that as a result, "a sense of Man's oneness with Nature, both animate and inanimate—the poetry of Nature—became the philosophy of my life."[66]

Unwilling to live with the injustice of military rule, Kropotkin resigned his commission and instead attended St. Petersburg University. He subsequently participated in several further scientific expeditions supported by the Imperial Russian Geographical Society and published a number of articles on his findings. Recognized for the quality of his work, he was formally admitted as a member of the Imperial Geographical Society and in 1868 was awarded their gold medal for his work on the Olekmin-Vitim expedition to discover a direct communication route between the gold mines of the Yakútsk province and Transbaikália. His growing political convictions led him to Europe to seek out the International Workingmen's Association in Geneva, which he had heard so much about in the radical press in his own country. However, the eruption of the Commune in Paris in 1871 revealed to him the problems of even a revolutionary centralized organization, and adopting the pseudonym "Mr. Borodin"—it was dangerous to be a revolutionary—he returned to Russia to advocate for anarchist revolution among the serfs.[67]

Although Kropotkin hoped to write a full monograph account of what he considered to be his greatest achievement in science—his discovery that the main structural lines of the geology of Asia run from the southwest to the northeast—with his increasing political involvement, he realized that it would most likely not be long before he would be arrested, and so contented himself with drawing up a new map of the region and writing up a brief description of his conclusions. His brother would see these through to publication in 1873 after Kropotkin had indeed been imprisoned.[68]

Kropotkin was aware that the tsarist spies and secret police monitored his every movement, although it later transpired that they were unsure until they actually had him in their custody that Borodin and Kropotkin were one and the same man. Aware that they were closing in, Kropotkin delayed his escape to give one last lecture to the Geographical Society—contrary to prevailing opinion, he was certain that glaciers had carved the topography of much of Finland and Russia. He returned home briefly to pack, but too late. A servant urged him to take the service staircase to make good his escape; he made it to the street and jumped into a cab, but the police were soon upon him. A second cab pulled alongside and his coachman reined in the horse—the game was up.[69]

Kropotkin was confined to the imposing fortress of St. Peter and St. Paul, which rises from the Nevá in front of the Winter Palace in St. Petersburg. Its history was one "of murder and torture, of men buried alive, condemned to slow death, or driven to insanity in the loneliness of the dark and damp dungeons." "To witness the destruction of a man's mind, under such conditions was terrible," he wrote.[70] It was only through the interventions of his brother and the Geographical Society that Kropotkin was allowed pen and paper and access to the books he would need to complete his geological work. This eventually stretched to two long volumes, the first of which was duly published, as mentioned, in 1873; the manuscript for the second lay unfinished in his cell on the day he escaped. It was a daring and elaborate plan worthy of the best prison-break drama. Comrades had signaled their plan to liberate him, lookouts were posted, and Kropotkin made a mad dash for the gate. Outpacing a prison guard who made several desperate attempts to impale the fleeing anarchist upon his bayonet, Kropotkin jumped into a waiting coach that raced him from the scene before the prison guards could mount an effective pursuit. Eluding the hunt for him that ensued, he traveled through Finland to Sweden and onward by ship to the port city of Hull in the northeast of England.[71]

In England, Kropotkin continued his scientific work and had several articles published in *Nature*, but he also could not abandon his life as a revolutionary. Traveling Europe, he hoped to sow the seeds of revolution. He was arrested again in France in 1882 for his continued political agitation, and sentenced to five years imprisonment, but was released in 1886 on the condition that he leave the country. By the time he returned to England, London had become the center of the nascent British socialist revival. It was also a refuge for so many exiles that the steps of the British Library became a veritable "Who's Who" of Europe's political renegades—Marx and Engels were by no means the only émigré revolutionaries in town. Kropotkin immersed himself in the movement. With the help of the English anarchist and Fabian Society member Charlotte Wilson, he founded Freedom Press in 1886. Situated in Angel Alley off Whitechapel High Street, the press was dedicated to the publication and distribution of revolutionary anarchist literature, including their journal, *Freedom*.[72] As Shaw has noted, they were influential across the broader socialist movement. Wilson and others with anarchist sympathies left the Fabians in 1887, shortly after Webb made their permeationist strategy explicit.[73]

Given Kropotkin's views on both man and nature, it is hardly surprising that he thought Huxley's article was dreadful. What was the worst of it though, was that he believed Huxley was fundamentally misrepresenting Darwin in the process. Kropotkin was aware that there were many blatant misrepresentations of Darwin's work doing the rounds, each claiming to be the one true application of evolution to humanity, but Huxley was the most prominent of Darwin's inner circle, the most famed scientist of his day, and the man who had made his name as "Darwin's Bulldog." The popularity of the *Nineteenth Century* under Knowles's editorship meant that Huxley could not go unanswered, and Kropotkin felt himself to be in a good position to respond. He had studied Darwin's work not only on the printed page, but had gone into the field to test its validity against the empirical data to be drawn from the Siberian steppes. He had also spent years in the silence of solitary confinement, reflecting upon the nature of man, his history, and his natural history. In the obituary he had written of Darwin in 1882, he had already raised the issue of the tendency of English Darwinians to overemphasize competition, and with encouragement from Knowles—there was nothing like a good argument to sell papers—he commenced a refutation of Huxley that began as an article but which grew to become his lifetime's occupation.[74] The entomologist Henry Walter Bates cheered him on from the wings: "It is horrible what 'they' have made of Darwin."[75]

Kropotkin continued his campaign even after Huxley's death in 1895. Huxley had given his own final words on the subject in his powerful essay "Evolution and Ethics," published with a prologue in 1894; Kropotkin had responded with a lecture on "Justice and Morality," which he delivered to the South Place Ethical Society in London and the Ancoats Brotherhood in Manchester during the autumn of 1893. But from this point onward he was outlining his own grand plan. The essays and lectures continued at a regular pace from September 1890 until June 1896; many were translated and published across Europe and America, and many, of course, were summarized in *Freedom.* These were the basis of what became *Mutual Aid,* but he did not stop there. He had also written a series of articles on "Anarchist Morality" that appeared in the French anarchist journal *La Révolte* in 1890 and in *Freedom* in 1891–92; they were republished that year together as a pamphlet. *The Conquest of Bread* (1892) and *Fields, Factories, and Workshops* (1899) followed—this last in the same year as he published his *Memoirs of a Revolutionist.* And then came the long essay "Modern Science and Anarchism" (1901).

Reflecting on the discussions he had had with his brother Alexander when they had first read *Origin,* Kropotkin also wrote a number of articles as part contribution, part commentary upon the debate that had erupted among evolutionists in the 1890s about the inheritance of acquired characters.[76] Darwin had included a role for acquired characters in his theory of inheritance, and the mechanism was central to Spencer's evolutionary ideology. Others waded in too, filling the pages of the *Contemporary Review,* but the issue was far from decided even then.[77] Throughout the 1910s, Kropotkin would vehemently refute the neo-Darwinian conclusions of the German cytologist Friedrich Leopold August Weismann, again in the pages of *Nineteenth Century.*[78] Kropotkin had replaced Huxley as the writer of the journal's "Recent Science" column in 1892, and so his views on the matter must be considered far from marginal.[79] Kropotkin continued to write on these issues even as war raged across Europe. Papers on "Inherited Variation in Animals" (1915) and "The Direct Action of Environment and Evolution" (1919) might seem a peculiar distraction for one of Europe's most-committed anarchists to be working on as the nations of Europe fell at each other's throats, but to Kropotkin these were central issues relating to his hopes for the future of mankind. Following his return to Russia, in what should be seen as his final rejoinder to Huxley's "Evolution and Ethics," Kropotkin spent the last years of his life writing what he clearly intended to be his own authoritative statement on the origin and nature of the human moral sentiments—and what might be expected of them in the future. He died in February 1921 before he could finish it, however; the first volume was published, unfinished, in 1924 under the title *Ethics, Origin, and Development.*[80]

Kropotkin, of course, is most readily associated with the ideas of mutual aid as a factor in evolution. However, in his study of the relationship between Darwin and Wallace, *A Delicate Arrangement*, Arnold Brackman has suggested that Kropotkin took the idea of mutual aid from a Russian translation of Wallace without acknowledgment. This, despite the fact that Kropotkin quite openly stated that he first came upon the idea in the work of the Russian naturalist Karl Fiodorovic Kessler.[81] Kropotkin had first read Kessler's work while a prisoner in Clairveaux, and he stated quite clearly in "Mutual Aid among Animals," the first essay in the series for *Nineteenth Century*, his opinion that, "of the scientific followers of Darwin, the first, as far as I know, who understood the full purport of Mutual Aid *as a law of Nature and the chief factor of evolution*, was a well-known Russian zoologist, the late Dean of St. Petersburg University, Professor Kessler."[82] Kessler had delivered this address in January 1880 but died before he had a chance to elaborate his main idea. As Kropotkin duly noted, Kessler was by no means the only naturalist working in this area who recognized that mutual aid rather than mutual struggle was the dominant factor in evolution; many of Kessler's students and colleagues in St. Petersburg following his lead.[83] Where Kessler had noted that it was unfortunate that some naturalists invoked "the cruel, so-called law of the struggle for existence" for their own political ends, Kropotkin saw that mutual aid might in turn be extended to explain human social life as well.[84]

Like Kessler, Kropotkin certainly did not deny that competition took place in nature or that many organisms died as a result of it. However, what he did deny was the almost exclusive emphasis that English Darwinians gave to this aspect of the struggle for existence—Huxley, whom he otherwise considered to be among the greatest of naturalists, more so than anyone.[85] Over and above this though, he was determined to show that cooperation and mutual aid were by far the most important factors in the progressive evolution of species. While Kropotkin believed that Huxley had willfully misrepresented Darwin's work, he did not deny Darwin's general conclusions—although, of course, with Darwin already six years dead, the differences across the corpus of his works left room for debate about exactly what Darwin's final conclusions really were. Needless to say, Kropotkin argued that his own reading was far closer to Darwin's last words on the matter than Huxley's account—even if he thought that Darwin had taken one or two of his own metaphors a little too literally at times, too.

While he differed from Huxley's conclusions, Kropotkin was not interested in making purely ad hominen attacks. Indeed, practically the first thing he drew to the attention of his readers was the fact that he had deeply appreciated *Origin of Species* and that from the moment of its publication he had been eager to take these new insights into the field with him on his Siberian expedition. It was here though that he saw that there were some discrepancies between the aspects of Darwin's theory that Huxley championed and what he observed in nature. Kropotkin had gone into the field in the company of his good friend and fellow naturalist, I. S. Poliakov. "We were both under the fresh impression of the *Origin of Species*," he later recalled, "but we vainly looked for the keen competition between animals of the same species which the reading of Darwin's work had prepared us to expect."[86] Indeed, in explaining his "theory of divergence," which was central to his understanding of speciation, Darwin had emphasized not only that individuals would compete with one another to the death for limited resources, but that the competition would be the most intense between those individuals who were most similar. "For it should be remembered," Darwin had written, "that the competition will generally be most severe between those forms which are most nearly related to each other in habits, constitution and structure." As a result, Darwin concluded, "there will be a constant tendency in the improved descendants of any one species to supplant and exterminate in each stage of descent their predecessors and their original parent."[87]

In this instance, and a point that Huxley had clearly taken to heart, internecine struggle was the driving force of evolution. Kropotkin acknowledged that there might occasionally be circumstances in which individuals of the same species, and even of the same group, might compete with one another. He maintained that it was cooperation rather than competition that led to the progressive development not only of individuals but of animal communities and thus of species. Conflict could only weaken those subject to it. Further, though, and as Kropotkin and Poliakov's observations bore out, the bulk of the struggle for life did not appear to go on between one individual and another, but rather tended most often to occur in terms of either an individual organism or a group struggling to survive in the context of a harsh and inhospitable environment. In the Siberian steppes, life was much more frugal than in the tropical regions that had been the experience of both Darwin and Wallace—a point that several historians of science have emphasized in their efforts to explain Kropotkin's conclusions—"but even in the Amur and Usuri regions where animal life swarms in abundance," Kropotkin continued, "facts of real competition and struggle between higher animals of the same species came very seldom under my notice, though I eagerly searched for them."[88] Indeed, and in stark contrast to the vision of nature as Huxley's gladiatorial arena, he noted that in fact, "wherever I saw animal life in abundance . . . I saw Mutual Aid and Mutual Support carried on to an extent which made me suspect in it a feature of the greatest importance for the maintenance of life, the preservation of each species, and its further evolution."[89]

Laying aside, at least for the moment, Kropotkin's suggestion that mutual aid might work for the "preservation of each species" rather than for the individual, or even the particular herd or group in question, there remain significant differences between Kropotkin's view of nature and that which Darwin presented in *Origin*, let alone that which Huxley had so recently portrayed—most notably, the emphasis that Darwin placed upon Malthus.[90] Indeed, given his anti-Malthusian views, some historians of science have asked whether it is really legitimate to classify Kropotkin as a Darwinian at all.[91] However, Kropotkin certainly thought of himself as a Darwinian, and what is more, in this instance he thought himself a better one than Huxley.

As Kropotkin was keen to point out, Huxley had made the mistake of taking Darwin's Malthusian references literally, when Darwin had been quite clear that he used the term in only a metaphorical sense. Indeed, regarding the use of the phrase "the struggle for existence," Kropotkin noted that "at the very beginning of his memorable work" Darwin had "insisted upon the term being taken in its 'large and metaphorical sense including dependence of one being upon another, and including (which is more important) not only the life of the individual, but success in leaving progeny.'"[92] Further, Kropotkin went on, Darwin was also aware that while his metaphor might work in some circumstances, in others it was clearly a stretch. Darwin had noted shortly after having first introduced the concept of the "struggle for existence," that "a plant on the edge of a desert is said to struggle for life against the drought, though more properly it should be said to be dependent on the moisture."[93] "In such cases," Kropotkin stated, "what is described as competition may be no competition at all." And this was exactly what he and Poliakov found on their Siberian expedition: "One species succumbs, not because it is exterminated or starved out by the other species, but because it does not well accommodate itself to new conditions."[94] While Darwin had certainly intended his Malthusian reference to be more metaphorical than literal, Kropotkin argued that as one works through *Origin* it is clear that Darwin had quickly forgotten himself and been carried away by the force of his own metaphor. When it came to his discussion of the competition between individuals of the same species, for instance, Kropotkin pointed out that here the weight of empirical examples with which Darwin was want to illustrate the other points of his theory were surprisingly thin by comparison. "The struggle between individuals of the same species is not illustrated under that heading by even one single instance," he noted. Rather, "it is taken as granted," and thus the examples that Darwin offered in support of competition between even closely allied forms, Kropotkin thought questionable.[95]

This weakness had further implications for other aspects of Darwin's theory. In seeking to explain the dynamics of divergent speciation as resulting from the intensity of competition between the most similar individuals of any one species, Darwin had invoked the idea that nature, like the breeder of exotic pigeons, preferred extremes—that is, because the most-extreme variations would tend to be the best adapted to divergent niches, they would tend to survive and prosper at the expense of any intermediates. These intermediate forms, he had concluded, would thus be exterminated.[96]

As Kropotkin pointed out, Darwin had relied upon the extermination of intermediate varieties not only as the mechanism by which he might account for the wide diversity of life that had its origins in a few or even a single form, but also to explain the lack of intermediate forms between one species and another in the present. As Kropotkin quite correctly pointed out, "Darwin was worried by the difficulty which he saw in the absence of a long line of intermediate forms between closely allied species, and that he found the solution of this difficulty in the supposed extermination of the intermediate forms."[97] But, and as Darwin's critics were keenly aware, this only deferred the problem, for now the absence of evidence was shifted to the fossil record. If evolution worked by such gradual and uniform processes, it might be asked (and it was asked), then even if the intermediates that we might expect to otherwise see in the present had been exterminated, would not we surely expect to find them in fossil form? Of course the problem here was that fossilized organisms showing a gradual evolution from one form to another in a nicely linear organization was exactly not what was being dug out of the ground. Hostile witnesses were not the only ones to point out this deficit in Darwin's argument. In addition to the objections raised by Henry Charles Fleeming Jenkin and Samuel Wilberforce, Huxley had also expressed doubts about this question in his own early reviews of Darwin's work, going as far as to suggest that Darwin might later be embarrassed by repeating so insistently that *Natura non facit saltum.*[98] Wallace had repeated Darwin's position on this though, Kropotkin noted, adding to its appearance as orthodoxy.

In contrast, Kropotkin suggested that "an attentive reading of the different chapters in which Darwin and Wallace speak of this subject soon brings one to the conclusion that the word 'extermination' does not mean real extermination"; rather, this too "can by no means be understood in its direct sense, but must be taken 'in its metaphoric sense'" as well. This was not a fatal criticism of Darwin's theory, though, for as Kropotkin went on to point out, "both he and Wallace knew Nature too well not to perceive that this [Malthusianism] is by no means the only possible and necessary course of affairs."[99]

If both the struggle between individuals of the same species and of the extermination of intermediate forms were both mere metaphors, the question remains as to what Kropotkin thought they were metaphors for. That is, by what means did he think that the many and diverse forms in nature had come into being, and if not exterminated by competition, by what means had so many of them gone extinct? The harsh and frequently changing environmental conditions that he witnessed in Siberia gave him his lead, but he recognized too that Darwin had paved the way for an alternative view, paying significant attention to the ways in which organisms interacted with and adapted to their environment, even if, on balance, he had been carried away by his Malthusian thinking.

In a static world where environment, climate, and the habits of a given species remained unchanged, the sudden introduction of a new variation into this environment might well result in struggle and extermination, Kropotkin thought, just as Darwin's metaphor suggested. But this demonstrated the importance of taking any theory into the field, for "such a combination of conditions is precisely what we do not see in Nature." Rather, and here clearly under Spencer's influence, organisms were continually adapting to an ever changing environment—they altered their habits, took to new sorts of food, or in many observed cases simply migrated to more favorable environments. "In all such cases," he pointed out, "there will be no extermination, even no competition—the new adaptation being a relief from competition if it ever existed." Further, of course, and what Kropotkin believed made this a stronger hypothesis than that which Huxley was touting, was that in his own case there was no need to invent competition to explain the absence of intermediate forms in the present, or to offer explanations about the paucity of the fossil record that many simply found unconvincing. Given time, adaptation, migration, and isolation alone might account for the divergent character of organisms without the problem of "missing" intermediate forms. Kropotkin concluded, "It need hardly be added that if we admit, with Spencer, all the Lamarckians, and Darwin himself, the modifying influence of the surroundings upon the species, there remains still less necessity for the extermination of the intermediate forms."[100]

Huxley, of course, had had no truck with Lamarck, and this fact had had as much of an impact upon his politics as did his Malthusian view of the world. By Kropotkin's analysis though, where Darwin had erred, Huxley had failed. In his eagerness to refute Spencer's individualism on the one hand and to deny Georgite radicals and socialists on the other, Huxley had portrayed a nature that was at best a sick caricature of reality. Huxley had his own saltationary answers to "gaps" in the fossil record of course, but that was beside the point.

Kropotkin argued that if nature was not a record of individual struggle, then this had two important implications. The first of these was that what it meant to be "fit" needed to be redefined. The vast majority of the higher mammals, at least those that thrived, were those that led very social lives and that helped those within their group. There was struggle in life, certainly: "No naturalist would doubt that. . . . Life *is* struggle; and in that struggle the fittest survive." But, he continued, the important questions to be asked were these: "'By which arms is this struggle chiefly carried on?' and 'Who are the fittest in the struggle?'" The answers were one and the same: the most sociable organisms, those that were most cooperative and that practiced mutual aid. It was these that were the most successful and that would survive to leave progeny; those that persisted in individualism were at a distinct disadvantage, were less prolific, and were on their way to extinction. Was it not the case that the vast majority of the higher mammals that lived in social groups were abundant? Were not the individualist predators to be found in ever fewer numbers?[101]

If one was to take a message of how to behave from observations of the natural world—and Kropotkin clearly believed that one could do so quite legitimately—the answer was clear. Those organisms that were successful and which were undergoing a progressive development in their evolution were those that were sociable and practiced mutual aid. Those that did not were on the road to extinction. Thus, he appealed to his readers, take this lesson from nature: "'Don't compete!—competition is always injurious to the species, and you have plenty of resources to avoid it!' That is the tendency of nature. . . . Therefore combine—practice mutual aid! That is . . . the best guarantee of existence and progress, bodily intellectual and moral."[102]

Across his first two articles for *Nineteenth Century*, the second of which appeared in November 1890, Kropotkin had thus made several important criticisms of the predominant interpretation of Darwinism in Victorian England. Where Darwin, and even Wallace, had perhaps been carried away with the individualism and competition at the heart of the Malthusian metaphor, Huxley celebrated it. Indeed, he had championed it, and what was worse, had made this the basis of his ringing endorsement of the low wages and devil-take-the-hindmost competitive markets of industrial capitalism. Kropotkin certainly agreed that competition did occur in nature, but it tended to be the exception rather than the rule, and where it did take place—and here in marked contrast to Walter Bagehot's opinion—it tended to be to the detriment of the individuals involved rather than to their advantage. What is more, Kropotkin's account had significant implications for his understanding of the natural condition of human life and thus of "human nature." Recall that in Huxley's account man had only become social in his most recent history. In doing so, he had turned his back on the harsh competition of the natural world of which he was a product. It was this that, by Huxley's reckoning, accounted for the dual nature of humankind: on the one hand, humanity was competitive and selfish and assertive, made so by long generations of struggle, competition, and conquest; on the other, man was sociable, sympathetic, and moral. Each of these latter qualities, however, was of a much more recent evolution and was thus more fragile. As he had explained in his Romanes Lecture of 1893, later published as "Evolution and Ethics," man's ethical nature was certainly a product of evolution, but it was but a thin veneer beneath which lay the natural man that betrayed his savage origins.

By Kropotkin's account, however, and here he also differed from Wallace as well as Huxley, there was no disjuncture between man and the rest of nature. In his 1864 paper on man, Wallace had made the point that whereas the natural world was dominated by individualism, humans combined together into social units for mutual advantage. Huxley had echoed this same point. However, by Kropotkin's analysis of the natural world, natural selection had functioned to advance ever more socialistic traits from the very origins of life. Cooperation and mutual aid were to be found among even the most primitive of life forms. Indeed, he told his readers, "we must be prepared to learn someday, from the students of microscopical pond-life, facts of unconscious mutual support, even from the life of micro-organisms"—and this was increasingly so the higher one looked in the animal kingdom.[103]

Kropotkin recognized differences between his own understanding of the relative merits of competition and cooperation compared to that which Darwin had made central to *Origin*; however, when it came to accounting for the evolution of man—and the evolution of morality in particular—Kropotkin knew that he was much closer to Darwin than Huxley was. He saw the tensions between the individualism that dominated *Origin* and the sociability and mutualism that were the unifying themes of *Descent of Man*, and he staked his claim to be a good Darwinian on the similarities between his own work and what Darwin had had to say in *Descent*.

Darwin's first concern in *Descent* had been to establish that the difference between humans and non-humans was one of degree and not kind. Given that even Wallace, formerly his closest ally on the issue of natural selection, had jumped ship, this was more important than ever. While in large part due to Huxley's work in comparative anatomy, the vast majority of naturalists admitted that, physiologically speaking at least, man might reasonably look for his ancestry among the animals, the main debate had shifted to mind and morals. Intelligence and ethics were deemed to be the mark of human exceptionalism and beyond the power of natural selection to explain. In *Descent*, Darwin had argued that it was from the instinctive sociability of animals that these most human faculties had arisen, and he had made the case that any sociable animal of an intelligence comparable to man "would inevitably acquire a moral sense or conscience."[104] Having made this point, he went on to show that sociability was rife throughout the animal kingdom and that as a consequence one might also find there the insipient stages of intellect, reason, and conscience. This was certainly the case among the higher mammals. Following the lead that Wallace had laid down in his 1864 paper, Darwin had made good use of the suggestion that natural selection would act upon the intellect and other psychological attributes just as readily as it had done on physiological characteristics. The crucial point, though, and what brought Kropotkin's position closer to Darwin's than was that of either Huxley, Wallace, or Spencer, was in his recognition that sociability, reason, and conscience existed in the non-human, and thus the pre-human, world—at least in incipient form, in instinct. Echoing Darwin's response to the Duke of Argyll's doubts about human evolution, Kropotkin noted that "Darwin so well understood that isolately-living [*sic*] apes never could have developed into man-like beings," and that in consequence "he was inclined to consider man as descended from some comparatively weak but social species, like the chimpanzee, rather than from some stronger but unsociable species, like the gorilla." Again contrary to Huxley's opinion, Kropotkin argued that in fact evidence from paleontology further supported the view that primitive man had lived in large and social communities. John Lubbock's descriptions of the flint implements found in the Aurignac region of southern France as "without exaggeration . . . numberless" indicated as much.[105] Darwin had anticipated these conclusions in the letter he had written to Morley back in 1871 in which he had confessed, "I do not think that there is any evidence that man ever existed as a non-social animal."[106]

Kropotkin's views on the origin of the moral sentiments thus had a lot in common with those that Darwin had expressed in *Descent*. Both men held that morality originated in the social instincts, which they believed were seated in the deepest evolutionary history of humankind's forebears. In this much Kropotkin was right to argue that his own reading of human evolution was closer to Darwin's than either that of Huxley, Spencer, or even Wallace—each of them had suggested that social and thus ethical behavior had arisen only recently in human history. Despite Kropotkin's claims to be the most Darwinian of them all, however, he did believe that there was one further important point of difference between his own views and Darwin's: the emphasis that Darwin placed upon the parental and filial affections.

This was something that had long been on Kropotkin's mind, and he turned to it again at length in a 1905 essay entitled "The Morality of Nature." Despite their agreement that the social instincts were important in understanding the origin and nature of human morals, Kropotkin noted that Darwin had been less than consistent in his account of what were, after all, very complex issues. "Unfortunately, scientific animal psychology is still in its infancy," Kropotkin lamented, "and therefore it is extremely difficult to disentangle the complex relations which exist between the social instinct, properly so called, and the parental and filial instincts, as well as several other instincts and faculties, such as sympathy, reason, experience, and a tendency to imitation." Darwin, of course, had suggested that each of these had played some role in the development of the moral sentiments. However, in his quest to ground human morals in a truly other-regarding instinct rather than one that might have had its origin in the "low motive" of self-interest, Darwin had given pride of place to what he called the "parental and filial affections." Kropotkin noted, though, that Darwin had been less than precise in exactly how these affections sat in relation to "the social instinct properly so called." At one point, he noted, Darwin had suggested that the parental and filial affections "apparently lie at the base of the social instincts," but had elsewhere acknowledged that the social instinct was "*a separate instinct in itself*, different from the others . . . developed by natural selection for its own sake, as it was useful for the well-being and the preservation of the species." This last was Kropotkin's view of things—and for what he considered to be good reason. In fact, he argued, the precedence that Darwin gave to the parental and filial affections was at odds not only with his own observations in the field but with the latest findings of anthropologists.[107]

Kropotkin's views on the limited explanatory power of the parental and filial affections were not hastily arrived at. In fact, he had had cause to reflect on the role that parental affections might play in the evolution of ethics from the first. Kessler, his mentor in mutual aid, had also, like Darwin, argued for a significant role for what he had called "parental feeling" and the care of progeny. Even at the time, Kropotkin had found Kessler's emphasis upon parental feeling wanting because it failed to account for the many instances of mutual aid that he had observed between organisms that were not related in this way. Kropotkin was appreciative of Darwin's attempt to deny that morals were grounded in self-interest, but he had already considered and rejected the German physician Ludwig Büchner's attempt to ground morality in "love." Kropotkin admitted that he had found Büchner's *Liebe und Liebes-Leben in der Thierwelt* (1879) interesting, but he wrote of it that "I could not agree with its leading idea."[108] For while Darwin had reframed both love and sympathy as evolved sentiments, Büchner had not, demonstrating only that love and sympathy existed among animals. This much was clearly far too limited in its scope for Kropotkin's purposes, and besides, he argued, "to reduce animal sociability to *love* and *sympathy* means to reduce its generality and its importance, just as human ethics based upon love and personal sympathy only have contributed to narrow the comprehension of the moral feeling as a whole. It is not love to my neighbor—whom I often do not know at all—which induces me to seize a pail of water and to rush towards his house when I see it on fire, it is a far wider, even though more vague feeling or instinct of human solidarity and sociability which moves me."[109]

While Kropotkin could appreciate Darwin's efforts to find a ground for the social instincts that was wider than self-interest, he believed that sociability—a sentiment that had "been slowly developed among animals and men in the course of an extremely long evolution"[110]—was neither limited to those organisms that were bound by the parent-offspring relationship or to the love and sympathy that developed between individuals who were well known to one another. Darwin of course had suggested that the true and other-regarding love that a parent felt for its offspring could and had spread throughout the broader population not only by natural and sexual selection, but also by the other mechanisms Kropotkin cited above, such as through imitation and repetition. However, to Kropotkin's mind, this was to put the cart before the horse. Not only were there plenty of organisms in nature that exhibited mutual aid that did not enjoy a close parent-offspring relationship, but the evidence from anthropology suggested that the exclusive family unit was of recent origin in the history of human societies as well.

Following Darwin's style of reasoning in *Origin*, the essays that made up *Mutual Aid* include a vast accumulation of observations. Kropotkin piled up incidence upon incidence of mutual aid between animals of the same species in support of his thesis, but he could also cite several of Darwin's own examples in support of his belief that the most powerful social instincts were not the parental and filial affections. Indeed, he argued, the social instinct "is so fundamental that when it runs against another instinct, even one so strong as the attachment of the parents to their offspring, it often takes the upper hand." Perhaps the most telling as well as the most emotive case illustrating this fact was that in which Darwin had described a mother bird giving in to the migratory instinct despite the fact that she had young chicks still in the nest that were too young for the flight. As Kropotkin saw things, this was a contest between the parental instinct and one of comradeship that found its resolution in the mother bird, however reluctantly, accompanying her fellows. "Birds, when the time comes for their Autumn migration, will leave behind their tender young, not yet old enough for the prolonged flight, and follow their comrades," he wrote.[111]

Kropotkin noted that his own observations further testified that the primary social instinct had a separate origin to that of the parental and filial affections. "To this striking illustration," he noted, referring to the birds described above, "I may also add that the social instinct is strongly developed with many lower animals." He had in mind the several species of land crabs and fishes that he had observed extensively in his Siberian travels. In these cases, sociability could not possibly have developed "as an extension of the filial or parental feelings," for their reproductive behavior was such that there was no noticeable affection between parent and offspring and the social group consisted of many individuals that were not related. Kropotkin's explanation was significant. In these cases, and by extension also among those of man's earliest ancestors, sociability had arisen as a result of "a considerable number of young animals, having been hatched at a given place and at a given moment" and continuing to live together regardless of "whether they are with their parents or not."[112] A shared environment and the mutual association that would arise as a result were as important for Kropotkin as anything else—and certainly more so than any sentiments that might derive from the parent-offspring relationship.

Certainly, the tendency toward mutual aid improved the chances that those who exhibited it would survive, but its motivation was not self-interest—Darwin had been right about this much. Rather, mutualism had arisen through the sheer pleasure to be derived from associating with others. Kropotkin cited a number of notable naturalists who had recorded that many animals exhibit "play" behavior far beyond anything that might reasonably be explained in terms of practicing survival skills. Sociability was the offspring of "the joys" of social life and "the love of society for society's sake."[113] Beyond his many years as a naturalist in the field, Kropotkin had spent years in solitary confinement and could speak to this point with some personal experience.

This was not to say that familial relationships were irrelevant. Kropotkin firmly believed that both the social and the parental instincts were "*two* closely connected instincts," but it was wrong to suggest that the parental affections were either of stronger intensity or of greater significance for the evolution of society. Where parental and filial relations had their limits, however, those of fraternity and sorority did not. Speaking of the examples he had outlined above, he wrote, "In these cases it appears rather as an extension of the brotherly or sisterly relations, or feelings of comradeship" that would pertain in each case as a result of hatching together and living in close proximity.[114] Just as Darwin saw great hope for the extension of liberal politics in the extension of the parental and filial affections, so Kropotkin saw hope of anarchist-communism through the extension of brotherly and sisterly affections. He would later invoke exactly these kinds of associations—based upon the contingencies of time and place rather than direct kin relationships—in his account of the historical development of human social organization.[115]

Like many evolutionists, Kropotkin went on to consider the anthropological evidence that he believed further supported his own interpretation of the natural-historical evidence. Fraternal and sororal rather than parental and filial affections would seem to fit better with the latest understanding of human history, especially since anthropologists were by this time generally agreed that human societies had polygamous origins, which would only dilute the ties of any parental and filial affection.

Before going on to consider this, though, it is worth noting that Kropotkin's characterization of Darwin's views about kin relationships was based upon a slight, but significant, misconception. Certainly, in referring to the social instincts as derived from the "parental and filial affections," Darwin had indeed appeared to put the parent-offspring relationship front and center. However, and what Kropotkin appears to have overlooked, is that Darwin was insistent that they also included the fraternal and sororal relations as well, noting explicitly that the difficulty in explaining the several neuter castes of the social insects "disappears, when it is remembered that selection may be applied to the family, as well as to the individual, and may thus gain the desired end."[116] Darwin had also, of course, noted the significance of the wider family in the preservation of altruistic and "patriotic" behaviors among human social groups.[117] Although this was a clear oversight on Kropotkin's part, and a surprising one given the attention that he too had paid to the social insects, it does not necessarily invalidate Kropotkin's point—Darwin had, after all, suggested that brotherly and sisterly attachments were an extension of the parental and filial affections rather than a power that might act independently.[118] In light of the clear importance of social and familial relationships in the evolution of society, it is little wonder that Kropotkin, like many in science and in the socialist movement, came to believe that human ethics could only be truly understood by studying the origin of such relationships in light of the latest findings in social anthropology.

Anthropology had become more and more significant since the publication of *Origin*. Evolution, it seems, had indeed thrown new light on the origin of man and his history. On the *Beagle* voyage Darwin had been prompted to think deeply about the relationship "between savage and civilised man." His experience of the Fuegians, both those who had been educated in England and later with the tribes from which they had been taken, left a deep and lasting impression upon him. Darwin was not alone in turning to primitive or native tribes for evidence in support of his evolutionary ideas. In the 1860s, Spencer, Wallace, and John Lubbock had done so, and so too had Huxley, of course, in light of the Neanderthal skull that had been unearthed in the Dordogne in 1864.[119] Like Darwin, they extrapolated from what they assumed to be the way of life of these people to build theories about the kind of creature that early man was and of how the ancestors of modern Europeans had lived. In order to stress the evolutionary connections between man and animals, these accounts frequently cast early man in the most bestial light. Just as in *Origin* Darwin had shown that nature was dominated by a ruthless Malthusian struggle, most of his contemporaries portrayed the life of the primitive savage, the "natural man" of Huxley's essay, in just the same way.

As Kropotkin pointed out, though, such observations more often than not came fully packaged with the political prejudices of the observer. This had certainly been the case with Darwin, for instance, who could not help but think that the tendency of the Fuegians to share all they had equitably among the tribe was a barrier to their further advance.[120] Kropotkin argued, however, that observers who had not lived among these tribes could not hope to understand the complex social systems that were so different from their own. As a result, Kropotkin pointed out, they had failed to comprehend the true significance of infanticide, or parricide, or—in Darwin's case—of the refusal of private property. Time and again they had described each of these behaviors as a barrier to the evolutionary advance of the people who practiced them, rather than as contingent moral behaviors that had evolved and persisted precisely because they aided in the survival of the group. Indeed, from the "mostly unconscious" individualist perspective of Western observers, such actions could only appear as illogical and immoral and guaranteed to undermine any hope of a progressive evolution.[121]

From the reports of anthropologists who had actually lived among the tribes they wrote about, however, it was evident that some form of communism was the rule rather than the exception among native peoples, and thus, if they were indeed a legitimate indicator of Western ancestral life, then it would seem that mutual aid rather than mutual conflict was the dominant condition of the life of early man. Kropotkin's own experiences among the native peoples of Siberia and Manchuria had led him to appreciate the importance of mutual aid and communitarianism for survival in a harsh environment, but he also noted that others had recorded similar findings among Hottentots, Ostyaks, the Samoyedes, the Eskimo, the Dyaks, the Aleoutes, the Papuas, and more. What Darwin had witnessed, therefore, was one of the Fuegian's survival strategies—he had recognized as much in their occasional cannibalism and even in their reluctant infanticide—but clearly even his broad liberal imagination could not fathom the utility of any behavior that denied the right of individual property ownership.[122]

Kropotkin noted that "a whole science devoted to the embryology of human institutions" had developed among anthropologists, including, as he noted, Johan J. Bachofen, John F. MacLennan, Lewis H. Morgan, Edwin Tyler, Henry Maine, and Post and Maksim Kovalevsky, in addition to Lubbock, whose *Prehistoric Times* (1865) had been so influential.[123] He cited these and other authors, many of whom Darwin had also cited, to show that not only was private property a recent social institution, but that so too was the independent family unit. "As far as we can go back in the paleo-ethnology of mankind, we find man living in societies," he wrote, "in tribes similar to those of the highest mammals; and an extremely slow and long evolution was required to bring the societies to the gentile, or clan organization, which, in its turn, had to undergo another, also very long evolution, before the first germs of family, polygamous or monogamous, could appear. Societies, bands, or tribes—not families—were thus the primitive form of organisation of mankind."[124]

As Darwin had noted in *Descent of Man*, this social and polygamous origin of mankind, however much it might disturb the polite presumptions of Victorian society, had significant implications for any account of the origins of human morals. Huxley, after all, had framed the condition of man as dominated by a Hobbesian war of each against all, with the only respite to be found within the bonds of the family. However, in light of a vast swath of anthropological evidence, it seemed that this was a groundless presumption. Darwin had located the deepest form of morality—that which was truly other-regarding—in the parental and filial affections, but what anthropology was revealing was that these affections encompassed the entire community. Further, anthropologists were claiming that among modern native communities, those who acted in any way contrary to the social interests of the group would receive the censure of the whole community. "Primitive folk . . . so much identify their lives with that of the tribe, that each of their acts, however insignificant, is considered a tribal affair. Their whole behavior is regulated by an infinite series of unwritten rules of propriety which are the fruit of their common experience as to what is good or bad—that is, beneficial or harmful for their own tribe." So, too, it must have been for modern man's forbears. It is interesting to note here Kropotkin's observation that among many native populations, "self-restriction and self-sacrifice in the interest of the clan are of daily occurrence," and that if anyone infringed upon even "one of the smaller tribal rules," he was "prosecuted by the mockeries of the women."[125] Kropotkin did not comment upon the implications this might have for sexual selection, but it is indicative that he too was aware that female opinion might be a powerful moral censure with evolutionary implications.

Mutual aid had resulted from a standard of morality that was measured by the extent to which an action worked for the good of the community. However, and as Darwin had noted, this set no expectation of the extension of moral consideration of members of other communities. Kropotkin agreed. "The life of the savage is divided into two sets of actions, and appears under two different ethical aspects: the relations within the tribe, and the relations with the outsiders." Thus, and again echoing Darwin's account of human ethics, he acknowledged that "when it comes to war the most revolting cruelties may be considered as so many claims upon the admiration of the tribe." It is notable that despite the fact that Kropotkin had often talked in terms of mutual aid working "for the good of the species," here he clearly differentiated between different social groups, concluding that "this double conception of morality passes through the whole evolution of mankind, and maintains itself until now."[126] Tempering this judgment though, he reminded his readers that in light of these same anthropological studies, and from a thorough investigation of modern history, it was clear that war was far from a perpetual state of human existence. Even among primitive tribes, territorial boundaries were generally respected, and an equitable trade and intermarriage rather than war was the most likely outcome of one tribe seeking to meet its own needs through the resources held by another.[127]

Thus, in a similar if not identical vein to that which Darwin had outlined in *Descent*, Kropotkin offered an evolutionary-anthropological argument that mutual aid was not only a factor in evolution, but that it was the most important one. As a result, it was a deep and abiding sentiment in mankind that had become an ingrained instinct through long millennia. Significantly, this led Kropotkin to a conclusion that was the exact opposite of that which Huxley had made in both "The Struggle for Existence" and "Evolution and

Ethics." Natural man was instinctively a communist and had only recently become subject to the laws of private property, of the family, and the imposition of other individualistic social institutions.

If this picture of voluntary social cohesion seemed to go against the received wisdom of the history books, Kropotkin reminded his readers that writing history was not an innocent occupation. Certainly, with the most popular history books in his hands "the pessimist philosopher triumphantly concludes that warfare and oppression are the very essence of human nature; that the warlike and predatory instincts of man can only be restrained within certain limits by a strong authority which enforces peace." But a closer investigation of history revealed that this was a slander upon the truth worthy of only a Hobbes or a Huxley.[128] "History will have to be re-written on new lines," Kropotkin believed, and in his "Appeal to the Young" he called upon the rising young men of science to turn their attention to setting the historical record straight. "Do you not understand that history—which today is an old woman's tale about great kings, great statesmen and great parliaments—that history itself has to be written from the point of the view of the people in the long evolution of mankind?" Just as political economy—and all it had to say about the nature of humanity—had been revised in light of socialism, so too "anthropology, sociology, ethics, must be completely recast, and that the natural sciences themselves . . . must undergo a profound modification."[129] Written history must reflect the new anthropology and place mutual aid at the heart of the history, as well as the natural history, of mankind.

Kropotkin was not alone in making use of this kind of anthropological evidence, and many of the socialists I have considered in chapter 4 also wrote lengthy analyses of the historical development of humanity from prehistory to the present. Engels's *The Origin of the Family, Private Property, and the State* (1884) was only the most famous of these; Hyndman, Morris, and Belfort Bax each also wrote a number of such historical studies. Individualism and private property were of comparatively recent origin, they argued, and thus had no claim to being an essential part of the natural condition of humanity.

Certainly, the course of history had not run smoothly, and there had been major points of upheaval that had disrupted and occasionally set back the gradual development of the tendency toward an ever greater mutual aid. However, given history's significance as a factor in evolution—as a law of nature, not a mere contingency of time and place—Kropotkin believed, like many of his colleagues in the movement, that its ultimate fulfillment was inevitable. Exactly what had brought an end to the primitive communism of these early clans, tribes, and nations was the subject of some dispute among historians as well as among men of science. "One cause, however, is naturally suggested to the geographer," Kropotkin wrote. Contemplating the ruins of large cities in the midst of the deserts of Central Asia—the dried-up river beds and the evidence of vast lakes reduced to mere ponds—it seemed likely that it was "a quite recent desiccation" that had driven what historians would later describe as barbarian hoards from their homelands and toward Europe. The outcome of this mass migration had been the breakup and intermixing of previously insulated groups. It would have been no wonder if the traditional social institutions, based largely on the membership of these groups, had been totally wrecked as a result of such dislocation. But they were not. "They simply underwent the modification which was required by the new conditions of life."[130] As clans and tribes dissolved and reformed into new social groups incorporating peoples of different tribes and different races, the family unit became a significant and recognizable entity—holding property, but only at the discretion of the community. Even then, property might be held only for a set number of years—the idea that property might be bequeathed was an alien concept. The blood bonds that had previously united the clans were undermined by this migration and intermingling of peoples, but—and echoing what he had witnessed in his land crabs and fishes—they were replaced by the formation of a common identity built upon a common location. Again, individual families might be given leave to take possession of allotments of land, but only for a limited tenure. No man could own the land, after all.[131] The history that Kropotkin told resonated with deeply felt English radical sentiments.

What developed into the village community became the new embodiment of the tendency toward mutual aid. It was governed by the collective folkmote which sought to preserve the will of the people in the interests of the whole and was the "chief arm of the barbarians in their hard struggle against a hostile nature." In an aside that was clearly aimed at Huxley, Kropotkin added: "It was also the bond they opposed to oppression by the cunningest and the strongest which might so easily have developed during those disturbed times. Clearly, the imaginary barbarian—the man who fights and kills at his mere caprice—existed no more than the 'bloodthirsty' savage" that ignorant English sailors had conceived out of their own misunderstandings.[132] Kropotkin was well aware that Huxley had begun his career in the Navy. "Leaving aside the preconceived ideas of most historians and their pronounced predilection for the dramatic aspects of history," he wrote, "we see that the very documents they habitually peruse are such as to exaggerate the part of human life given to struggles and to underrate its peaceful moods. The bright and sunny days are lost sight of in the gales and storms."[133] Any dispute that arose between one member of the community and another would be resolved in public under the arbitration of the "common law" of the folkmote, as would any claim brought by outsiders.[134] Peace prevailed as the preferred state of existence. Kropotkin quoted the legal historian and sociologist Henry Maine on this point: "'Man has never been so ferocious or so stupid as to submit to such an evil as war without some kind of effort to prevent it.'"[135]

Kropotkin knew his history. He had had the opportunity to study it as few men of his time had. For years in the fortress of St. Peter and St. Paul he had had the silence of solitary confinement in which to study and to contemplate the history of the Russian peasantry, their social institutions, and their traditions. He read about the origins of serfdom and tsardom, his opposition to which had brought him there. Once in London, he found others who shared his interest in the politics of history—he read Engels's work of course, but he also discussed early and medieval English history as well as Icelandic folk community life with William Morris. Both subjects were within the latter's expertise—indeed, there were few who knew the material better. It had been a similar story across Europe: the founding of villages followed by the establishment of independent cities; then a flourishing of trade and markets; and with all this series of developments, the rise of a ruling class that established themselves on the back of the people either through the superstition of religion or by military might. The moral economy of the community was abrogated by the blight of private ownership and the rise of the state as an institution to affect the tyranny of the few. The folkmote and the guilds were disbanded and despoiled, the common lands enclosed, and—much as Hyndman and Morris had outlined in their own accounts of this process in England—the institutions of mutualism were subject to the corruption of the state and the state's church.[136]

It is telling that even though Kropotkin's account of human history had differences to that which Spencer relied upon—like Huxley, Spencer believed that man had previously existed in a Hobbesian state of nature that was "solitary, poore, nasty, brutish and short"—it was still Spencer's understanding of biology that continued to influence the anarchist.[137] Kropotkin, like Spencer, emphasized the inheritance of acquired characters that resulted from the use or disuse of particular morphological or behavioral characters, a fact which was to influence his interpretation of the historical degeneration of the mutual-aid instinct as well as to inform his hopes for their revival. As the social institutions that had been founded upon organic mutualist associations were disbanded and replaced by centralized institutions, so mutualism was no longer fostered. Indeed, without the frequent recourse to such autonomous associations, the capacity to do so withered, and with the decline of man's mutualist tendency, so correspondingly, individualism arose. "The absorption of all social functions by the State necessarily favored the development of individualism," Kropotkin wrote, and echoing Spencer's concerns about the negative outcomes of government overreach, he pointed out that "in proportion as the obligations to the State grew in numbers the citizens were evidently relieved from their obligations to each other." Given but a few generations of such arrangements and the present state of things would prevail and be accepted as if it were itself a natural and instinctive way of life. Indeed, Kropotkin was aware that in Darwinian England this was already the common belief: "The theory which maintains that men can, and must, seek their own happiness in a disregard for other people's wants is now triumphant all round—in law, in science, in religion . . . and to doubt of its efficacy is to be a dangerous Utopian."[138]

Although Kropotkin had no knowledge of the letter that Darwin had written to Morley in which he had confessed that he thought that man had always been a social creature, it was clear from *Descent* that this was Darwin's view and that it was quite contrary to Huxley's position. Indeed, as the historian Mark Borrello has pointed out, when Kropotkin reflected on these events several years later, and with the benefit of those parts of Darwin's correspondence that had been published in the meantime, he had good cause to argue that Darwin had in fact come around to share much of his own view of evolution.[139] Kropotkin certainly believed that Darwin had said enough in private correspondence to indicate that by the time he wrote *Descent* he had given up a great deal of what he had written in *Origin*. In an essay Kropotkin wrote entitled "The Theory of Evolution and Mutual Aid," which was published in 1910, he laid out the evidence.

In *Origin*, Darwin had argued the case that organisms were not immutable but rather were subject to transmutation over vast amounts of time, diverging to form the wide variety of different species that populated the world. Naturally occurring variations were subject to the forces of what he called a "natural selection," in which those organisms that best fitted the environment they were born into would be more likely to survive and reproduce, outcompeting other organisms for ever scarce resources in a bitter struggle for existence. This was Malthus writ large, as Kropotkin noted. However, while Darwin had mustered a wealth of facts in support of the variability of both wild and domestic organisms, Kropotkin reminded his readers that natural selection remained only a hypothesis—indeed, it lacked any inductive observations to make it anything more than this. Certainly, Darwin's hypothesis had a lot going for it: as a Whewellian "consilience of inductions" it explained the wonderful adaptations of organisms to their environment without appealing either to supernatural interventions or to woolly ideas such as were often associated with Lamarckism—an organism willing its own adaptation and all that unsubstantiated rubbish. Kropotkin rejected such notions as absurd and later ridiculed the popularity of the philosopher Henry Bergson's "creative evolution" that was based on this very idea. Indeed, Kropotkin argued that natural selection had appealed to a great many naturalists for the simple reason that it set the whole species question on scientific grounds, subjecting it to naturalistic and inductive inquiry. Darwin had recognized the importance of this for the advancement of science as well as for pushing back the Anglican-dominated social elite. Here, Kropotkin thought, Darwin had done real and good service, but once the species question had been opened up to inductive inquiry, doubts had quickly been raised about the efficacy of natural selection to account for all that Darwin claimed for it and other naturalistic mechanisms were suggested in its place.

Darwin appeared to have recognized as much in a letter to Hooker, Kropotkin noted. "Personally, of course, I care much about Natural Selection," Darwin had written, "but that seems to me utterly unimportant, compared to the question of Creation *or* Modification."[140] Among the alternatives that were suggested, either in place of natural selection or to shore up Darwin's theory where it was found wanting, were the effects of climate and diet, of the use and disuse of what Kropotkin called an organism's "direct adaptation" to its environment, a phrase he took from Spencer. Each of these, Kropotkin noted, had been a part of Darwin's thinking in his transmutation notebooks before he even happened upon Malthus. Kropotkin was clear to say that this criticism of Darwin's science did nothing to invalidate the significance of his contribution. Rather, it strengthened it. For although Darwin was reluctant to give up the Malthus-inspired element of his theory on the basis of feeling "a sort of paternal predilection" for natural selection, Kropotkin argued that he had ultimately done so in *Descent*. Further, he pointed out that Darwin's vacillations were evident from his correspondence with two of his closest confidants and personal friends, Asa Gray and Joseph Dalton Hooker.[141]

Kropotkin was aware from Darwin's notebooks that from 1837 he had embraced a whole range of ideas about how organisms might adapt to their surroundings—use and disuse, food, and climate, to name a few—and while reading Malthus gave Darwin an appreciation of the importance of the individual in any given population and the idea of individualistic competition that fit so well with his liberal Whig politics, Kropotkin suggested that it was not until Darwin's friends pressed upon him the similarities between his own ideas and those of Lamarck that he began to stress natural selection as "the main but not exclusive means of modification."[142] Kropotkin confessed that "in studying the letters from that period I cannot refrain from the idea that the more [Darwin] was told by his friends (especially since the appearance of the *Vestiges of the Natural History of Creation*) of the near resemblance between his own ideas and those of Lamarck popularized in that book, the more he insisted upon showing in what they differed."[143]

Indeed, by 1856, the year in which Lyell and Hooker had pressed him to publish, Darwin had evidently become more confident in selection, and aware of Lamarck's associations with bad politics as well as bad science, he narrowed his focus. As Kropotkin showed, writing to Hooker again, Darwin tried to distance himself from such Lamarckian associations, pointing out to his friend that "external conditions (to which naturalists so often appeal) do by themselves *very little*." Yet, and as Kropotkin noted, even here he lacked the full courage of his convictions, adding, "How much they do is the point of all others on which I feel myself very weak."[144] This remained Darwin's position in *Origin*. The lion's share of the action went to natural selection, but even in the first edition Darwin still found room for the inheritance of characters acquired through use and disuse, a role for diet, and the direct effects of the environment. Indeed, even in the midst of explaining his theory of divergence, Darwin had emphasized the role of isolation in preventing the blending-out of useful variation.

Darwin's doubts pervaded his correspondence. It was in a letter to Hooker that he first mooted what would later become his famous "stone house" analogy, in which he refuted Gray's teleological suggestion that evolution was quite compatible with Calvinist preordination.[145] The point of the analogy was to drive a wedge between the naturally occurring variations which, Darwin admitted, were at least in theory fully determinable, and the success or failure of these variations in the environment in which they appeared. The laws of variation were fixed and immutable, environmental circumstances much more contingent. It was here that he had described variation as the mere "handmaid" that offered up her fruits to natural selection, "the mistress." This was Darwin at his most hard-line on the subject, referring to natural selection as the "natural preservation" of the fittest variations.[146] Kropotkin noted that here "Darwin did not anywhere admit the suggestion—which for us is now an established fact—that under the influence of external conditions the variations themselves are produced chiefly in a certain definite direction, and therefore have already a positive character. . . . That direct action might be—to use Herbert Spencer's terminology—a *direct adaptation*."[147] Hooker had pressed Darwin on this, and as a result, Kropotkin noted, "we see that in 1862 a change began to take place in his mind in this respect." Hooker had given Darwin cause to think deeply about his earlier demarcation of natural selection as the "mistress" and the variations caused by direct adaptation as the "handmaid," and Darwin had been forced to confess that there were circumstances in which direct adaptations to the environment might "submit to Natural Selection variations so useful that little choice was left for the approval of 'the mistress.'" Indeed, might not the respective roles be reversed? Kropotkin noted that Darwin had confessed a little disappointment at this realization: "'I hardly know why I am a little sorry, but my present work [on variation] is leading me to believe rather more in the direct action of physical conditions. I presume I regret it, because it lessens the glory of Natural Selection, and it is so confoundedly doubtful. Perhaps I shall change again, when I get all the facts under one point of view, and a pretty hard job that will be.'"[148] However, he did not "change again." Far from it. Rather, by 1863 he had become more convinced than ever. "I have underrated this action in the *Origin*," he wrote.[149]

Kropotkin argued that this was a most significant admission on Darwin's part. In *Origin* Darwin was still of the individualist mindset and thinking of individual variations—whether by chance or in response to external conditions. But as Kropotkin pointed out, if all organisms in a particular environment were subject to the same stimuli, then the variations that might be produced by this mechanism would not be in single individuals, but rather would occur throughout the population. "It is a *group* variation," Kropotkin claimed, and in consequence, "the sharpest struggle for life goes on no longer between the individuals of the same group, but between the group and its competitors from other species."[150] Kropotkin recognized that Darwin's admission on this point got rid of Malthusian individualism at a stroke. The most significant competition was between species, not between individuals, and if variations might be produced in response to the environment in this way, then they would no longer be random, or "chanceful," as Darwin had suggested in his opposition to Gray, but would be directed and cumulative.

Extrapolating, Kropotkin asked, "Could not new species, better appropriated to new conditions, be produced in the same way as the function produces the organ—as had been indicated by Herbert Spencer?" Certainly, organisms that could not adapt to a given environment—entire species as well as individuals—would die, but if this was "natural selection," then it was of a very different character to the individual war of each against all that Darwin had implied in his most Malthusian of moods. Here, geographic isolation also took on a new tenor. No longer merely a mechanism to prevent intercrossing of rare variations, now it was the source of variation—and variations that were prompted as well as favored by the prevailing conditions. Here was a much more powerful theory of divergence at work, Kropotkin suggested.[151]

The recognition of the power of these elements of Darwin's theory and his own realization that he had overstated the Malthusian mechanism in *Origin* should be of the utmost significance to naturalists, Kropotkin argued. It would not only change their opinion on the mode of speciation, engendering a new appreciation not only of the role of the environment in the formation of new species, but of the role of sociability, mutual aid, and cooperation as well: "There is not the slightest doubt that the hesitation of many biologists to recognize sociability and mutual aid as a fundamental feature of animal life is due to the fundamental contradiction they see between such a recognition and the hard Malthusian struggle for life which they consider as the very foundation of the Darwinian theory of evolution. Even when they are reminded that Darwin himself, in the *Descent of Man*, recognized the dominating value of sociability and 'sympathetic' feelings for the preservation of species, they cannot reconcile this assertion with the part that Darwin and Wallace assigned to the *individual* Malthusian struggle for *individual* advantages in their theory of Natural Selection."[152]

Kropotkin did not deny that there was a contradiction between Malthus and mutualism. Indeed, if there was a struggle for existence, both for food and to leave progeny, going on within each animal group "to the extent admitted by most Darwinians (which *must* be admitted if the natural selection of individual variations plays the part that is attributed to it), then it excludes the possibility of association being a prevalent feature among animals." But, Kropotkin continued, the converse was also true. If association and mutual aid went on in nature to the extent that so many capable naturalists had observed in the field—and here he listed a great many of them—"then struggle for life cannot possibly have the aspect of an acute war within each tribe and group."[153] Harking back to Huxley again, he concluded, "Darwin was quite right when he saw in man's social qualities the chief factor for his further evolution. Darwin's vulgarizers are certainly wrong when they maintain the contrary."[154]

Kropotkin's emphasis upon direct adaptation to the environment and the heritability of characters that were acquired in this way had significant implications for his politics. Anarchist-communism was thus not only sound politics, but it had been proven to be the most successful mode of social organization across long eons of evolution. As a result, he could not but oppose the views of the German cell biologist Friedrich Leopold August Weismann, who had argued in a series of publications directed against Herbert Spencer that there were no grounds for belief in such Lamarckian mechanisms of heredity.[155] Weismann had first published his refutation of the inheritance of acquired characters in his 1883 essay "On Heredity," and he noted the significance of Spencer's proposition. "If these views . . . be correct," he had written, "all our ideas upon the transformation of species require thorough modification, for the whole principle of evolution by means of exercise (use and disuse), as proposed by Lamarck, and accepted in some cases by Darwin, entirely collapses."[156] The political implications of Weismann's work is my subject in the next chapter.

Before moving on though, it is important to note that throughout the rest of his life Kropotkin continued to repeat the message that evolution vindicated mutual aid because historically it had preserved those societies that had adopted it. In *The Conquest of Bread* (1912), Kropotkin echoed the views of many in the socialist movement when he argued that the productive output of modern industry was such that there was no reason for people to starve or to live and die amid squalor, filth, and disease—Malthus was an ideology, not an induction, and Huxley was wrong to present Darwin as an out-and-out Malthusian, especially where he considered man. While Kropotkin was certainly skeptical of the way in which science was being misrepresented by Huxley and other prominent men of science in England at the time, he recognized that if rightly directed, the insights of science might be applied to agriculture with great benefit. The inspiring technology that Paxton had demonstrated in the construction of the Crystal Palace suggested a future in which acres of glass houses would ensure that England might meet the needs of her own subsistence with ease, and all the more so with a just and equitable production and distribution of the goods thereby produced.[157] Similarly, and as the proponents of sanitary science had long since argued, the application of science to pollution would end the dreadful conditions endured by the working classes in the city slums. "If a Huxley spent only five hours in the sewers of London," Kropotkin argued, one might "rest assured that he would have found the means of making them as sanitary as his physiological laboratory." While it seemed to Kropotkin that the Huxleys of the world saw in the slums only useless mouths, he reminded his readers that it had not been that long since "somebody [had] said that dust is matter in the wrong place. The same definition applies to nine tenths of those called lazy."[158] It would not have been lost on his readers, of course, that this "somebody" was Huxley.

In challenging Huxley's construction of man's place in nature, Kropotkin was determined to show his readers that Huxley's credentials as "Darwin's Bulldog" did not give him the exclusive right to speak in Darwin's name. In fact, the extensive field research that Kropotkin had engaged in demonstrated that not only was Huxley misguided in stressing a Malthusian view of nature and society, but that Darwin—who had clearly stated his intention that Malthus should be taken metaphorically—had done so too. The nature of human nature was at stake, and thus so too was the kind of society that might be possible for mankind. For Huxley, mankind had only recently become sociable, as a result of a self-interested rational calculation. For Kropotkin, as was the case for Darwin too, humans were instinctively sociable and self-interest was a much more recent development. Where Darwin had grounded the origins of truly other-regarding social instincts in the parental and filial affections, Kropotkin thought otherwise. They originated in the fraternal and sororal instincts that predated not only the modern family, but the evolution of sexual reproduction as well. Kropotkin was certain that in the future scientists would find evidence that mutual aid was evident among microorganisms and thus that there was reason to believe that it had been a factor in evolution from the very origin of life.

Historians need to take Kropotkin more seriously than they have done to date: Stephen Gould was certainly correct when he stated that "Kropotkin was no Crackpot!" Dan Todes and, more recently, Mark Borrello, Oren Harman, and Lee Dugatkin, have incorporated Kropotkin into their histories of evolutionary ideas as well, but as Shaw pointed out, Kropotkin was responsible for an "anarchist influenza" that infected the early British socialist movement, one that was based on deeply evolutionary ideas. We thus need not only to acknowledge Kropotkin, but to acknowledge his influence. We need to recognize that the concept of "Darwin without Malthus," to borrow Todes's phrase, was not only a Russian phenomenon, and that Kropotkin was far from alone in advancing a non-Malthusian evolutionary politics in English politics.

As I have indicated in the previous chapter, the radical origins of English socialism meant that many socialists were deeply ambivalent to Malthus, and as a result there remained debate about the moral meaning of Malthus beyond the turn of the century. However, this was not the only debate in biology that had significant political ramifications. By the mid-1880s, the inheritance of acquired characters was also increasingly subject to question. As I shall show in the next chapter, the debates over Malthus and Lamarckism shaped the subsequent development of English socialism.

6

Of Mice and Men: Malthus, Weismann, and the Future of Socialism

> Were we free to have our untrammelled desire, I suppose we should follow Morris to his Nowhere, we should change the nature of man and the nature of things together; we should make the whole race wise, tolerant, perfect—wave our hands to a splendid anarchy, every man doing as it pleases him, and none pleased to do evil, in a world as good in its essential nature, as ripe and sunny, as the world before the Fall.
>
> H. G. WELLS, *1905*

The moral meaning of Malthus was central to nineteenth-century politics. From 1798, Malthus's *Essay on the Principle of Population* had been a vehicle to attack Godwinian radicalism, but by the 1830s a new breed of philosophical radicals led by Harriet Martineau had succeeded in turning the moral significance of Malthus upside down. Where previously Malthus's observations about nature had been articulated as setting limits to human improvement, Martineau almost singlehandedly succeeded in changing the public perception of Malthus as having set the conditions by which improvement might be obtained. Hard work, industry, and self-restraint were all good moral attributes as far as Whigs were concerned. Darwin was eager to acknowledge that it was this interpretation of Malthus that had been his inspiration when he was searching for a mechanism that might drive speciation. However, and as we have seen, it was precisely the Malthusian elements of Darwin's theory that remained controversial long after the religious concerns about human being descended from apes had been laid to rest. Now that the evolutionists had placed man well and truly among the animals, it was the processes of evolution that became the focus of debate because this issue spoke to the question of what kind of an animal a human being was, of what it meant to be human, and of what kind of society it might be possible for us to live in as a result. In this chapter, I will show that the moral meaning of Malthus remained an intensely political issue well into the twentieth century. Further though, from the 1880s this debate expanded beyond arguments about selection to include new views on the mechanisms of heredity. From the mid-1880s, but especially in the 1890s, the observations and experiments of the German cytologist and naturalist Friedrich Leopold August Weismann were deemed to have political significance. Contrary to popular belief and in light of his work on heredity, Weismann suggested that the characters an organism acquired in the course of its own lifetime were not transmitted to the next generation. This was certainly a controversial finding in science, but it was of no less significance to contemporary socialists, because it called into question the Lamarckian assumptions that informed the hopes that many of them had of mankind evolving into more socialist beings. Without the inheritance of acquired characters and with only the slow process of the natural selection of chance variation to work with, evolution was apparently a much slower process than many had previously presumed. Thus, by the turn of the century, not only did a stance on Malthus divide the socialist movement, but so too did a stance on the inheritance of acquired characters.

The historiography on the debate over the inheritance of acquired characters has largely focused upon the debate that was carried on in the pages of the *Contemporary Review* between 1893 and 1895. Weismann and Herbert Spencer were the main protagonists, but the physiologist George J. Romanes also weighed in on the issue. In his important article "The Weismann-Spencer Controversy over the Inheritance of Acquired Characters," the historian Frederick B. Churchill has argued that although Weismann ultimately won the debate, this was not because he had the better evidence, for he did not. Indeed, the majority of Weismann's contemporaries in biology thought his germ-plasm theory overly speculative and his views on cell differentiation problematic. Weismann won out, Churchill argues, because he fit better with the prevailing naturalist tradition that was popular in English science at the time. Spencer's claims, on the other hand, were grounded in the older, physiological tradition. As Churchill points out, although we tend to think of Weismann today as a cytologist, recognition of the significance of his contribution to cell biology only came later, and Churchill reminds us that for a decade after 1864 Weismann was forced to abandon his microscope as a result of hyperemia of the retinas, a disease associated with eyestrain from extensive periods of fine and detailed work. The total rest of the eye that treatment demanded turned Weismann toward work on caterpillars and butterflies—natural history, even if in a museum setting.[1]

By 1895, when the debate came to an end, at least in the *Contemporary Review*, the old guard was gone. Huxley was dead and Hooker had given up the directorship of Kew to his son-in-law William Turner Thiselton-Dyer a decade earlier. Only Wallace, now aged sixty-six, was still actively publishing, and since he had never given any weight to the inheritance of acquired characters, he backed Weismann all the way. With both Darwin and Huxley gone, Wallace's was a powerful voice indeed. Churchill points out, too, that many of the new generation of Darwinian scientists were, like Weismann, also well versed in the naturalist tradition of natural history. Thiselton-Dyer and E. Ray Lankester were just two of the most notable; others included the new Linacre Professor of Human and Comparative Anatomy, H. N. Moseley, and the Hope Professor of Zoology at Oxford, E. B. Poulton. Each of them was also happy to see Lamarckism undermined.[2]

Churchill's argument from natural history is generally convincing, but he hints that there is more at work here than mere resonance of styles of work. As he points out, there remained problems with Weismann's explanations, and the fact that Weismann's views were considered speculative also suggests that there might have been other agendas at work. "The issue of heredity, then as now, had implications which stretched beyond the scientific claims," he adds. Churchill does not flesh out this last point, but I believe it is an important one, and I will attempt to develop it in this chapter. As I shall show here, Weismann's work had significant political as well as scientific implications.

As I have already shown in chapter 4, the inheritance of acquired characters was fundamental to the strategy of "making socialists," which dominated the early years of the English socialist revival. If Weismann was right, then what might be achieved through this strategy was clearly much less than its advocates had claimed for it. There were many people in the socialist movement who took part in, or at least took a stance on, this issue. We have already seen, for instance, that David Ritchie was aware that Lamarckism was being debated in biology at the time he wrote *Darwinism in Politics* in 1889; at the same time, Peter Kropotkin was also aware of the challenge to the inheritance of acquired characters, but he remained convinced of the heritability of both physical and behavioral characters that had been developed through use and disuse as a means of an organism adapting to environmental change. In this much, and despite their other differences, both Ritchie and Kropotkin were in agreement, and in this much too their beliefs were broadly representative of those held across the socialist movement.[3] In this chapter, however, I focus on three figures who were deeply involved in the intellectual development of British socialism: the craftsman, artist, and founder of the Socialist League, William Morris; the science writer, aspiring science-fiction writer, and sometime Fabian, H. G. Wells; and the Dublin-born literary critic and playwright, George Bernard Shaw. Like Wells, Shaw was a Fabian too, but one of long standing; however, his views were unique enough to merit the label "Shavian," to take account of his idiosyncrasies!

Morris was a friend and colleague of Kropotkin and was just as opposed to Malthusian politics. He was also an ardent advocate of making socialists as a strategy for bringing about socialism, and remained so until his death in

FIGURE 6.1. William Morris, 1834–1896. Photograph by London Stereoscopic & Photographic Company, 14 March 1877. The author of *News from Nowhere,* Morris was an influential figure in the English socialist revival of the 1880s; influenced by Kropotkin, Morris argued for a non-Malthusian evolutionary socialism. (NPG x3728; © National Portrait Gallery, London)

1896. His socialist-utopian novel *News from Nowhere* (1891), founded upon his anti-Malthusian and Lamarckian presumptions about social evolution, was immensely popular. Wells and Shaw debated the insights that Darwinism might have for Morris's hopes for the future; both found Malthus convincing but were divided over Weismann. To Wells, Weismann's observations radically altered what might be possible for humanity to achieve in terms of the evolution of a new humanity; Shaw, on the other hand, never gave up his faith in Lamarck and found Weismann's arguments less than convincing.

William Morris had been converted to radicalism by the artist and art critic John Ruskin, whom he had met in the mid-1850s while a student at Oxford. He was particularly moved by Ruskin's passion for social justice and the way in which he characterized the awful and demoralizing effects of industrial capitalism upon the nation's workingmen. Ruskin became a lifelong influence upon Morris; his *Unto This Last* (1862) and his letters to the workingmen of England, which were published under the title *Fors Calvigera* in the 1870s, proved particularly important in shaping Morris's views on the importance of work in shaping humanity.[4] Morris's favorite work by Ruskin, however, was the essay "The Nature of Gothic." Ostensibly about architecture, it was also and most deeply about the impact of living and laboring in a soulless, shoddy, and ersatz environment, and the consequences of ignoring the fact that the nature of a man's work was important for his humanity.[5] Ruskin lamented that "the great cry that rises from all our manufacturing cities, louder than the furnace blast, is all in very deed for this—that we manufacture everything there except men; we blanch cotton, and strengthen steel and refine sugar, and shape pottery; but to brighten, to strengthen, to refine or to form a single living spirit never enters into our estimate of advantages."[6] It was recognition of the naked truth of Ruskin's words that compelled Morris to throw over his efforts at liberal reform. Eager to "push forward matters," he declared, "I am going in for socialism: I have given up these Radicals."[7] Although Ruskin was no evolutionist, by the time that Morris turned to socialism, the "making of men" was something that had distinctly Darwinian associations.

As I have made clear in chapter 4, the nascent socialist movement that Morris gave up liberalism to join in 1883 was very much in its infancy. Indeed, beyond a few small groups of individuals, there really was no "movement" to speak of. In fact, when Morris first attended a meeting of the Democratic Federation, one of the London radical organizations, it had yet to declare itself for socialism, although Morris clearly anticipated that it would surely shortly do so.

When it did, it did not take long for Morris to become a towering figure in the English socialist movement, and he did so out of all proportion to the meager size of the various socialist organizations he either founded or to which he belonged throughout his life. By the time of his death in 1896, Morris was among the most influential men in English socialism. Indeed, as the historian Stephen Yeo has noted, Morris's utopian novel *News from Nowhere*, which was published in 1890 in *Commonweal*, the journal of the Socialist League, became "a central text in this period of British Socialism."[8] While Morris lectured on labor, art, and socialism from the moment he joined the movement, it was in *Nowhere* that he most clearly outlined his non-Malthusian and Lamarckian account of the evolution of society toward socialism. The novel was published serially from January to October 1890 and issued as a book the following year. In it, Morris portrayed England as he hoped it might become; it was a vision that many people found appealing—and many still do.

Set in an imagined London of 2102 following a socialist revolution, the world Morris portrayed had undergone substantial change from the one his readers knew. Gone were the belching industrial chimneys, the ugly and oppressive architecture, and the dirt and disease of the industrial age. In their place Morris described a world in which humanity lived in harmony with nature and where unskilled and polluting industrial practices had largely been replaced by skilled craft labor. Beautiful architecture, built to harmonize with the landscape, was commonplace, and salmon swam in the Thames once more. It was a vision that resonated with the radical agrarian ideals of those who thought that perhaps England had taken a wrong turn when it embraced industrialism and set out to become the world's workshop.[9] Morris portrayed a world at peace with itself; there is neither waste nor want, no driving commerce or economic competition, and no Malthusian struggle for existence. The book is subtitled "An Epoch of Rest." Perhaps the biggest difference between the world that Morris portrayed in his utopia and that which he lived in was the changes that he depicted as having been effected among the inhabitants of Nowhere. Not only were they both good-natured and selfless, both they and their children had also become much more physically beautiful as well.[10]

It is not the narrative of Morris's tale that interests us here so much as his rejection of Malthus and his conception of how the transformation of both individuals and society were to be effected. Morris's views reflected many of the radical assumptions about Malthus and evolution. Like many other radicals, Morris had seen Huxley champion the Malthusianism in Darwin's theory as a "Whitworth gun in the armoury of liberalism" and had not been impressed. He accused Huxley of presiding over "a Whig committee dealing out champagne to the rich and margarine to the poor."[11] Indeed, many of the older members of the socialist community, those who had cut their political teeth as radicals, had watched as Huxley waged a holy war for secularism in an Anglican age. Many who had been impressed with his attacks on the establishment soon changed their opinion in the wake of his essay "The Struggle for Existence."[12] As Kropotkin made abundantly clear, not only had Darwin intended Malthus to be read as a metaphor, he had also ultimately realized that it was not a particularly good one. Huxley, however, was carrying on as if his distorted view of life was the truth of the matter. Of course, Marx had also pointed out the political implications of Darwin's adoption of Malthus, and thus it is not surprising that in the same month as Huxley's article appeared, one contributor to *Justice*, the paper of the Social Democratic Federation, anticipated Kropotkin's point: "Every discovery of Science, every invention of mankind, has been seized upon by the bourgeoisie to delude and exploit the proletariat. . . . In a like manner the bourgeoisie accept the teachings of Malthus and pervert those of Darwin to bolster up the tottering fabric of society today, and they steal from the armoury of the evolutionist weapons which they use in their own defence."[13]

It is not that socialists refuted Malthusian conclusions in the natural world, but they did reject the argument that they necessarily applied to humanity. Humanity could raise itself out of the struggle for existence through labor, cooperation, the application of technology to the natural resources that lay all around them, and the fair and equitable distribution of the goods that they produced as a result. Alfred Russel Wallace, who had converted to socialism in 1889, also made the case that female emancipation would mitigate fertility rates. Wallace, who like many in the socialist movement had come to socialism from an earlier commitment to radicalism, also believed that however true Malthusian predictions might be for the rest of the natural world, they had no necessary hold over humanity. To believe as much, he said, was "the greatest of all delusions."[14] Not only had Malthus admitted—at Godwin's promptings—that moral restraint might fend off the threat of population growth, but also that under socialism the Malthusian assumption that the divergent ratio between production and consumption might be quite easily subverted. Scarcity in society was the result of injustice, not the inevitable working out of natural laws, and Wallace argued that in addition to the appropriation of the means of production, distribution, and exchange as a means to meeting the needs of all, the emancipation and education of women would discourage early marriage and its consequences. Basing his argument upon his experience of the many and diverse cultures he had encountered in his travels, Wallace was keen to point out that there was a direct correlation between female education and fertility rates.[15] Of course, given that the primary concern of most socialists was with labor, the most frequently articulated reason to deny Malthusian pessimism was the only too evident productive capacity of modern industry. The mastery of nature that had made Britain the workshop of the world might easily meet the subsistence needs of the growing populous.[16]

Anti-Malthusian evolutionary politics were rife across the socialist movement. William Morris popularized these ideas, doing more than most to shape the character of English socialism. From his earliest days as a socialist, Morris attacked the Malthusian arguments that were supposed to show that the poor were poor as a result of the inevitable workings of natural law. In the brief year that he was a member of the Democratic Federation, Morris lectured incessantly and wrote a number of pamphlets arguing this point. Most pointed, perhaps, was the pamphlet he co-wrote with Henry Hyndman, *A Summary of the Principles of Socialism*, in 1884. In it, they made the point that a reorganization of labor would secure adequate food, clothing, and shelter for all. "This foolish Malthusian craze is itself bred out of our anarchical competitive system and those who are smitten with it cannot see that the power over nature is such that, if his [*sic*] labour were properly organised, he would produce in food or its equivalent at least four times the amount of wealth which he would require, if he lived in absolute comfort, provided he worked only six hours a day."[17]

Morris believed that such was the power of mankind over nature that not only could the present population be adequately provided for, but that the guarantee of adequate provisions would also remove the incentive for large families as a form of insurance against the inability to provide for oneself in old age. Also, and echoing Wallace's views, the women in Morris's *News from Nowhere* are the equals of the men in every way. Indeed, Morris was so deeply convinced that the Malthusian "population problem" was an artificial construct that once social justice has been achieved in *Nowhere*, the struggle for existence plays no further part in his imagined future.

Although Morris rejected the conclusions that many drew from Malthus's work, it is clear that the utopian vision he portrayed in *News from Nowhere* relied upon neo-Lamarckian mechanisms of inheritance. Morris was not alone in assuming that the humanity of the future would be different to that inhabiting late nineteenth-century England, that the coming of socialism and the bringing of the world "to its second birth" would effect substantial changes upon humanity as well as in the political and economic arrangements of society.[18] These beliefs were not merely hypothetical, but found overt expression in the strategy of "making socialists." It is significant too that while Morris found no place for natural selection—at last in terms of the competition between one individual and another—he was quite open to the significance of sexual selection—and of the role of female choice in that process in particular.

Morris was by no means ignorant of evolutionary science. He subscribed to the popular journals in which the implications of evolution for humanity were published, and it was a regular topic of discussion in the socialist press as well. Further, from 1886, Morris and Kropotkin became quite close. The two had met at a dinner that March to commemorate the Paris Commune shortly after Kropotkin's release from Clairvaux, and thereafter Kropotkin became an occasional visitor and dinner guest at Morris's home, becoming particularly close to Morris's daughter, May.[19] While Kropotkin never quite convinced Morris of anarchism in their discussions, the two men shared an abhorrence of state power and a faith in the ingenuity, creativity, and free agency of humanity. And despite the fact that Morris frequently stated that he had no head for science, several of his associates attributed this to modesty.[20] Despite Morris's reservations about anarchism, Kropotkin had no qualms about endorsing *News from Nowhere.* In a review for the anarchist journal *Freedom* he described it as "perhaps the most thoroughly and deeply Anarchistic conception of future society that has ever been written."[21]

Like Kropotkin, Morris was particularly concerned with individual liberty, and both male and female agency were central to his conception of socialism, as well as to the strategy by which he hoped to see it achieved. Indicative of the sort of direct democracy that both Morris and Kropotkin envisioned, the people of Nowhere have converted the House of Commons into a manure store, making their own decisions by collective consensus reminiscent of the medieval folkmote.[22] Morris was aware that an emphasis upon agency was being squeezed out of both contemporary socialist theory and practice as Marxism and Fabianism gained ground, and although he was no anarchist, it was this that brought him to sympathize so deeply with Kropotkin. Without free agency, Morris believed, socialism could only be imposed upon the population, something that could only become a "slavery . . . far more hopeless than the older class-slavery" of capitalism.[23] It was only with a socialist people that one might have a socialist society worthy of the name, he wrote.[24]

Morris voiced a common opinion when he asserted that socialism was "emphatically not merely a 'system of property-holding,' but a complete theory of human life . . . including a distinct system of religion, ethics and conduct"—a view that reflected his faith in the political strategy of "making socialists."[25] Those who had been degraded, demoralized, and disfigured by capitalism needed to be made anew—in mind, in body, and in spirit. It was this commitment to "making socialists" that increasingly became mere rhetoric in the mouths of those who sought either to seize state power in a Marxist coup or to gradually infiltrate government in order to legislate their way to a socialist future.

The point of any utopia is to persuade its readers that things can be different than they are, and in this respect Morris's *News from Nowhere* was no different. He drew on his extensive knowledge of the medieval era to demonstrate that the life of the past had been very different from the present, and that thus it might be different again in the future. Morris aimed to make his readers aware of the fact that there was nothing natural or immutable about the lives they led or the society they lived in. What most Morris scholars have missed, however, is the fact that there is an evolutionary component to Morris's utopia and his politics. Like the vast majority of his contemporaries, Morris believed that men, like any other organism, adapted to the prevailing environment, and that those adaptations were heritable. As a result, he believed that it was not only the conditions of life that had changed for the worse under capitalism, but that so too had the people.[26] In short, nineteenth-century society, with its embrace of laissez-faire and refusal to ensure better conditions for the people, acquiesced in the unmediated adaptation of its citizens to groveling competition and squalor, and worse, it ignored the fact that future generations would inherit the characteristics that were acquired as a result of such degrading conditions. Because the average workingman had been so degraded by the world he lived in and was too mentally and spiritually degraded to imagine that life might be different, Morris argued that it was left to "those who worked for change because they could see further than other people" to show that there might be more to life than their common experience taught them. As the old man, named Hammond, in Morris's story remarked, clearly on Morris's behalf: "Contrast is necessary for this explanation."[27] Morris sought to illustrate this contrast in *News from Nowhere*, juxtaposing the clean, healthy, and vigorous lives of the citizens of Nowhere with the degraded, broken, and dispirited workers under capitalism. Morris underlines the message of his tale in the last lines of the book. Having awoken back in the nineteenth century, the visitor to Nowhere imagines what the inhabitants of the future want to say to him: "'Be the happier for having seen us, for having added a little hope to your struggle. Go on living while you may, striving, with whatever pain and labour needs must be, to build up little by little the new day of fellowship, and rest, and happiness.'" Then Morris concludes: "Yes, surely! and if others can see it as I have seen it, then it may be called a vision rather than a dream."[28] This was exactly the aim of making socialists, to build up little by little the fellowship of the new life of socialism. By actually experiencing how radically different life might be, and by being physically immersed in and subject to the influences of a more healthful, social, and "natural" environment, Morris sought to awaken in them the all-important "longing for freedom and equality" that was the "great motive-power of the change."[29]

As the character Hammond went on to recall: "In times past, it is clear that the 'Society' of the day helped its Judaic god, and the 'Man of Science' of the time, in visiting the sins of the father upon the children. How to reverse this process, how to take the sting out of heredity, has for long been one of the most constant cares of the thoughtful men amongst us."[30] It is clear from Morris's description that Morris thought that an altered environment and the adaptation to it through the use and disuse of characters and their subsequent inheritance would be important ways in which humanity might "take the sting out of heredity."

Like Kropotkin—and indeed, like many radicals and old Owenite socialists—Morris emphasized the environment and the conditions of existence that people lived under as an important influence upon the development of their character, and thus it is no surprise that Morris's vision of what socialism might be like in the England of 2102 is set in a very different environment than existed in Morris's own time.[31] Morris leaves it to Hammond to note the potential of even a relatively subtle difference in climate and environment to effect a significant adaptive change upon its inhabitants. Pointing out the differences between the neighbors of the Thames Valley and those who live farther north, he observes that "'there are parts of these islands which are rougher and rainier than we are here, and there people are rougher in their dress; and they themselves are tougher and more hard-bitten than we are to look at.'" That these are adaptive differences and are intended to be read as having hereditary significance is emphasized by Hammond's rejoinder that for all their differences, "'the cross between us and them generally turns out well.'"[32] Morris made it clear that he believed that human nature was malleable and should not to be limited by contemporary notions of self-interest and the war of each against all. Rather, he believed that men were the product of their environment and could be shaped by circumstance: "'I have been told that political strife was a necessary result of human nature,' the Guest remarked to Hammond. 'Human nature!' cried the old boy, impetuously: 'What human nature? The human nature of paupers, of slaves, of slaveholders, or the human nature of wealthy freemen? Which? Come, tell me that!'"[33]

Of equal importance for Morris's strategy for making socialists was the role of use or disuse in acquiring (or losing) characters, which could then be passed on to and further emphasized (or deemphasized) across subsequent generations, and it is significant, if unsurprising, that Morris thought that labor was of paramount importance in this regard. Morris believed that the divided and alienating practices of commercial labor under capital forced the workers into a position of "real inferiority . . . involving a degradation both of mind and body," and thus, far from being a pleasure, labor under capitalism was actually the means by which the working class was oppressed and alienated.[34] Indeed, Morris brought his evolutionary aspirations to bear on the concerns of his mentor, John Ruskin, and the historian and social critic, Thomas Carlyle. Ruskin had taught Morris more about labor and art than any man, and Carlyle, too, had written at length about what he called the "gospel of work" in his social critique *Past and Present.* There was something divine about true hand labor, he had argued. "A man perfects himself by working."[35]

Thus, Morris was not only concerned about the negative effects of capitalist labor, he believed that a lack of wholesome labor—a lack of the kind of work described by Ruskin and Carlyle—was detrimental to physical and mental vitality as well. The resulting maladies not only affected the unemployed, but the idle rich of the middle class as well. In contrast to the Fabian view that the middle class performed a function that was vital to the health of the social organism, Morris believed that as a class they produced nothing of real value. Not only was the typical bourgeois male parasitic, effeminate, and weak, but the idleness that was habitual to his class had, over time, become so culturally ingrained as to have become biologically heritable. As Hammond recalled: "'It is said that in the early days of our epoch there were a good many people who were hereditarily afflicted with a disease called Idleness, because they were the direct descendants of those who in the bad times used to force others to work for them—the people, you know, who are called slave-holders or employers of labour in the history books.'"[36]

The most disconcerting effects of this disease, however, were those inflicted upon middle-class females, who bound by Victorian social convention—not to mention Victorian corsetry—were kept from exerting themselves. They did no useful work and were thus prevented from ever exercising their bodies and minds or their creative spirit. As a result, they were typically ugly, pinched, and susceptible to the "mully-grubs" of mental weakness and hypochondria.[37] Subject to such middle-class ailments, only bourgeois men could find these women attractive, an element of Darwinian sexual selection that only served to compound these symptoms in their offspring. As Morris had the old man who traveled with Dick and William Guest for a short while say of the middle-class women of the pre-revolutionary days: "'They were as little like young women as might be, they had hands like bunches of skewers, and wretched little arms like sticks; and waists like hour glasses, and thin lips and peaked noses and pale cheeks. . . . No wonder they bore such ugly children for no-one except men like them could be in love with them—poor things!'"[38]

In line with Morris's conception of the restorative potential of unalienated hand labor and the central place it had in his reconception of society under socialism, it is unsurprising that work played a central role in his hope to redeem the degenerate nineteenth-century individual and transform him or her into the healthy, vital, and creative human being. Again, Morris left it for Hammond to recall the stern, if paternal, measures that had to be taken in order to remedy the negative hereditary effects of disuse: "'I believe that at one time they were actually *compelled* to do some work, because they, especially the women, got so ugly and produced such ugly children if the disease was not treated sharply, that the neighbours couldn't stand it.'"[39] Morris believed that labor under socialism would have not only physical benefits for humanity, but would also reclaim them at a deeply psychological and even spiritual level.

On a physical level, of course, the production and just distribution of the things that people actually needed, rather than the waste of labor expended on so-called luxury items for the wealthy and shoddy tat for the consumption of the workers, would have its obvious rewards. The end of starvation levels of poverty, of malnourishment, and of destitution would be a straightforward outcome of the changes that Morris proposed. In addition, though, there would be no place under socialism for the sweatshops of the nineteenth-century factory system. The practice of cramming men, women, and children into dank, cramped, and ill-ventilated buildings where they were hurried and harried by relentless machinery into putting quantitative output above all else, would come to an end. "Temples of over-crowding and adulteration, and overwork," Morris called them. They were instrumental in the degradation and degeneration not only in the immediate sense of the worker's everyday experience of life, but it is also clear that he thought that the dire effects of these conditions would be hereditary.

And yet, as Morris repeatedly pointed out, things did not have to be this way. Without the pretended urgency of increasing output regardless of the quality of the items produced, Morris argued that it was a very real possibility—he believed it a necessity—that all labor might become a joy and that it might be undertaken in pleasant and healthy surroundings. As he wrote in *Justice* in the spring of 1884, it was quite possible for the factory to be made a "pleasant place," situated in pleasant gardens in which the workers might enjoy "open air relaxation from their factory work." And of course, he added, "our factory must make no solid litter, befoul no water, nor poison the air with smoke."[40] As for the buildings themselves, they too would be beautiful, with a "simple beauty" that characterized Morris's preference for a workshop. However, because the factory would not have production of wares as its sole end, and making Ruskin's concerns his own, Morris imagined that the factory would have other buildings too, devoted to the refinement, the education, and the refreshment of the laborers. These buildings, "which may carry ornament further, might include dining halls, libraries, and schools, as well as facilities to study knowledge of various kinds—the sciences as well as the arts."[41] In short, the factory was to become a beautiful place where workers would find both leisure and learning and remake themselves in the process. Socialist labor, Morris believed, would allow humankind to realize their full and true potential as social beings.

Socialist labor and socialist life exercised "the energies" of a man's "mind and soul as well as of his body," while at the same time, as Morris had long asserted, "it helps the healthiness of both body and soul to live among beautiful things." Given Morris's acceptance of the inheritance of acquired characters and the importance of adaptation to the physical environment, the benefits that accrued (both physiologically and culturally) to the laborer during his lifetime in these ways could be transmitted to and compounded across subsequent generations.[42] In this way, Morris could well imagine a substantial change, from the degraded worker of the capitalist era of commerce to the "well-knit" men and women he portrayed in *Nowhere* as the living embodiment of socialism. Clearly, the inheritance of acquired characters was as important to Morris's strategy to bring about socialism as his rejection of Malthus was for his critique of capitalism. However, neither of these positions went uncontested. Indeed, the young science writer and aspiring science-fiction writer H. G. Wells, who deeply admired Morris in the 1880s, later became one of his most outspoken critics on exactly these points.

It is well known that H. G. Wells studied under Huxley in 1884 during his first year at the Normal School of Science at South Kensington.[43] He had been admitted on a meager scholarship of a guinea a week to train as one of the first generation of nationally accredited science teachers that were to bring about a transformation in science education across the country. In part this was testimony to the success that Huxley had had in advancing the cause of science education; in part it was the legacy of the "payment by results" that had been the outcome of the 1862 "revised code" in education. The code rewarded schools and students who scored highly in exams.[44] A career as a science teacher had been far from Wells's parents' intentions for him. Coming from a lower-middle-class family of somewhat fallen fortunes—his father was a failed shopkeeper and sometime professional cricket player, his mother a lady's housekeeper—it had long been his mother's intent to see him indentured as a draper's assistant. However, through the combination of a determined refusal to shine in the calling that his mother had chosen for him and an extraordinary proficiency in passing examinations, Wells found himself at the age of eighteen signing on "at the entrance to that burly red-brick and terra-cotta building" of the Normal School. Walking across Kensington Gardens from his lodgings in Westbourne Park on the first morning of classes was an act full of significance for the young Wells. He had long been a religious skeptic and convert to Darwinian ideas, and was in awe of the prospect of walking the same halls as the great T. H. Huxley. He later recalled that taking the elevator to the upper floor and first setting foot into the biological laboratory was "one of the great days of my life." "Here I was under the shadow of Huxley, the acutest observer, the ablest generalizer, the great teacher, the most lucid and valiant of controversialists."[45] He was enthralled to be assigned to Huxley's classes in elementary biology and zoology. "The study of zoology . . . was an acute, delicate, rigorous and sweepingly magnificent series of exercises. It was a grammar of form and a criticism of fact," wrote Wells.[46]

The sense that he was not only embarking on a new stage in his own career but making a contribution to a new view of life is palpable from Wells's description of the school and the great man who had brought it into being. The laboratory was a long and narrow room at the very top of the building, and "on the tables were our microscopes, reagents, dissecting dishes or dissected animals as the case might be. In our notebooks we fixed our knowledge." Despite advancing years and declining health, Huxley still made a deep and lasting impression on his students—and on this young student in particular.[47] "Huxley himself lectured in a little lecture theatre adjacent to the laboratory, a square room, surrounded by black shelves bearing mammalian skeletons and skulls displayed to show their homologies, a series of wax models of a developing chick." Huxley was "a yellow-faced, square-faced old man, with bright little eyes, lurking as it were in caves under his heavy grey eyebrows, and a mane of grey hair brushed back from his wall of forehead. He lectured in a clear firm voice without hurry and without delay, turning to the blackboard behind him to sketch some diagram, and always dusting the chalk from his fingers rather fastidiously before he resumed."[48] Wells's portrait of the aging naturalist was as clear as that painted by Huxley's son-in-law, the artist John Collier.[49] Clearly enamored, Wells was one of only three students to finish the first year of instruction with a first.[50]

FIGURE 6.2. Thomas Huxley, 1825–1895. Painted by Thomas Hamilton Crawford, 1922; mezzotint, after the original by John Collier, 1883. (NPG D36430; © National Portrait Gallery, London)

Huxley restricted his teaching to the basics of anatomy and zoology, avoiding discussion of the then ongoing debate about the various mechanisms of evolution, which Wells recalled was regarded at the time as "a field for almost irresponsible speculation."[51] However, and unsurprisingly, it was exactly this field of speculation that excited the students, and they could not help but be aware that the outcome of these discussions would have implications for humanity. Already familiar with Darwin's work, Wells also read E. Ray Lankester's 1880 work *Degeneration: A Chapter in Darwinism*. Lankester had worked alongside Huxley at South Kensington earlier in his own career, but had moved on to become Jodrell Professor of Zoology at University

College by the time of Wells's enrollment. Lankester had translated many of Weismann's works into English and continued to lecture at the Normal School, where his own and Weismann's works were evidently the subject of lively discussion.[52]

Whatever the source of his own initial encounter with Weismann's work, Wells recalled in his autobiography that although Weismann had his critics, in the Normal School at least, "Weismann and his denial of the inheritance of acquired characters was in the ascendant," a fact that is hardly surprising given the teachers there.[53] Wells lamented that "there was only one Huxley in the Normal School of Science." He had ensured his enrollment for a second year of study by his early academic achievement only to find that classes in physics and geology had "none of the stimulation and enlargement of that opening year."[54] There were extenuating circumstances as to why his tutor, Professor Guthrie, was distracted; ill and suffering with an undiagnosed cancer, he died the following year, in October 1886.[55] The immediate result of Guthrie's illness, at least as far as Wells was concerned, however, was a growing indifference to his prescribed course of study, which led him to seek out other interests. With some friends he founded the *Science Schools Journal*, in which he published his first short story, "The Chronic Argonauts."[56] It was later to form the basis of *Time Machine*, the science-fiction story that was to make his name. *Time Machine*, written across 1894 and 1895, reveals just how deeply Wells was influenced by Huxley and Lankester, and later by Weismann, but all this would come later. For now, biology was just great inspiration for writing stories.

In addition to taking his first serious steps into writing fiction, Wells also turned to politics as a worthy distraction from the studies that he found increasingly tedious. Throughout his youth, Wells had read widely, and in addition to Darwin's works, among much else he had also imbibed Plato's *Republic* and Henry George's *Progress and Poverty*. While Plato—and his own sudden rise from draper's assistant to undergraduate student—had opened Wells's mind to the idea that there was nothing fixed or immutable about the social hierarchies of Victorian Britain, George fired his interest in the many radical clubs that proliferated in the capital as well as in the nascent socialist organizations that were coming into being just as he was starting student life. Wells thus came to seek out some of the early socialist meetings. Always an avid reader and with the intellectual pretensions one might expect of a young man of his obvious talents, he took himself to the library of the British Museum, where like many of the key players in the nascent socialist movement he eagerly immersed himself in history, sociology, and economics.

As I have outlined already, the years during which Wells was enrolled at the Normal School were also the years that saw a significant resurgence of socialist politics in Britain. Discussion of the social and political implications of evolution—and of Darwin's work in particular—was rife among these groups, albeit often with a considerably different emphasis than that which Huxley and Lankester had given it. As a result, when Wells began to immerse himself in the politics of social change, he had cause to reflect upon the politics of the biology he was learning.

It was through these excursions into the socialist community that Wells first saw William Morris speak. In the company of a small cadre of like-minded friends, he would attend the open meetings of the Fabian Society and "went on Sunday evenings to Kelmscott House, Hammersmith, where William Morris held meetings in a sort of conservatory beside his house."[57] It was here too that Wells also first encountered the "raw aggressive Dubliner with a thin flame-colored beard," George Bernard Shaw, who was later to become a close friend—and amiable thorn in his side—for the rest of his life. Other eminent speakers in the coach house of Morris's Hammersmith residence that year included Peter Kropotkin, Ernst Belfort Bax, and the atheist, socialist, Darwinist, and partner of Eleanor Marx, Edward Aveling, as well as Sidney Webb, who was to become so central to the subsequent development of the Fabian Society, and Annie Besant, who spoke on the "Evolutionary Aspect of Socialism."[58]

It is testament to the freshness and vitality of Wells's prose that even though his recollections were written down some fifty years after the event, the enthusiasm and excitement that he felt at these meetings is evident. Morris was never comfortable with public speaking, but he forced himself to it as a matter of duty to the cause. As Wells recalled, he would "stand up with his back to the wall, with his hands behind him when he spoke, leaning forward as he unfolded each sentence and punctuating with a bump back to position."[59] It is clear from Wells's descriptions of these meetings—"None of our little group had the confidence to speak . . . [although] our applause was abundant"—as well as in his singling out Morris in his recollection, that Wells held Morris in some esteem.[60] Indeed, Wells wrote a short semi-autobiographical story of his life at the Normal School called "A Slip under the Microscope," which he published in 1896, the year that Morris died. As he is setting the scene of the laboratory in this story, Wells notes that among the possessions that the students had left about the room was "a prettily bound copy of News from Nowhere." It is significant and further testimony to Wells's high opinion of Morris during his own college days that the book's owner, a student named Hill, who is clearly based upon Wells, is noted as a socialist and materialist who troubles himself to lend Morris's book "to everyone in the lab." By 1896, however, and as a result of his studies, Wells would come to be much less enamoured by Morris. Mirroring both his own disaffection with Morris and Morris's disaffection with Huxley, he notes in the story that *News from Nowhere* was "a book oddly at variance with its surroundings."[61]

During these first forays into socialism and prior to his disillusionment with Morris, Wells confessed "a certain amount of subconscious antagonism towards science, or at least towards men of science," apparently falling quite readily into the prevailing outlook of his contemporaries in the socialist movement. Morris and Kropotkin were particularly skeptical of the politicization of contemporary science as it had grown to become a part of the establishment it had once criticized.[62] It had long been Morris's view that science was "so much in the pay of the counting-house, the counting-house and the drill sergeant," that without significant redirection it could not be of much use in bettering the lot of the mass of mankind.[63] Kropotkin echoed this opinion, arguing in a series of articles written for *Justice* that the scientific establishment was "only an appendage of luxury," an assessment he repeated in his popular pamphlet *An Appeal to the Young*.[64] Notably, and perhaps in light of his later enthusiasm for Huxley's views, Wells did not divulge how his confessed ambivalence to "men of science" affected his opinion of his former tutor at the time, but he clearly recognized that science, and evolutionary biology in particular, had political implications.

It is unclear exactly when Wells began to combine the insights of his scientific training under Huxley with his socialism, but as a result of his encounters at Kelmscott House, Wells found himself in an intellectual environment in which discussion of the political importance of evolution was an integral part of what it meant to call oneself a socialist. Thus, it is probable that Wells would have been given to think about such connections as a matter of course. It was doubtless in the company of Morris, Kropotkin, and Shaw, among others, that Wells initially became critical of "men of science," and only later—perhaps even only after reading Huxley's "Evolution and Ethics" in 1893—that he found himself at odds with the anti-Malthusian views at the heart of Morris's socialism. Indeed, it is quite possible that this was the turning point, for as Wells noted in his biography, Huxley did not lecture on evolution at the Normal School, confining himself to anatomy and physiology. Further, the Malthusian theme of "Evolution and Ethics" runs throughout *Time Machine*: throughout 1893 and 1894 Wells had been reworking "Argonauts" as a way to explore all the excitement. As Wells reimmersed himself in biology in the 1890s—he had taken to reviewing the latest developments in science for the popular weeklies to supplement his meager income—it is evident that his ambivalence toward Morris and the kind of socialism he imagined only increased.

It was in his capacity as a reviewer that Wells was led to revisit the work of August Weismann in 1894—probably for the first time since his college days. This time around, however, he read it carefully, and even though he was initially skeptical of Weismann's germ-plasm theory, he immediately saw the political implications of Weismann's arguments, not only against the inheritance of acquired characters—he had this from Huxley and Lankester, after all—but of "panmixia." Over the years, he had heard many arguments for the evolution of socialism, but now rereading Morris's idyll in light of the biology he had learned both at the Normal School and subsequently, he found it wanting in a number of ways. Morris's hopes for the future of mankind were not only incompatible with the Malthusian Darwinism that he took from Huxley, they were also at odds with Weismann's apparent experimental proofs against Lamarck and the degeneration that he forecast would result from panmixia.

Portraying the future as "an epoch of rest," Morris had pictured a world free from the Malthusian struggle for existence. Like many of his socialist colleagues, Morris believed that Malthus's conclusions in his famous essay were the consequence of injustice, not of nature. Indeed, like many a radical, Morris envisioned that with the overthrow of capitalism humanity might once more be reunited with nature. Socialism would liberate humanity from want and enable the rethinking of all aspects of life—the human relationship with nature, the organization of labor, and the woman question in particular. *Nowhere* was thus an act of imagination that opened up all sorts of possibilities for the future under socialism. It is to these three aspects of Morris's vision that Wells took exception.

In *Anticipations of the Reaction of Mechanical and Scientific Progress upon Human Life and Thought* (1900) Wells stated his conversion in no uncertain terms. He now believed, quite contrary to Morris, that "probably no more shattering book than the *Essay on Population* has ever been, or ever will be, written. . . . It made as clear as daylight that all forms of social reconstruction, all dreams of earthly golden ages must be either futile or insincere, or both, until the problems of human increase were manfully faced."[64] Wells's political *volte face* was total. The ambivalence toward Huxley and other men of science that he had hinted at as a result of his first encounters with the socialist movement evaporated in an instant and he subsequently took every opportunity to find fault with Morris's socialism in his writing—in his fiction as well as his more obviously sociological work.[65] As a recently graduated science student—Wells eventually graduated with a science degree from

London University in 1890—and making as much as he could of his Huxleyan associations, Wells attacked Morris with all the scientific authority he could muster. Modern science ruled out all possibility of "the Nowheres and Utopias men planned before Darwin quickened the thought of the world."[66]

Morris's vision of the future rested upon and celebrated the reintegration of the rural and the urban environments. The cities would empty and the people would throw themselves upon the land, revitalizing themselves and their society in the process. *Nowhere* was lauded by the "back to the land" movement for this very reason.[67] Contrary to the sentiments of Marx and Engels in the *Manifesto*, Morris saw no "idiocy" in rural life—and certainly not once it had been invigorated by the influx of a creative and productive population. Contrary to Huxley's views and fully consonant with the aims and ambitions of English radical agrarianism, Morris idealized the reunification of humanity with nature. Morris believed that the Huxleys of the world had been misled by capitalism. They were "always looking upon everything, except mankind, animate and inanimate—'nature,' as people used to call it—as one thing, and mankind as another. It was natural to people thinking in this way, that they should try to make 'nature' their slave, since they thought 'nature' was something outside them," he had argued.[68]

To Wells though, who now followed Huxley in all things, this was beyond naive. The idea of living in harmony with nature was, as Huxley had put it in 1888, "little better than a libel upon possibility."[69] "There is no justice in Nature," Wells wrote. "The method of Nature 'red in tooth and claw' is to degrade, thwart, torture and kill the weakest and least adapted members of every species in existence in each generation, so as to keep the average rising . . . using the stronger and more cunning as her weapon."[70] Thus, even if there was any merit to Morris's hopes of people adapting to fit their new natural environment, this was hardly the kind of influence that would help make socialists. Following Huxley's lead, the best hope for man was to put his efforts into opposing the methods and mores of nature rather than aspiring to mimic its qualities.

The fact that nature was at best indifferent, if not actually hostile to the aims of man meant that Morris's ideas of labor were also wide of any practicable mark. Society would have to win its subsistence by the hard toil of subordinating nature to man's will. She would not yield sufficient food for all by some harebrained scheme of allotment gardening and hand labor. Certainly, Wells acknowledged, arts and crafts might find their place under socialism, but they were not the bread, but the roses. Morris's contention that people might work for free as a result of some newly acquired altruism was improbable at best, too, even if labor could be reduced to the six hours a day that Morris had optimistically talked about.[71]

In his 1905 book *Modern Utopia*, a work that is rife with anti-Morris sentiment, Wells argued that "there is, as in Morris and the outright Return-to-Nature utopians—a bold make believe that all toil may be made a joy, and with that, a levelling down of all society to an equal participation of labour. But indeed this is against all the observed behaviour of mankind. It needed the Olympian unworldliness of an irresponsible rich man of the shareholding type, a Ruskin or a Morris, playing at life, to imagine as much."[72] The steady application of industry, science, and technology to nature would be imperative if humanity was to win out in the struggle to win from nature the necessary resources for all.[73]

What Wells accepted to be the Malthusian realities of existence thus ruled out much of Morris's vision, leaving his utopian reconception of the human relationship with nature and his hopes for labor deeply flawed. What Wells had learned from Lankester, and from Weismann in particular, only further problematized Morris's presumptions.

Even though he remained skeptical about all of Weismann's claims about the germ-plasm theory, in a review he wrote of Weismann's *On Heredity* (1883) for the *Pall Mall Gazette* early in 1895, Wells did acknowledge that "Professor Weissmann [*sic*] has at least convinced scientific people of this: that the characteristics acquired by a parent are rarely, if ever, transmitted to its offspring."[74] Weismann had first reported his experiments that were designed to test the inheritance of acquired characters in 1888, but they only later became notorious—in part because of the attention that the likes of Wells and, later, Shaw gave to them. In these experiments, Weismann had cut the tails off a quantity of mice and then bred the mice to see if the acquired mutilation was passed on. Unsurprisingly, the offspring had been born with fully developed tails, and so Weismann had repeated the amputations on five successive generations. He later increased this to twenty-two generations in a subsequent experiment, and again with no noticeable shortening of the mice's tails. Weismann took the fact that there was no evident diminution of the tail length in any of the mice as demonstrated proof that acquired characters—and certainly acquired mutilations—were not hereditable.[75]

Weismann hypothesized that if acquired characters were not heritable, then the material that was passed from one generation to the next must somehow be isolated from the somatic cells. His conviction over this point was the basis of his germ-plasm theory. Thus, with the inheritance of acquired characters ruled out, by default, natural selection became the sole mechanism by which new species came into existence. The physiologist George John Romanes had dubbed Weismann's theory "neo-Darwinism" in 1895 on the basis that this emphasis upon natural selection was to return to the emphasis that Darwin had given it in the first edition of *Origin*.[76] In fact, of course, if this was "Darwinism," Weismann was considerably more Darwinian than Darwin had ever been. The label stuck, however, and gave Weismann the cultural authority of being Darwin's heir apparent in the further development of evolutionary theory.

The following year, this time in a review for the *Fortnightly*, Wells was prepared to be even more explicit: "Assuming the truth of the Theory of Natural Selection, and having regard to Professor Weismann's destructive criticisms of the evidence for the inheritance of acquired characters, there are satisfactory grounds for believing that man . . . is still mentally, morally and physically, what he was during the later Palaeolithic period, and that the race is likely to remain, for a vast period of time, at the level of the Stone age."[77]

The implications were clear. In a neo-Darwinian world there was to be no changing human nature, no newfound altruism, and above all, no "making socialists." Even if science and technology might conquer the Malthusian dilemma, humankind would forever bear the scars of his long evolutionary history, and as Huxley had already pointed out, like every other species, mankind had been molded through eons of struggle and competition and consequently bore the hallmark of the self-asserting individualist.

In light of Weismann's insights, any schemes for a future socialist society had to be premised upon the assumption that people would remain, even into the distant future, pretty much what they are now. No newly evolved presumptions of universal altruism could be made, and thus although Wells admitted that even though his own modern utopia might be less appealing than that which Morris had portrayed in *News from Nowhere*, his was at least grounded in reality. "Were we free to have our untrammelled desire," he wrote, "I suppose we should follow Morris to his Nowhere, we should change the nature of man and the nature of things together; we should make the whole race wise, tolerant, perfect—wave our hands to a splendid anarchy, every man doing as it pleases him, and none pleased to do evil, in a world as good in its essential nature, as ripe and sunny, as the world before the Fall."[78]

To Wells's mind, therefore, biology had thus destroyed Morris's hope for a reunification with nature, of the reconception of labor, and now of any hope that humanity might evolve to become more socialistic than they currently were in any meaningful time scale. This was not all, however, and if advocates of the strategy of making socialists were thinking that things could hardly get worse for them, they were to be sorely disappointed.

FIGURE 6.3. Friedrich Leopold August Weismann, 1834–1914. Weismann's work convinced many naturalists that the inheritance of acquired characters was untenable. (From Romanes, *An Examination of Weismannism*; courtesy of History of Science Collections, University of Oklahoma Libraries)

In "On Heredity," Weismann had described what he called "panmixia," which meant literally the "all-mixing" of available variations. Weismann used the term to describe the specific circumstance in which natural selection had ceased to select a particular character as a result of a change or changes in the environment. Natural selection, he argued, not only worked to improve the level of perfection of the attributes of a species, but to keep them up to that standard. The slightest relaxation in the conditions that had created an organ—an eye, for instance—would result in its degeneration, not through its "disuse," as had long been associated with Lamarckian beliefs about inheritance, but through the free intercrossing of those individuals with the most-perfect eyesight and those who were less well endowed. Under previously existing conditions, these latter of course would have been exterminated by natural selection. "This suspension of the preserving influence of natural selection may be termed Panmixia," Weismann wrote, concluding that "the greater number of those variations which are usually attributed to the direct influence of external conditions of life, are to be ascribed to panmixia."[70]

Lankester benefited from Weismann's work as he fleshed out his own ideas on degeneration, and he was one of the first real converts to Weismann's germ-plasm theory. If Weismann was right, then panmixia showed the inevitability of degeneration in the instance that the Malthusian stimulus of necessity was removed. This was particularly significant for any understanding of evolution given Weismann's argument against the inheritance of acquired characters. Weismann and Lankester later collaborated to publicize these neo-Darwinian ideas, jointly leading a seminar at the 1887 meeting of the British Association for the Advancement of Science in Manchester.[80] Even while Wells remained skeptical of Weismann's germ-plasm theory in its totality, Weismann and Lankester between them had done enough to convince him that panmixia had some very real and disturbing political implications. In light of panmixia, any "epoch of rest" would assuredly herald mankind's decline.

As I have suggested above, Wells explored the concerns that Huxley, Lankester, and Weismann had raised in his first popular novel, *Time Machine*, the first drafts of which he had written a decade earlier while still a student in South Kensington. He wrote it in 1894 just as he was reading "Evolution and Ethics," and revised it in 1895 having read Weismann. The novel was at once a harsh indictment of capitalism and a rejection of the "epoch of rest" to which Morris aspired. Wells opens the narrative of the tale with a man known only as the Time Traveller regaling his dinner guests with tales of the possibility of time travel, revealing to their initial disbelief that he has invented a machine that can accomplish such a feat. Over a series of subsequent dinners attended by a variety of guests, Wells unfolds a macabre tale of the future evolution of mankind.[81]

In humanity's distant future, the Time Traveller had hoped to find mankind's much-improved progeny and learn from them the secrets of millennia of advancement. But instead of finding a race with superhuman intelligence and advanced technologies, he is shocked and dismayed to find that the inhabitants are inferior in every way to present-day man. Lacking in intellect, the creatures, soon to be identified as the "Eloi," are simpletons. Effete and simpering, they exist on a diet only of fruit and are no more sophisticated than children. According to the Time Traveller's first impressions, the Eloi live in a land with no predators and with a ready supply of food. There is no evidence of industry or commerce, and just as Morris had outlined in *Nowhere*, there is no need to struggle and no competition. Secure in their subsistence, they pass their lives in casual and playful decadence. However, the Traveller is soon brought to have to reconsider this first analysis of the state of things.

Originally, he had thought that this future society had found some way of banishing want, old age, and the struggle for life, but the more he became acquainted with this delicate and frivolous race, the less he found this a convincing hypothesis. Rather, it quickly becomes apparent that these future descendents of mankind have degenerated as a result of the suspension of the struggle for existence—panmixia had taken its toll. Set in the same area of the Thames Valley that Morris had chosen for his own evolutionary narrative, the Eloi live amid ruins that are the last testament to the civilization of their ancestors. The Time Traveller also realizes that despite his initial assumptions, fear has not been entirely banished from the world—or if it had been at some point in the past, it must have been reawakened.

Further exploration reveals an increasingly grim series of discoveries. The future, it seems, is peopled not by one race of hominids, but two. Hidden in the bowels of the earth are the "Morlocks," a subterranean and nocturnal race who are industrious, carnivorous, and—worst of all—who have taken the time machine! The Time Traveller quickly surmises that these two races of men are the divergent evolutionary descendents of nineteenth-century society. The Eloi are the descendents of a parasitic middle class who did no labor themselves but relied instead upon the work of others; the Morlocks are the degenerate offspring of the workers. Now deformed—and bleached, just like the denizens of the Kentucky caves that Darwin had described in *Origin*—they are also malign, and the source of the Eloi's fear. It dawns upon the Time Traveller that the reality of this utopia is the worst of nightmares. The Eloi, through long and habitual ease, had fallen victim to the inevitable panmixia that was its reward: "The too-perfect security of the Upperworlders had led them to a slow movement of degeneration, a dwindling in size, strength, and intelligence."[82] As they had become more and more incompetent, so they had lost dominion over the workers, who through long-endured habit continued to live underground and provide for their former masters. The relationship was entirely turned, however, for though the Eloi remained at ease in the hours of daylight, at night, like cattle, they were harvested in the darkness.

Indicative of the impression that Weismann's arguments against the inheritance of acquired characters had made upon Wells is the fact that in between the two editions of the story, written in 1894 and 1895, respectively, he substantially pushed back the date at which the Time Traveller encounters the Morlocks and the Eloi. In the 1894 version, the action takes place in the year 12,203. In light of reviewing Weismann's work, however, in the 1895 version, the encounter does not occur until the year 802,701. Wells's recognition that speciation by natural selection alone would take much longer than might be expected with the aid of the inheritance of acquired characters was to have a profound impact upon his politics as well as his fiction.

Acknowledging his debt to his schooling, and to his tutor, Wells sent Huxley a presentation copy of *Time Machine* upon its publication in May of 1895. Huxley was wracked with fever and gravely ill, and with less than a month to live it is unlikely that he read the book. It is probable, however, that the following letter, which Wells had enclosed, was read to him:

> May 1895
> Dear Sir,
>
> I am sending you a little book that I fancy may be of interest to you. The central idea—of degeneration following security—was the outcome of a certain amount of biological study. I daresay your position subjects you to a good many such displays of a range of authors but I have this much excuse, I was one of your pupils at the Royal College of Science and finally: the book is a very little one,
> I am Dr. Sir
>
> Very Faithfully yours
> H. G. Wells[83]

Huxley died on the afternoon of 29 June 1895.

Following his rejection of the inheritance of acquired characters early in 1895, Wells did not give up on shaping the future, but rather than looking to biology alone, he was led to follow Huxley in seeking his hopes for the future development of humanity through education. While he still championed the necessity of individual liberties, he increasingly brought these freedoms within the compass of his own utopia of a world-state government, which he described in his book fittingly entitled *A Modern Utopia* (1905). Given the biological realities of the world, education—particularly in the sciences—

combined with the state oversight of production and reproduction were now the means by which Wells believed socialism might be achieved.

Wells rededicated himself to the cause of educational reform—and to the advance of science education in particular. As he now argued in the essay "Human Evolution, an Artificial Process," written in October of the following year, he believed that natural selection was not responsible for recent human ethical advance. Rather, and here clearly echoing Huxley's "Evolution and Ethics," he argued that "the evolutionary process now operating in the social body is one essentially different from that which has differentiated species in the past and raised man to his ascendancy among the animals." Weismann had convinced him of the main point of Huxley's essay, that the amount of time demanded by natural selection meant that human social development was much more the result of culture than it was of biology. And it was Weismann too who had convinced him that the two were different orders of things. Without the inheritance of acquired characters, cultural change had no effect on biology. The biological material that man had inherited from his forebears was to all intents and purposes unchanged and unchangeable. Over the same period, however, man's cultural education had undergone a momentous development. The only considerable evolution that had occurred in man since the later Paleolithic, he argued, "has been . . . an evolution of suggestions and ideas," but these were not insignificant. It was ideas and suggestions that had driven technological advances in industry, medicine, and all other aspects of human knowledge, and even though Weismann had ruled out the inheritance of acquired characters on a biological level, as was evidenced in every library and every bookshelf, *cultural* characters might still be transmitted and compounded across the generations. This culturally acquired character of humanity was, he wrote, "the padding . . . necessary to keep the round Palaeolithic savage in the square hole of the civilised state." Thus, although Wells retained "making socialists" in his political vocabulary following his full appreciation of Weismann's work, he used it to signify something significantly different than Morris had intended by the phrase.[84] The learned and social changes in behavior that might be brought about through education now trumped any that might be hoped for from biology.

Conclusions such as these were perhaps unsurprising for someone who had trained as a teacher. Nevertheless, this was a significant moment in the evolution of Wells's socialism. From this point onward, Wells's speculations as to the biological evolution of humanity were confined to the remote future in his fiction, and from this point too his political concerns focused more and more upon the importance of education, and later, in the 1930s, upon developments in behavioralist psychology.[85] As Richard Barnett has shown, this concern with education was what brought Wells into close collaboration with Lankester.[86]

Lankester had been well impressed with *Time Machine*, and he and Wells became collaborators and increasingly close acquaintances from 1900 as a result. Lankester shared Wells's deep admiration for Huxley, and he had also been one of Wells's examiners for his Bachelor of Science degree. In light of their shared diagnoses of social decline, Lankester and Wells collaborated to advance education as a remedy. They believed that science education in particular would give people the knowledge they would need to avoid the consequences of panmictic nonselective breeding. A thorough knowledge of science would equip people to place themselves in relation to the inevitable working out of nature's laws. The imminent possibility of decline that might result if the appropriate education was not forthcoming—even for such an evolutionarily advanced race as the English—is evident in Lankester's argument that "the full and earnest cultivation of Science—the Knowledge of Causes—is that to which we have to look for the protection of our race—even of this English branch of it—from relapse and degeneration."[87]

Education in biology in particular, Wells believed, would show people the naiveté of Morris's vision, which much to his frustration remained as influential as ever even after Morris's death. Just as had been the case in his own experience, he thought, an education in biology would reveal the futility not only of Morris's ideas about production, but would also highlight the importance of tackling the other half of Malthus's famous equation: reproduction. Historians have been ambivalent in how they characterize Wells's views on reproduction. A number of scholars have drawn attention to Wells's vocal endorsement of the "sterilisation of failures," linking him to the contemporary eugenics movement. However, as John S. Partington has pointed out, to assume from such a statement that Wells was uncritical of contemporary eugenics would be to neglect the complexities and ambivalences that Wells felt toward the movement.[88] As G. R. Searle has shown, for instance, by 1910, G. K. Chesterton, the Catholic critic of eugenics, confidently listed both Shaw and Wells among those who he believed had "chucked the notion."[89] That said, however, Wells was certainly concerned about reproduction, and however much he was critical of simplistic notions that the state might quietly do away with the unfit or instigate state breeding programs, he did see a role for the state in the management of reproduction.[90]

As early as 1902, Wells had rejected such popular "positive" eugenic breeding strategies: "Let us set aside at once all nonsense of the sort one hears in certain quarters about the human stud farm," he had written in *Mankind in the Making.* Indeed, it was the persistence of popular enthusiasm for eugenic selection that was his chief motivation for having this work reprinted in 1914.[91] Although Darwin had made much of the analogy between the artificial selection of plant and animal breeders and selection in nature, famously devoting the first chapter of *Origin* to the subject of pigeon breeding, Wells, like Huxley, concluded that it would be erroneous to try and extrapolate from the practices of animal breeders back to human populations. Neither had faith in what Huxley denigrated as the "pigeon-fanciers' polity" of the eugenicists.[92] While eugenics enthusiasts often repeated the call for the same discernment to be applied to human reproduction that was applied to the breeding of livestock, as Wells pointed out, the chief difficulty was that however rational and desirable the intended outcomes of such schemes, they overlooked one important difference: where the stockman bred for beef, "we are . . . not a bit clear what points to breed for."[93] The many and diverse qualities required for a vigorous and vibrant society were much less easy to decide upon—or to recognize.

Wells clearly found the whole notion of a state-controlled breeding program problematic on a number of levels, and again the intuitions that Huxley had expressed in "Evolution and Ethics" were his guide. Huxley had pointed out that even the most skilled of observers would have difficulty discerning which young boys and girls might later make a success of their lives, stressing that the qualities that determined a successful life might not make themselves known until the individual was in their forties, or even later—by which time, of course, they would undoubtedly have already passed on their innate qualities for good or ill to their offspring.[94] Further, and true to the character of Darwinian selection, Huxley was conscious of the fact that much of how one might evaluate one particular trait or another must necessarily be contingent upon the environment in which it was expressed. The qualities that made a successful villain might under altered circumstances lead the same person to acts of great heroism and social worth. It was for these reasons that Huxley had rejected the aims and intentions of contemporary eugenicists, and Wells was similarly reticent.[95]

Like Huxley, Wells recognized the implications of the contingent nature of fitness inherent to Darwinian selection. Fitness was not an innate or essential quality, but rather was merely whatever attribute happened to give an organism a selective advantage in the context of its own given ecology. Thus, socialism was not merely a matter of improving the lot of those who struggled under capitalism, but of changing the environment so as to ensure that those with the most socially desirable attributes fared the best. However, Wells recognized that just as evolution was contingent, so too it was also "chanceful." Certainly, many of those who fared the best under capitalism did so as a result of their individualistic natures, but so too there were those at the top of the social tree simply by virtue of having either inherited (in the fiscal rather than the biological sense) the trappings of a successful life or of having made good out of the roulette of the stock market—neither of which were the result of any quality that might denote fitness by any objective measure.

Awareness of such social contingencies at the top of society led Wells to a similar conclusion regarding capitalism's failures, which is perhaps surprising given his noted lack of sympathy for the working class, the poor, the oppressed, and the apparently degenerate. Where many of Wells's contemporaries decried the threat of "outcast London," "the residuum," and the "submerged sixth"—who were referred to by many in the eugenics movement as the "social problem group"—Wells was reluctant to write them off.[96] Just as Huxley had suggested that in many cases it could be said that "dirt is riches in the wrong place," so Wells refused to condemn anyone who might simply have fallen through the cracks of capitalism.[97] After all, not only had Huxley's own personal history involved modest origins, but so too had Wells's own. There was no necessary indication that those at the bottom had failed to achieve success as a result of their biology, and in light of this fact Wells echoed the calls of his sometime fellow Fabians who argued for slum clearance, sanitary reform, and, most importantly, educational reform as a remedy. However, and in contrast to the majority of the Fabian reformers, Wells also argued the necessity of an economic revolution.[98] Only once a minimum standard of living, of environment, and of education had leveled the playing field, so to speak, would it be possible to tell who had the aptitude, the character, and indeed the innate biological qualities to apply themselves and make a success of their lives—and this was important.[99] In spite of his leanings toward state socialism, Wells was deeply concerned to protect individualism, the factor that he believed was the source of the variation that was so necessary to any evolution. The individual should be allowed to make his experiment and succeed or fail accordingly. This was something that the hand of the state could not replicate, but only facilitate, through education and the maintenance of an "equality of opportunity" that was real rather than just the ideological cant of capitalism.[100]

Further illuminating Huxley's concerns, increasing recognition of the complexity of the genetic mechanisms of heredity showed that the relationship between the germ and the soma, or what we would now call the genotype and the phenotype, was turning out to be a much more complex affair than had previously been imagined.[101] Wells was well placed to appreciate this fact given his close working friendship with Julian Huxley, who kept him abreast of the latest developments in genetics. Thus, the hope of the eugenicists that it might be possible for the state to accurately select the "fit" from the multitude was, to Wells's mind, not only beyond the realms of current scientific knowledge, but indicative of the level of the ignorance of many in the eugenics movement when it came to science.[102]

Despite these problems with the eugenics movement, however, as Malthus had made only too clear, some intervention to control spiraling fertility rates was still essential if society was to avoid panmixia and degeneration. And in this matter Wells was willing to move beyond Thomas Huxley's clearly articulated reservations. While Wells sought to distance himself from the uninformed and unscientific aspirations of many eugenicists, he had little time for those who imagined that neither population nor degeneration were an issue. This had been Morris's position in *News from Nowhere*, and Wells was still concerned to refute the basic assumptions of the book. As might be expected, given his belief that "the problems of human increase [must be] manfully faced," Wells did not balk at taking seriously the problem posed by those who were proven to be beyond the reach of even the most far-reaching social reforms—those who actually *were* held back by their biology. One might appeal to the working classes to restrain their reproductive urges in the common interest, but what of the very lowest orders of humanity? "It is our business," Wells wrote, "to ask what Utopia will do with its congenital invalids, its idiots and madmen, its drunkards and men of vicious mind, its cruel and furtive souls, its stupid people—too stupid to be of use to the community, its lumpish, unteachable and unimaginative people?"[103] Certainly, "there would be no killing, no lethal chambers," talk of which (not least by Shaw) was fodder for the inflammatory imagination of contemporary writers in the press, but he did envisage the need for the state to intervene and provide the inducements necessary to encourage these people to give up their fertility—by voluntary sterilization if not by segregation.[104] Wells hoped that in light of such proposals the majority of those so afflicted might be encouraged to surrender their reproductive capacity as a small price to pay in exchange for the right to remain at liberty in society.[105]

Wells's concern to restrict his points of advocacy to areas of scientific certainty made him much more conservative in his endorsement of the sterilization of the "unfit" than the vast majority of eugenicists, however, and he carefully made the case that even these "negative" eugenic interventions should only be employed in the case of those who were proven beyond doubt to have some dangerous congenital disease. Even in these extreme cases, he was reluctant to advocate the forcible sterilization that later characterized

North American eugenics. While the great tome *Science of Life*, the result of his collaboration with Julian Huxley and his own son, G. P. Wells, expressed admiration for the advances of the American movement, there were significant differences between Wells and Huxley on this issue. Even by February 1930, for instance, and despite his own reservations about mainline eugenics, Huxley had felt compelled to counter Wells's claims regarding environmental contingency and education. Complaining about Wells's copy in drafting the book, Huxley wrote: "As they stand, the remarks about different social classes are to me untenable. You make sweeping assertions about the absence of difference between them which I really can't pass."[106]

A fundamental part of Wells's scheme to clear away the biases of capitalism so as to avoid the degeneration of society was the resolution of the "woman question." Wells, like Morris, took the question of female emancipation seriously—it was not something that could be left until after the revolution.[107] Wells argued that as long as women were restricted to the role that they had under the capitalist sexual division of labor, in which they were molded to be little more than the ineffectual "parlour borders" stereotyped in so many Victorian novels, then the sexual selection that was so important to Darwinian evolution would always be distorted. Wells set out his hopes for the future in this regard not only in *Modern Utopia* but also in his fiction—scandalizing reviewers in the process![108]

Wells recognized that just as men would need to be freed from the capitalist economic constraints that distorted their environment and stymied their potential, the same must necessarily apply to women. However, whereas men needed liberation from their economic and social conditions, women also needed emancipation from the domestic tyranny that their economic dependence upon their husbands dictated. To this end, in Wells's *Modern Utopia* women were to be afforded the same education and employment opportunities as were accorded to men; indeed, Wells was adamant that women should not be barred from even the highest offices in society. That said—and perhaps unsurprisingly given his Darwinian preoccupations—and despite himself, Wells could not help but think of women's position in any society in terms of their biology. Rather than recognizing women's sexual difference as a justification for their subordinate role in society, however, socialism not only required their political and social emancipation, but also—and typically controversial given Wells's own sexual profligacy—their sexual emancipation as well. Only by the emancipation of women in the realm of reproductive choice would it be possible to halt the degeneration that Wells perceived to be increasingly apparent under capitalism.[109]

Under capitalism, it was men who monopolized the marriage market.

Under Wells's vision of socialism, however, the emancipation of women was to alter things considerably: men would have to compete to attract a wife. Those who were the most successful—and exhibited those phenotypic traits deemed most desirable in a husband—would be accepted, while those devoid of such traits would not. Wells's utopia was to be a biological meritocracy, not the plutocracy that presided over such matters under capitalism, and significantly, it is clear that Wells believed that the female of the species should have at the very least an equal role in the selection process. The restoration of a healthy sexual selection unhindered by the false inequalities of capitalist accumulation would provide the necessary incentive for men to strive to make successes of themselves. Women, as the future mothers of the race, would be ideally positioned to make discerning choices as to who might make a suitable father for their offspring—there was no need for some state-run scheme of selective breeding. Nevertheless, the state would not be unimportant in the ultimate success of Wells's hopes for the future of sexual selection.[110]

Although Wells clearly believed that women quite naturally had a discerning eye, he also believed it would be necessary for the state to facilitate female choice and help to place women in the best possible position to make the most-informed of decisions. This was to take two forms. First, and as Wells suggested in his *Modern Utopia*, the state was to enact a policy of full disclosure regarding all information relevant to reproductive health, including one's financial, employment, and marriage history as well as one's medical records. Physicians and healthcare providers would be required to log all such information with the central authority, which would then not only be given due consideration by the state should the individual in question apply for a marriage license, but would be made available to the individual's intended partner as well.[111]

Wells readily acknowledged that "this question of marriage is the most complicated and difficult in the whole range of Utopian problems," and surmised that as a minimum it should be stipulated that both of "the contracting parties be in health and condition, free from specific transmissible taints, above a certain minimum age, and sufficiently intelligent and energetic to have acquired a minimum education." The man at least "must be in receipt of a net income above the minimum wage. . . . All this much it is surely reasonable to insist upon before the State becomes responsible for the prospective children," he concluded.[112]

The second measure was that all citizens should receive a full education in the science of heredity—women as well as men. Education was going to be fundamental in making effective reproductive choices. It is no coincidence that the title character of Wells's "new woman" novel *Ann Veronica* (1909) is led to think about her position as a woman, about heredity, about socialism, and about feminism, all as a result of the science education she receives under a character named Russell—who was quite clearly modeled upon Huxley, whereas his able demonstrator in comparative anatomy, Capes, is most likely a fictitious rendition of Lankester.[113]

In *Ann Veronica*, Wells controversially discussed the important connections between socialism, biology, education, and sexual emancipation. Just as Wells and Lankester believed a knowledge of scientific laws would give one an advantage in the struggle for life, so too the education of women in the mechanisms of heredity would enable them to make more-prudent choices in their choice of mate. Wells later confirmed his view in his 1923 novel *Men Like Gods*, in which he suggested that even though biological science might not yet be in a position to judge a man on the hereditary quality of his germ plasm, an independent and scientifically educated woman would certainly be in a position to judge of his achievements and his character—and both, Wells presumed, would give a fair indication of an individual's biological worth. The idle and the unfit, he supposed, would simply not be chosen and thus would not pass on any negative hereditary tendency to subsequent generations. "If the individual is indolent there is no great loss, there is plenty for all in Utopia, but then it [*sic*] will find no lovers, nor will it ever bear children, because no one in utopia loves those who have neither energy nor distinction. There is much pride in a mate in Utopian love," he wrote.[114] The praise and blame that Darwin had discussed in relation to sexual selection in *Descent* were not absent from the modern utopia, it seems. As Wells put into the mouth of his future utopians as they looked back on the path that their own development had taken, however, "The supreme need of our time was education."[115]

Although it is to move beyond the chronological scope of this book, it is worth noting that the connections that Wells drew between education and biology were strengthened in the 1930s by the work of others in the newly developing disciplines of psychology and physiology—particularly among the behaviorist school of the former. Wells was particularly interested in the conditioned responses of Pavlov's dogs, for instance, and was clearly excited about the potential implications of this line of research for controlling human behavior—perhaps a lot of the behaviors thought to be hard-wired in the germ line might turn out to be learned and thus malleable after all.[116] The new sciences of mind, psychology, and psychiatry could thus not only guard against the irrationalities of tradition and religion, but might also be used to suppress the individualism and self-assertiveness that had served humanity in an earlier, pre-social epoch of its evolution. Wells argued, and in some contradiction to his earlier determination to preserve individualism, that "this mental modification is steadily in the direction of the subordination of egotism and the suppression of extremes of uncorrelated individual activity."[117] Such hopes reignited his passion for education as a means by which to inaugurate a new era in the political history of the world under one overarching administration. Wells remained ambivalent about the long-term prospects for humankind, though, and was by turns optimistic and then deeply pessimistic about the future. Despite his self-identification with the socialist movement, by the 1930s Wells's hopes for a strong centralized state run by an elite of superior, clean-cut, industrious middle-class men had become closely akin to the fascist politics that appealed to many across Europe in the early decades of the twentieth century. The fact that in the 1933 film *Things to Come*, which Wells had a hand in directing, the Airmen's uniform closely resembled the black shirts favored by the British Union of Fascists did not go unnoticed.[118] Wells, whose increasing political influence gained him access to a great many world leaders, including Theodore and Franklin D. Roosevelt, also met with Oswald Moseley, Lenin, and later with Stalin, although he was little taken with either Moseley or Mussolini, describing the former as "our own little black-head" and the latter as "highly distended." However, it is clear that rather than disagreeing with their aspirations, Wells rejected them as individuals on the basis that they were representative of "a spotty stage in the adolescence of mankind."[119]

By contrast, in his last work, *Mind at the End of its Tether* (1945), notably written in the aftermath of the Second World War, Wells despaired of even this political avenue. Echoing the evolutionary possibilities he had explored in his more-fictitious accounts of the future, he pondered the possibility that humanity was simply incapable of adapting to the political demands of the modern world. "Homo Sapiens," he wrote, "is in his present form played out. The stars in their courses have turned against him and he has to give place to some other animal better adapted to face the fate that closes in more and more swiftly upon mankind."[120]

Wells was not alone in his struggle to make sense of the political and social implications of biology for humanity. His friend and former Fabian colleague, George Bernard Shaw, wrestled with the same demons, and argued with him on this account until Wells's death in 1946. Prior to Weismann's experiments on mice, it is clear that Shaw and Wells recognized each other as fellow "world betterers." However, Weismann effectively divorced them—not only scientifically, but, and as a result, politically as well. Shaw, who was committed to the agency and purpose inherent in the inheritance of acquired characters, was dismayed at the backlash against all things Lamarckian, lamenting the fact that by 1906 "the neo-Darwinians were practically running current Science."[121] Shaw was one of those who, like Morris, stressed agency as central to human evolution—the will to change, to exercise certain faculties and suppress others, he believed, was central to his hopes for the future of humanity. He was thus dismayed by what he perceived to be the moral vacuity of neo-Darwinism, a judgment in which he was by no means alone.[122] His own progressive and consciously Lamarckian evolutionism, which he called "creative evolution," had many influences, including Friedrich Nietzsche, Arthur Schopenhauer, and Heinrich Ibsen as well as his most keenly acknowledged mentor in biology, the novelist Samuel Butler, whose *Erewhon* and *Life and Habit* had been published in the 1870s. It is worth noting that despite the many similarities between their respective theories, Shaw repeatedly claimed independence from the French philosopher Henri Bergson, who in 1907 had written of his own *élan vital* that provided an internal guidance to a progressive evolution.[123]

Shaw is usually pegged as an anti-Darwinian, but he initially saw in the "idiocy" of Weismann's mice and subsequent germ-plasm theory a perversion of Darwinism rather than its advance, and blamed what he believed to be Weismann's error on the mechanistic logic of neo-Darwinism. After all, he conceded, Weismann "was not a born imbecile."[124] However, as the neo-Darwinian assault on Lamarckism became more intense—Weismann's polemic served to polarize a debate among evolutionists on this issue—Shaw responded by rejecting natural selection as relevant to evolutionary progress. In *Man and Superman* (1903) Shaw had clearly articulated a neo-Lamarckian combination of selection and the inheritance of acquired characters, but by the time he came to write *Back to Methuselah* (1921), reflecting upon the chance nature of Darwinian selection and what he now saw as the insidious permeation of Darwinism into the politics that had led Europe into the horrors of the trenches, Shaw had given up on selection as anything other than a chanceful irrelevance. "Neo-Darwinism in politics had produced a European catastrophe of a magnitude so appalling, and a scope so unpredictable, that as I write these lines in 1920, it is still far from certain whether our civilization will survive it." He lamented that neo-Darwinism had "a hideous fatalism about it," and he ultimately rejected the term "natural selection" as an adequate description at all. Arguing that there was "nothing natural about an accident," Shaw instead preferred to call the mechanism at the heart of Darwin's theory "circumstantial selection," believing that this was more indicative of the role it actually played in evolution.[125] Emphasizing the role of the will combined with the inheritance of acquired characters, Shaw had no need to invoke natural selection at all, concluding that "we have here a routine which, given time enough for it to operate will finally produce the most elaborate forms of organized life on Lamarckian lines without the intervention of Circumstantial Selection at all."[126]

Given Shaw's moral and metaphysical concerns, it is perhaps unsurprising that Wells could only put Shaw's continued faith in the inheritance of acquired characters and belief in vital forces down to what he described as his "fine and sympathetic nature" that found it "unchivalrous and vile for science to recognise that the weakest go to the wall," and Wells alternately reasoned with and ridiculed his friend in their correspondence.[127] In response, Shaw argued the logic of Lamarckism, pointing out that Weismann's experiments in no way invalidated the popular ideas of the French naturalist. Lamarck had made "many ingenious suggestions as to the reaction of external causes on life and habit, such as changes in food supply, climate, geological upheavals and so forth," he pointed out. But what Lamarck "really held as his fundamental proposition [was] that living organisms changed because they wanted to."[128]

Shaw's position regarding the role of the will in evolution is worth considering in more detail, not least because while the extent to which he emphasized it set him apart from many Lamarckians, he was not alone in his views. As the historian Peter Bowler has pointed out, the growth of Idealist thought in the last decades of the nineteenth century echoed a more popular desire to retain a prominent role for mind and purpose in evolution.[129] The popularity of Bergson's work, which was translated into English in 1911, as well as that of the philosopher Alfred North Whitehead and the Jesuit priest and paleontologist Pierre Tielhard de Chardin, are testament to the fact that Shaw's views were not as beyond the pale as Wells and other mechanists hoped to portray them. Indeed, Wells's collaborator on *Science of Life*, the grandson of T. H. Huxley, Julian Huxley, was also fascinated by the progressive potential of de Chardin's vitalist thought, even if he later tried to undermine Shaw's continued influence.[130] Further, it was in 1923 that the psychologist Conway Lloyd Morgan proposed the idea of "emergent evolution" to explain the evolution of mind at a certain point in the development of physiological complexity.[131] The appeal of Lamarckism, of the inheritance of acquired characters and of the willing of evolutionary change, was understandable because it allowed a place for human agency and moral effort in the evolutionary process that was missing from the more determinist account of life's development that Weismann and his followers like Wells offered.

Shaw was surprised that so many biologists had been taken in by what he believed to be the erroneous nature of Weismann's experiments, but more than anything else he was shocked that Wells was among them. "This won't

FIGURE 6.4. George Bernard Shaw, 1856–1950. Photograph by Frederick Henry Evans, 1896. Shaw remained an ardent Lamarckian throughout his life. (NPG P113; © National Portrait Gallery, London)

do," he wrote to his friend. "Why did you, who have put your finger with ridiculous ease on dozens of political absurdities that have duped generations of Englishmen, never put your finger on the absurdity of that experiment of Weismann's with the mice's tails?"[132] No matter how many generations of mice Weismann might mutilate, Shaw argued, such an experiment made no test of the will to change that he believed to be at the heart of Lamarckian inheritance. A committed anti-vivisectionist, Shaw could only bemoan the fact that "a vital conception of evolution would have taught Weismann that biological problems are not to be solved by assaults on mice" and that he should instead have thought to devise an experiment that tested the ability of an organism to adapt through its own desire to change; or, better still, to have realized that evidence for the inheritance of acquired characters already existed "within the personal experience of all of us."[133]

While, given his anti-vivisectionist predilections, the subject would have equally disgusted him, there was significant scientific backing for the idea that Weismann's experiments on mice were not the whole story. Even the most ardent supporters of the inheritance of acquired characters were becoming aware that much of the evidence in its favor was speculative and anecdotal at best. However, there remained one compelling piece of experimental evidence, from the laboratory of the highly regarded physician and physiologist Charles-Édouard Brown-Séquard. Indeed, from the third edition of *Origin*, which had been published in 1861, Darwin had made mention of "the remarkable cases observed by Brown-Séquard in guinea-pigs," which he expanded upon in subsequent editions. Certainly, what appeared to be the "inherited effects of operations, should make us cautious in denying this tendency."[134] Brown-Séquard had induced epilepsy in his experimental animals by cutting the sciatic nerve, and he reported that the offspring of his guinea pigs had also exhibited various epileptic traits. Further, in some of his guinea pigs the operation had rendered their hind feet insensible to pain, resulting in the somewhat disturbing fact that they had gnawed off their own toes. Most surprising, though, was the fact that the offspring of these guinea pigs appeared to be born without toes![135]

What made these experiments so telling was not only the fact that Brown-Séquard was a respected authority and known for his careful experiments, but his investigations were not intended as a test of the inheritance of acquired characters. He was thus a truly objective witness. Kropotkin recognized the importance of these findings and marshaled them against neo-Darwinian conclusions. "Altogether, these experiments were conducted so carefully, during a long succession of years, that the opinion which has prevailed among specialists is, that they really prove the hereditary transmission of certain abnormal states of different organs, provoked by certain lesions," he argued.[136] While Brown-Séquard's guinea pigs had indeed been significant pause for thought in 1861, Kropotkin was writing in 1912 and there had been no further credible evidence produced. Weismann had claimed that these experiments were inconclusive and doubtless indicated a microbial infection; indeed, it was Brown-Séquard's experiments that motivated his own investigation of the issue.[137]

Regardless of what the vivisectionists might say, however, Shaw argued that common experience taught us that it was through just such an exercise of will that humans acquired characters—either abilities or physical attributes—and subsequently passed them to their offspring. In an argument that mirrored Bergson's notion of explosive evolution interspersed with periods of stasis, Shaw suggested that "this process [of the acquisition and subsequent inheritance of characters] is not continuous, [and gradual] as it would be if mere practice had anything to do with it," but rather occurred through a series of efforts and relapses.[138] Indeed, Shaw wrote to his colleague Gilbert Murray, the man he occasionally consulted as a thesaurus while writing, asking him for "a good word for this phenomenon of relapse—something that will sound Weismannic, like panmixia." In reply, Murray suggested "metaneisis," meaning "an intermittent loss of hold or strength"; or "metasphalsis," "the process of tripping or falling in between."[139] Employing a cultural analogy which would no doubt have appealed to Wells—both men were keen cyclists—Shaw suggested that the acquisition and inheritance of characters operated through exactly the same mechanisms by which one might learn to ride a bicycle: "Though you may improve at each bicycling lesson *during* the lesson, when you begin your next lesson you do not begin at the point at which you left off: you relapse apparently to the beginning. Finally, you succeed quite suddenly, and do not relapse again. More miraculous still, you at once exercise the new power unconsciously."[140]

Just as this process occurred in the individual, so it might be transmitted across generations: "When your son tries to . . . bicycle in his turn, he does not pick up the accomplishment where you left it. . . . The set back that occurred between your lessons occurs again. . . . Your son relapses, not to the very beginning, but to a point which no mortal method of measurement can distinguish from the beginning."[141] Nevertheless, Shaw believed that even if the level of inheritance of a particular characteristic from one generation to the next appeared immeasurable, the accumulated effort put into the attempt was inherited in full and was compounded across the generations. Eventually, the buildup of this "life force" and creative desire to change would be irresistible and a child would finally be born without relapse with a fully inherited acquired aptitude. Significantly, in contrast to Wells, who had reluctantly resigned himself to the infinitely slow and gradual process of natural selection, Shaw's theory of creative evolution demanded substantial saltationary leaps in the evolutionary process as a matter of course.[142] Indeed, he hypothesized that this was a probable explanation of those who were born with freak "natural" talents—a Newton or a Mozart. Such gifted children were the embodiment of the life force of creative evolution that had finally succeeded in breaking through the physical shell that proscribed its previous limitations.[143] It was in this sense, for instance, that Shaw believed Morris to be a "readymade poet and decorative draughtsman."[144]

Shaw, like Wells, remained concerned about the amount of time that the evolution of a socialist humanity might take in its attempts to address the obvious "political inadequacy of the human animal." After all, people could only achieve so much in their allotted "three-score-years-and-ten," and as a result were "for all the purposes of high civilization, mere children when they die; and our Prime Ministers, though rated as mature, divide their time between the golf course and the Treasury Bench in parliament."[145] Unlike Wells, however, Shaw had little faith in the prospects for improving humanity through education or mere social reform, but rather confined his hopes to creative evolutionary biology.[146]

Creative evolution offered two potential answers to the apparently intractable limits of human longevity. First, he believed that the popular theory of embryological recapitulation suggested that in the future people might be born with the mental and physical faculties already fully developed that presently took them a lifetime of effort to achieve. Second, he thought that there were grounds for believing that, just as humanity had acquired their present lifespan in answer to the needs of an earlier stage in their evolutionary history, so they might—through the same evolutionary process—expand their longevity to meet their present and future needs.[147]

The fact that during the course of their development embryos pass through stages that appear to correspond to the developmental stages of the history of life on earth had long been observed, and neither was it unusual for these facts to be appropriated as an endorsement in nature of progressive political ends.[148] Further, Chambers, Spencer, Darwin, and Huxley had all counted on it. The embryological development of an organism proceeded in sequence, beginning with stages that represent its most remote and lowest ancestors, proceeding through stages that correspond to its closer and higher kin.[149] Before the work of the talented Estonian embryologist Karl Ernst von Baer became widely known—for a popular audience largely through Huxley's *Man's Place in Nature*, Darwin's *Descent of Man*, and of course, Robert Chambers's *Vestiges of the Natural History of Creation*—it was generally believed that this apparent correspondence was not merely indicative that humans had diverged from the common ancestor that they shared with other species, but rather that humans literally passed through these earlier stages in the history of life in the course of their own development.[150] This led to a conception of life as a linear progression from lower and simpler forms upward and onward through ever higher and more complex forms, with humanity at its peak.[151] Despite the fact that Huxley, Spencer, Darwin, and others made it clear that embryology in fact indicated the divergence of different organisms from a common ancestral form, Shaw clearly subscribed to the earlier view. This evidence of humanity's developmental history informed the mechanism of creative evolution through which Shaw believed mankind had the opportunity to develop into better beings. It was through exactly this process of embryological recapitulation that Shaw imagined that his own supermen—at least at one stage in their evolution—would pass through a stage akin to the Samurai that Wells so eagerly anticipated in *Modern Utopia*: Shaw referred to Emry Straker, one of the principle characters of *Man and Superman*, as "the contemporary embryo of Mr. H. G. Wells' anticipation of the efficient engineering class."[152]

Somewhat bizarrely, Shaw's take on the theory of embryological recapitulation emphasized the role of individual effort even in prenatal development. Before entering the world, he wrote, each person in every new generation "had to go back and begin as a speck of protoplasm, and to struggle through an embryonic lifetime, during part of which he was indistinguishable from an embryonic dog, and had neither skull nor backbone. When he at last acquired these articles, he was for some time doubtful whether he was a bird or a fish. He had to compress untold centuries of development into nine months before he was human enough to break loose as an independent being." Given that all the previous stages in the evolutionary history of humans had become compressed into a much shortened prenatal development, it seemed logical to Shaw to suppose that "the time may come when the same force that compressed the development of millions of years into nine months may pack many more millions into even a shorter space."[153] If this was indeed to be the case, then it was reasonable to suppose that future generations might be born both mentally and physically more advanced than those of today; the development that present-day humanity struggled with over the course of a lifetime being pushed back into but a prenatal moment.

In Shaw's opinion, the significance of this realization could not be overstated: "Nothing is so astonishing and significant in the discoveries of the embryologists, nor anything so absurdly little appreciated, as this recapitulation, as it is now called: this power of hurrying up into months a process which was once so long and tedious that the mere contemplation of it is unendurable by men whose span of life is three-score-and ten."[154] Indeed, Shaw suggested that rather than being concerned that evolution had taken many more millions of years than had originally been thought, a problem which had taxed the minds of Darwinians as well as their critics, in fact the opposite might well be the case. Shaw argued that whereas in the years that followed the publication of Darwin's uniformitarian theory of natural selection his followers had striven to prove the antiquity of both the earth and of man, the realization of the saltations possible under Shavian creative evolution suggested that "acquirements can be assimilated and stored as congenital qualifications in a shorter time than we think; so that, as between Lyell and the Archbishop Ussher, the laugh may not be with Lyell quite so uproariously as it seemed fifty years ago."[155]

In light of his conversion to Weismannism, across the 1894 and 1895 editions of *Time Machine* Wells had moved the date of his Time Traveller's encounter with the Morlocks and Eloi forward from the year 12,203 to 802,701. In comparison, and indicative of the evolutionary acceleration allowed under creative evolution, in *Back to Methuselah* Shaw portrayed a number of scenes spread at intervals to illustrate the evolution of humanity, the last and most advanced of which he set in only 31,920. However, while indicative of the comparative pace of Lamarckism compared to neo-Darwinism's much more stayed conclusions, the fulfillment of Lamarck's evolutionary promise for humanity—at least by Shaw's reckoning—was clearly not just around the corner either.

Shaw's second cause for optimism, also tied to creative evolution and the inheritance of acquired characters, was his belief that death itself was an acquired and thus mutable characteristic, a belief that led him to consider the possibility of what he called "voluntary longevity."[156] Although it appears somewhat outlandish today, voluntary longevity was actually based upon Weismann's early work and was endorsed by Alfred Russel Wallace, among others. In an essay entitled "The Duration of Life" (1881), Weismann had argued that "the duration of life is forced upon the organism by causes outside itself . . . governed by the needs of the species, and that it is determined by precisely the same mechanical process of regulation as that by which the structure and functions of an organism are adapted to its environment."[157] The "origin of death" thus became something that could be explained in terms of historic utility at a species level.[158] Weismann continued: "I consider that death is not a primary necessity, but that it has been secondarily acquired as an adaptation. . . . Death is to be looked upon as an occurrence which is advantageous to the species as a concession to the outer conditions of life, and not as an absolute necessity, essentially inherent in life itself."[159] Writing to Wells, Shaw was not only surprised at Wells accepting the validity of Weismann's experiments on mice, but could not believe that his friend had not "spotted that great hit of his: that death is an evolved expedient and not an eternal condition of life."[160]

Given that death was not, as it were, a fact of life, and that through "creative evolution" humanity could will its own development, Shaw surmised that there was every reason to believe that the duration of life could be extended by force of will alone. Whereas a lifespan of seventy years had been acquired as the age most suited to the advance of the species at some earlier point in mankind's evolutionary history, it was clear to Shaw that humanity was now at a point where an increase in longevity had become an evolutionary necessity. In a conclusion that has led many historians of science to question whether Shaw was in earnest, he contended that "if on opportunistic grounds Man now fixes the term of his life at three score and ten years, he can equally fix it at three hundred or three thousand."[161] Shaw explored the implications of this belief in his epic play *Back to Methuselah*, reminding his audience that "the legend of Methuselah is neither incredible nor unscientific. Life has lengthened considerably since I was born; and there is no reason why it should not lengthen ten times as much after my death."[162] Apparently in anticipation of an incredulous response, Shaw added, "This is not fantastic speculation: it is deductive biology."[163]

In *Back to Methuselah* Shaw had worked out his own metabiological theory through which he might anticipate a "democracy of supermen," which in many respects mirrored Wells's own hopes for an ideal governing elite. Indeed, Shaw's faith in the progressive meanderings of creative evolution led him to hold a similarly skeptical position as Wells vis-à-vis the eugenics movement. However, although as skeptical of positive eugenics strategies as his colleague, Shaw was much less apologetic when it came to negative eugenics.[164] Indeed, even Francis Galton, who was in many ways the founding father of eugenics, regarded Shaw's sometime support with unease. In February 1910, for example, Galton had written to the statistician and eugenicist Karl Pearson, who was Shaw's friend, expressing concern about what Shaw might come out with in his scheduled address to the Eugenics Education Society. "It is to be hoped that he will be under self-control and not be too extravagant," he wrote. As Michael Freeden has noted, however, Galton's fears were well founded, and headlines followed about Shaw's advocacy of "free love and lethal chambers."[165]

Where Wells was skeptical of the possibility of contemporary positive eugenic strategies in light of the emerging complexity of heredity, Shaw entertained similar doubts for altogether different reasons. Foremost among these was his belief that the aspirations of eugenicists were as nothing compared to the inevitable power of the life force that was continually and irresistibly probing onward and upward—ever seeking the most perfect form of expression. Further to this, of course, was that given man's imperfect state and limited knowledge, it was unlikely that mere humanity might second-guess the needs of the life force anyway—this was especially the case regarding the majority of scientific men who, obsessed by neo-Darwinian mechanism, denied the very existence of such a force.[166]

In 1913, when British enthusiasm for eugenics was at its height, Shaw addressed the Political and Economic Circle of the National Liberal Club and in similar terms to Wells's views on the limitations of positive eugenics, echoed concerns about the inappropriateness of the common eugenics analogy between the breeding of livestock and human reproductive choices. Certainly, it was possible to breed horses for speed or for strength, if that was the desired characteristic; however, things were less clear in terms of breeding humanity. The former, Shaw argued, "is quite simple because you know what sort of horse you want. But do you know what sort of man you want? You do not. You have not the slightest idea. You do not even know how to begin. You say: 'Well, after all, we do not want an epileptic. We do not want an alcoholic.' . . . [But as to positive characters] I tell you that you really do not know. I think the first thing you have to do is to face the fact that you do not know, and that in the nature of things you never can know . . . and therefore, you are thrown back on the clue that Nature gives you.[167]

What nature gave them was a wide variety of persons with different attributes and dispositions, however. Echoing common socialist concerns, Shaw believed that such natural differences were complicated and often obscured by the imposition of artificial social hierarchies of social distinction, wealth, and manners. Thus, the first and only truly positive eugenics measure that could be taken was the removal of the artificial distinctions imposed upon humanity by capitalism. It was to this end that Shaw advocated the "free love" that was so eagerly pounced upon by journalists; Shaw's intention with the phrase, however, was to indicate a love free from the artificial barriers of social distinction, bourgeois marriage, and inequality of income. However, given that Shaw, like Wells, was seen in the company of many different women, the journalists can hardly be blamed for mistaking his meaning.[168] The conditions of equality and liberty that would result from such an emancipation would allow the truly gifted to shine and thus for the natural progressive flow of the life force to proceed unhindered. However, as is implicit in Shaw's determination that alcoholics and epileptics were undesirable, he shared Wells's opinion that the most-clearly diseased and deranged could only inhibit human progress, and he made no bones about voicing his opinion that incurable criminals should be "painlessly liquidated," and the "half-created" "idiots" in mental asylums "sensibly and mercifully killed."[169]

Shaw's views too became more pessimistic in light of the desperate carnage of the Great War. Although he tried to see the outbreak of war in 1914 as a purgative renewal of European politics, it is clear that it was a humanitarian tragedy that ever after colored his vision. "Every promising young man I know has been blown to bits lately," he wrote to Lady Mary Murray. Alan Campbell, Robert Gregory, and J. M. Barrie's much-loved godson George Llewellyn Davies had all lost their lives. Shaw wept bitterly when he heard. The list of friends and acquaintances who would not be coming home went on inexorably; the extent of the slaughter pressed home to him "the fundamental waste and folly of the whole business."[170] He laid the blame for war on the permeation of Darwinian ideas into politics and became disillusioned with the liberal democracy he believed had allowed it to happen. After the debacle of Versailles, which both Shaw and Wells prophetically viewed as a set-up for a second global conflict, both were only too ready to sympathize with the range of alternative experiments in government that dominated the 1920s and 1930s. The elitist ideal inherent to the idea of the superman in particular made it easy for Shaw to accept the singular effectiveness of Mussolini, Hitler, and Stalin, in each of whom he at one time or another saw the mark of the life force at work. Thus, although Shaw had turned from Morris's revolutionary purism following the easy dispersal of demonstrators in the infamous "Bloody Sunday" demonstrations on 13 November 1887, in which thousands of marchers had been quickly routed by a small but well-trained and well-armed force, after the 1892 general election he became more sanguine.[171] Sidney Webb had succeeded in convincing the Fabian Society that their efforts were best exerted in trying to bring about social change change by lobbying the existing political administration rather than attempting to overthrow capitalism and replace it with socialism at one fell swoop. But by 1903 Shaw had given up hope that even this "permeation strategy" might bear fruit worth eating. Following the war, however, Shaw became convinced that a revolution would indeed be necessary to transform society—as not only Morris, but as Wells too had acknowledged. For each of them, revolution was as much to do with biology as it was to do with politics.

Looking back from the perspective of 1931, Shaw reflected somewhat ruefully upon his earlier political compromises and the position they had led him to. The Easter Rising as well as events in Russia had demonstrated that the evolution of socialism need not be mere administrative Fabian tinkering, and he lamented that he had had a hand in attempting to talk Morris down the parliamentary road, clearly pondering a few "what ifs" of his own: "When the greatest Socialist of that day, the poet and craftsman, William Morris, told the workers that there was no hope for them save in revolution, we said that if that were true there was no hope at all for them, and urged them to save themselves through parliament, the municipalities and the franchise.

Without, perhaps, quite convincing Morris, we convinced him that things would probably go our way. It is not so certain today as it seemed in the eighties that Morris was not right."[172]

The moral and political implications of evolution only became more pronounced as rival political organizations sought to appropriate Darwin's name and work and to fit their own politics to evolution. In the process, the debate about the moral meaning of Malthus remained as much a point of contention as ever, but by the mid-1880s it was clear too that Weismann's work on heredity was also politically significant. Morris's *News from Nowhere* was one of the most influential texts in the early years of the British socialist movement. In it, he had described a future socialist utopia as a means to encourage others to see that another world was possible beyond the drudgery of industrial capitalism. Like many socialists of his day, Morris was deeply influenced by Kropotkin, and it is unsurprising that he described his conception of social and political change in terms of a Lamarckian adaptation to altered circumstances and the inheritance of acquired characters. Morris was one of the many who, like Kropotkin, denied that Malthus had presented an accurate description of the world or that those who had rearticulated Malthus in defense of competition and Whiggism were any nearer to the true nature of things. He acknowledged that it was necessary for men to labor to meet their material needs, but given the technology available to them, this might be easily accomplished. Indeed, Morris imagined that the socialist future would be an "epoch of rest."

It was this connection between socialism and an anti-Malthusian and Lamarckian biology that H. G. Wells came to find deeply problematic. Not only had his studies under Huxley and Lankester convinced him of the veracity of Malthusian population dynamics, but by 1895 he had also been convinced by August Weismann that the Lamarckian mechanisms that Morris relied upon were untenable. Wells's friend and colleague, George Bernard Shaw, debated this. Although he too was convinced that Malthus did have significant implications for society—a conclusion that led him to think seriously about the necessity for at least some limited form of eugenics to prevent social degeneration—he could not believe that Wells had been taken in by Weismann's experiments on mice.

Wells had not only taken up Weismann's argument against the inheritance of acquired characters, however, but also his articulation of "panmixia," or the tendency for organisms to degenerate once they were removed from selective pressures. To Wells's mind, Morris's "epoch of rest" spelled out a future of certain social and biological decline. He explored these ideas in *Time Machine*, a counter to Morris's utopia—both books were set in the Thames

Valley. While not everyone in the various English socialist communities was convinced by Weismann's work on heredity, Wells was certainly influential. Further, the arguments about panmixia and its implications for contemporary understandings of evolution and heredity remained prominent, among scientists as well as among socialists.

7

Fear of Falling: Evolutionary Degeneration and the Politics of Panmixia

> By aid of a mysterious and novel principle termed panmixia, added by Weismann to the Darwinian theory, it is said to follow, not only that progress is impossible without natural selection, but that without natural selection degeneration must set in as certainly as death follows life.
>
> KARL PEARSON, *"Socialism and Natural Selection,"* 1894

By the last decade of the nineteenth century, the moral and political significance of Malthus's essay on population had become as problematic for socialism as it had been for radicalism. By invoking the incommensurable ratio between population growth and available resources, Malthus had sought to describe the natural limits within which humanity might live; his message was that the kind of unlimited human improvement that radical Enlightenment thinkers had imagined was impossible. Mid-century philosophical radicals like Harriet Martineau and John Stuart Mill had reinterpreted the moral meaning of Malthus in such a way as to conclude that scarcity prompted hard work and moral restraint as the means of achieving precisely the human improvement that Malthus thought impossible. However, the fact that their opinions echoed Whig political economy meant that socialists found this revised version of Malthus no less contentious. Thus, even as old-school radicals and socialists embraced evolution, they struggled to come to terms with the Malthusian political economy that Darwin had made central to natural selection. As we have seen, many socialists reconfigured their understanding of evolution around a conception of Darwin without Malthus. However, for the majority of evolutionary theorists—those who were Whigs or liberals rather than socialists—the presumption that competition was necessary to develop moral character as well as to drive social and industrial progress was self-evident. As a result, they believed that evolution without competition and struggle—Darwin without Malthus—could only lead to an evolutionary degeneration. The struggle of competition was necessary not only to keep evolution going forward, but to prevent it from falling off, and this was as true for social as for biological progress.

Ideas of evolutionary degeneration first surfaced in the last decades of the nineteenth century. Although Darwin had been well aware that every evolutionary development was not progressive, as Michael Ruse has pointed out, the first English naturalist of note who embraced the idea that degeneration might be a widespread phenomenon that explained the life history of many lower organisms, and of marine invertebrates in particular, was E. Ray Lankester. In this respect Lankester was influenced by the founder and director of the Naples Marine Biological Station, the German biologist Felix Anton Dohrn, and subsequently developed his own ideas on the subject in an address to the British Association for the Advancement of Science in 1879, which he published the following year as *Degeneration: A Chapter in Darwinism.*[1] Acknowledging the important role that Malthus had played in the development of Darwin's theory, Lankester pointed out that "naturalists have hitherto assumed that the process of natural selection and survival of the fittest has invariably acted so as either to improve and elaborate the structure of *all* the organisms subject to it, or else has left them unchanged, exactly fitted to their conditions."[2] Defining degeneration as "a loss of organisation making the descendent *simpler* or *lower* in structure than its ancestor," he acknowledged that prior to his own work on the subject, "only one naturalist—Dr. Dohrn, of Naples—has put forward the hypothesis of Degeneration as capable of wide application to the explanation of existing forms of life."[3] However, while Lankester raised the issue of degeneration in biology and used rich social analogies to make his point clear, it was the publicity afforded Weismann's theory of panmixia that really drove home the full implications of degeneration as a widespread possibility in the event that selective pressures either weakened or were removed. Panmixia was the term that Weismann had introduced to describe the degeneration of a biological character in the absence of a selective pressure that had previously ensured its preservation. His panmixia idea was part and parcel of Weismann's attack on Lamarckism. Just a few years after the publication of Edward Poulton's translation of Weismann's *Essays upon Heredity and Kindred Biological Problems* (1889), Havelock Ellis had overseen the translation and publication of Weismann's *Keimplasm* (1892), which appeared in English as *The Germ-Plasm: A Theory of Heredity* in 1893.[4] As I have shown in the previous chapter, it was this idea of degeneration following security that H. G. Wells, who was familiar with both Lankester's and Weismann's work, made central not only to his chilling science-fiction story *Time Machine*, but to his critique of the peaceful and abundant socialist future that William Morris had portrayed in his utopian novel *News from Nowhere.* Wells took biology seriously and revised his hopes for socialism in light of Weismann's work. As I shall show in this chapter, Wells was by no means alone in marshaling the prospect of a panmictic evolutionary degeneration as an argument against the kind of socialist future that Morris had made popular. Indeed, degeneration through panmixia not only became a major subject of debate regarding the processes of evolutionary development in the scientific journal *Nature* in the mid-1890s, it also became prominent in the wider public discussion of the implications of evolution for human political, moral, and social development.

Historians have long since noted that August Weismann's work on heredity raised serious questions about the biological inheritance of acquired characters. Indeed, Weismann had joined Lankester and Oxford zoologist Edward Poulton at the British Association for the Advancement of Science in 1887 to present a session entitled "Are Acquired Characters Inherited?"[5] However, historians have paid comparatively little attention to his theory of panmixia or to the debate over its broader social implications.[6] As Frederick B. Churchill has documented in his article "The Weismann-Spencer Controversy over the Inheritance of Acquired Characters," what Weismann termed "the All-Sufficiency of Natural Selection" was hotly debated in the pages of the popular journal *Contemporary Review* in the early 1890s. The issues were significant enough for the entire series of articles to be reprinted for a North American audience in *The American Naturalist*.[7] In his own assessment of the debate, Churchill focused on Weismann's place in the naturalist tradition as a significant factor in the ultimate success of his theory, giving little attention to panmixia per se. Further, although he hinted that there were social and political as well as biological aspects to the debate over the inheritance of acquired characters, he chose not to expand on this point in his essay. The *Contemporary Review* covered prominent intellectual debates on all manner of issues, and from 1882, under the editorship of Percy Bunting it had become increasingly liberal, giving more coverage to politics and social reform issues than had previously been the case. As I have shown, although the debate over the mechanisms of evolution was ostensibly one of biological importance, it was seen as having very clear social and political implications; like socialists and radicals, liberals also embraced the Lamarckian idea of the inheritance of acquired characters as a central element in their various strategies to bring about social change. To Weismann, Alfred Russel Wallace, and other "neo-Darwinians," the rejection of Lamarckism left only the default presumption that any hope of an evolutionary development in human nature was necessarily limited to what might be achieved through the slow accumulation of chance variations by natural selection. In consequence, those who accepted such neo-Darwinian views had to accept that any anticipation of significant human evolution lay only in the distant future. Just one consequence of this development, for example, was that the utopias that Wells created after his conversion to Weismannism were all set significantly farther into the future than those he had written earlier.

Here, I extend my consideration of the political implications that were taken to follow from Weismann's theory of heredity to focus upon the debate that surrounded panmixia. Not only had Weismann attempted to show through his memorable experiments on mice that acquired characters were not heritable, but he also invoked panmixia to explain those instances of the atrophy of one character or another that his contemporaries in biology explained as resulting from the heritable effects of disuse. Indeed, the appeal to the heritable effects of disuse had become orthodox in both the scientific literature and in more popular understandings of evolutionary processes. Darwin had made good use of the heritable effects of disuse in *Origin* to explain the loss of the eyes in organisms that had taken up subterranean niches, for example, and this was also something that Wells had incorporated into his story of the degeneration of half of humanity into the short-sighted cave-dwelling Morlocks.[8] Weismann proposed that in light of panmixia the atrophy of characters across generations could be explained by selection alone, further undermining prevailing Lamarckian ideas.[9] For example, and referring to the very case that Darwin had cited in *Origin*, Weismann argued that given that there was no longer any selection for good eyesight among cave-dwellers, any individuals with poor eyesight would not be eradicated from the breeding population. As a result of the indiscriminate interbreeding of the population, the average level of eyesight would fall off across the generations. The debate about panmixia was thus an important one, and this was true regardless of whether the focus was on the debate in the scientific journals or on the popularization of science in the context of the more public discussion of the politics of evolution. It is therefore surprising that this debate has been given little explicit attention in the literature. Historians have instead tended to focus upon the aspects of the debate that went on in the pages of the *Contemporary Review* and that Churchill highlighted in his article, despite the fact that the debate about panmixia and its consequences raged in *Nature* and involved some of the most prominent biologists of the day.[10]

In contrast to the exchange in the *Contemporary Review*, the debate in *Nature* took the form of a series of letters to the editor between 1890 and 1896. At issue was disagreement over the priority as well as the veracity of panmixia, and while as Churchill has pointed out, the broader context of the debate was the relative merits of neo-Lamarckian and neo-Darwinian explanations of speciation, in *Nature* the central point of contention was the extent to which evolutionary degeneration might occur. Although prompted by Weismann's work, the starting point of the debate was Francis Galton's statistical work on biological regression. In addition to Spencer and Weismann, E. Ray Lankester, George John Romanes, and Alfred Russel Wallace were just a few of the protagonists. Given the relevance of Galton's statistical analysis to the debate, both the biometrician W. F. Raphael Weldon and the mathematician Karl Pearson also weighed in. While Weldon was keen to underline the importance of statistical analysis in resolving the biological questions about selection and degeneration, it was Wallace's review of the popular book *Social Evolution* (1894) by Benjamin Kidd, which he published under the title of "The Future of Civilisation," that prompted Pearson to put pen to paper. Kidd, a previously little-known writer, had argued that Weismann's views undermined the possibility of socialism, using similar arguments to those that Wells had marshaled against Morris to do so. Both Kidd and Wells believed an end to the individualistic struggle for existence through socialism would bring about the degeneration of mankind. What set Kidd apart, however, was his belief that the natural evolution of religion would intervene to prevent this outcome. He believed that it was not the imposition of socialism that would guarantee the future peace and unity of mankind, but the natural development of Christianity. Wallace disagreed with Kidd's judgment of socialism, but given that he also saw a role for the supernatural in evolution, his review was broadly sympathetic despite his measured criticism. The same cannot be said of Pearson's reaction. Realizing that Kidd's views were becoming widespread because of the popularity of *Social Evolution,* Pearson chose to publish his response in the *Fortnightly Review* rather than in *Nature.* It had a much-wider circulation and gave dedicated coverage to the political and social issues of the day. In "Socialism and Natural Selection" he accused Kidd of not having read Darwin, of misunderstanding the basics of evolution, and of blindly accepting and exacerbating the errors in Weismann's work. Quite contrary to Kidd's conclusions, and as Darwin had made clear in *Descent of Man,* Pearson pointed out that while religious beliefs might play a role in human evolution, their significance was that just like any other human ethical belief they were shaped by natural selection so as to facilitate the good of society. Religion was thus a product of evolution, not a separate guiding factor as both Kidd and Wallace presumed.

In terms of positioning the debate over panmixia and heredity in the broader context of the evolutionary biology of the day, it is important to note that while much of the debate in the public sphere—among popularizers such as Wells, Shaw, and Kidd—centered upon the implications of natural selection for the future evolution of humanity; the scientific community was much more skeptical of the role that natural selection played in the formation of new species. Churchill points out that "the decades between 1880 and 1900 present a period of an extraordinary profusion of explanations of evolution."[11] Indeed, Peter Bowler has referred to the entire period between the publication of *Origin* and the modern evolutionary synthesis of the 1930s and 1940s as the "non-Darwinian revolution," or using Julian Huxley's phrase, as the period that witnessed "the Eclipse of Darwinism."[12] Instead of highlighting the role of selection in evolution, the vast majority of men of science who voiced an opinion on the matter appealed to a variety of alternatives including the inheritance of acquired characters or the existence of various types of "orthogenesis," or otherwise argued that speciation most likely came about not through the gradual incremental changes that Darwin had favored in *Origin* but by the stochastic occurrence of "sports" or "saltations"—significant "freak" variations—in a population.

I have already discussed Lamarckism at length, but these other mechanisms warrant a brief explanation. "Orthogenesis" was a term popularized by the German biologist Gustav Heinrich Theodor Eimer to signify his belief that evolution occurred along specific pathways determined by whatever variations were available. Eimer's work became well known in England in the last decade of the century as a result of Joseph T. Cunningham's translation, which was published under the title *Organic Evolution as the Result of the Inheritance of Acquired Characters according to the Laws of Organic Growth* (1890). It was predominantly a polemic against Weismann and found echoes in Shaw's neo-Lamarckism as well as in the evolutionary "life force" of Henri Bergson's *Creative Evolution* (1911). As we shall see, even at this point, there were few biologists who totally ruled out the inheritance of acquired characters.

The belief that speciation originated by saltations had a much older pedigree, and it remained prominent throughout the second half of the nineteenth century and into the twentieth century as well. Ever since the publication of *Origin*, in which Darwin had articulated his theory of natural selection as the primary means by which new species came into being, even some of Darwin's most ardent supporters were skeptical of the gradualism that Darwin had insisted was inherent to the process. Huxley was foremost among them. He had warned Darwin from the first that he risked tying himself to "an unnecessary difficulty in adopting *Natura non facit saltum* so unreservedly." Even a year after *Origin*'s publication, Huxley remained so clearly convinced that nature "does make small jumps," or "saltations" as he called them, that Darwin appealed to him not to come out in opposition. "For Heaven sake," Darwin wrote, "don't write an anti-Darwinian article; you would do it so confoundedly well."[13] Francis Galton was another Darwinian who doubted the continuity of the evolutionary process. Like Huxley, he thought that new species were most likely to be formed through the freak appearance of individuals with large variations, although he referred to them as "sports" rather than as "saltations." As William Provine has pointed out in his *Origins of Theoretical Population Genetics*, there were two reasons for this. First, Galton's statistical analysis of variation in a population revealed a normal Gaussian distribution in which he felt that any small anomalies in one generation would quickly be negated through a regression toward the mean across subsequent generations. Second, he believed that a significantly novel "sport" might found a new breeding population that would establish itself around its own mean.[14] Galton had first stated his beliefs about the discontinuous nature of speciation in *Hereditary Genius* (1869), but he developed his ideas in more detail in his influential 1889 work *Natural Inheritance*. There, he made the case that "sometimes a sport may occur of such marked peculiarity and stability as to rank as a new type, capable of becoming the origin of a new race with very little assistance on the part of natural selection."[15] The presumption of saltationary or discontinuous evolution became particularly prominent after 1900, following its popularization by the Cambridge biologist William Bateson and the subsequent rediscovery of Mendel's work on heredity by the botanists Hugo de Vries and Carl Correns.[16] This emphasis upon discontinuity meant that there was little need to appeal to natural selection to explain the formation of new species. Moreover, as Provine has demonstrated, the fact that a great deal of personal animosity developed between the principal biometricians, Weldon and Pearson, on the one hand, and Bateson, who led the Mendelian camp, on the other, meant that the two theories were initially viewed as being mutually exclusive.[17]

While Churchill, Bowler, Provine, and other historians have emphasized the diversity of biological explanation that was prevalent prior to the reconciliation of selection and Mendelism in population genetics beginning in the mid-1920s and continuing into the modern evolutionary synthesis of the 1930s and 1940s, the debate did not become polarized until Weismann pressed the issue, both in his *Essays upon Heredity* and *The Germ-Plasm* and in his article for the *Contemporary Review*, "The All-Sufficiency of Natural Selection." What is clear from analyzing the debate over panmixia in *Nature* is that the disagreement that followed forged a divergence of opinion that was only exacerbated by the insistence by the biometricians Weldon and Pearson that their mathematical and statistical methods were the only trustworthy tools to evaluate what was really going on in evolution, and by the equally dogmatic insistence by the Mendelians that biometrics proved nothing. Pearson did not help matters by dismissing the work of the vast majority of his contemporaries rather than patiently attempting to win them over to his view. The mathematical approach he used was new and beyond the competence of most biologists of the day as well as being a departure from the sorts of morphological studies and attempts to reconstruct phylogenies that occupied the older generation of evolutionists. As a result, even those of Pearson's contemporaries in biology who did not have a Mendelian axe to grind saw little in the grand claims that he made for his mathematical approach to biological questions. Pearson's manner, his departure from the working norm, and his training as a mathematician all made it easy to dismiss him as an uninformed outsider, and many did so. Indeed, in several ways Pearson was an interloper who sought to change the underlying presumptions of the participants in an already ongoing conversation. Pearson wanted to talk about how to measure variation across a population and how it changed over time. Up until he and Weldon intervened in the discussion, the focus had very much been upon the ways in which presumed evolutionary forces acted upon individual organisms.

The debate about panmixia in the pages of *Nature* took the form of a protracted exchange of letters to the editor. Those who took part did so not solely to argue the merits of the case, however, but also to debate Weismann's claim to priority. They recognized that if the doctrine of panmixia was not only true, but was as novel as Weismann claimed, then it was a significant contribution to prevailing Darwinian theory. While a number of prominent men of science argued that Weismann's views were either unproven or that the conclusions that he drew were faulty, there were also those who claimed that although Weismann had coined the term "panmixia" to describe the atrophy of organs in the absence of selection, this was a phenomenon that was nothing new. Lankester was the first to broach the subject in a letter under the heading "The Transmission of Acquired Characters and Panmixia" that appeared in the March 1890 edition of the journal. Although Lankester appreciated Weismann's attacks on the Lamarckian belief in the inheritance of acquired characters, he argued that what Weismann had described as panmixia was in fact nothing more than just one aspect of the degeneration that resulted from "disuse," a concept that Darwin had already described at length in the sixth edition of *Origin* and which he, Lankester, had further discussed in his own work, *Degeneration.* George John Romanes, on the other hand, had little time for any of Weismann's ideas. For the eight years prior to Darwin's death in 1882, Romanes had, in practice if not in name, acted as Darwin's research associate. Whereas Darwin's son Francis had helped his father with his researches involving plants, Romanes had aided Darwin's research upon animals.[18] Perhaps unsurprisingly, therefore, he thought Dar-

FIGURE 7.1. Karl Pearson, 1857–1936. Photograph by Elliott & Fry, May 1890. Pearson believed that evolution was a statistical issue; based upon his assessment of the relative strengths of inter-group and intra-group selection, he argued that the evolution of other-regarding ethical sentiments in society would ultimately lead to socialism. (NPG x12708; © National Portrait Gallery, London)

win's theory of pangenesis—which had been an attempt to describe a means by which acquired characters might be inherited—more convincing than Weismann's germ-plasm theory, and while he accepted that what Weismann had identified as panmixia was indeed a significantly different phenomenon to that which either Darwin or Lankester had described, he argued that it was not Weismann who had first described it, and further that there was a wealth of difference between the "reversal of selection," which Lankester had described, and the "cessation of selection" of panmixia.[19] However, not only was it not original to Weismann, Romanes claimed, but it was he who had first described the phenomenon as long ago as 1874, and he had done so in more detail and with greater subtlety than Weismann's attempt to do so.[20] Unlike Weismann or Lankester, Romanes followed Darwin in not rejecting the inheritance of acquired characters, and he cited Herbert Spencer's essay, which had begun the debate in the *Contemporary Review*, as having raised some significant examples that Weismann seemed unable to account for.[21]

That Alfred Russel Wallace was among Weismann's most significant critics in the *Nature* debate might appear surprising given what they had in common. Both men were as opposed to the inheritance of acquired characters as they were convinced of the sufficiency of natural selection to effect the origin of new species; both were among the most outspoken neo-Darwinians; and Wallace had already endorsed some of Weismann's earlier work in print.[22] However, Wallace had both scientific and political reservations about panmixia. He believed that panmixia was a misrepresentation of natural selection and was concerned that it was being erroneously used as an argument to undermine socialism. Even though Wallace had reined in his confidence in natural selection when it came to the evolution of certain aspects of human mind, morals, and even morphology, he criticized Weismann for thinking that natural selection ever ceased its vigilance. If circumstances changed so that selection ceased to operate on one aspect of an organism, then it would immediately act upon another, he argued—here was Darwin's ten thousand wedges at work. Wallace argued that the degeneration that Darwin had attempted to explain by appealing to use and disuse, and which Weismann, in rejecting such Lamarckian ideas, now sought to explain by the cessation of selection, could quite easily be accounted for by natural selection alone without having to appeal to supplemental factors. Just as selection could enhance an organ or character if it was advantageous in a given environment, so too it could diminish those attributes to mere vestiges, or even eradicate them entirely, if they became in any way injurious as a result of a change in the environment. To think that there could ever be a circumstance in nature where natural selection ceased to operate at all was to misunderstand natural selection, he argued.[23]

This was Wallace's main criticism of Weismann's theory, and he said as much in private correspondence to the entomologist Edward Poulton. Poulton, appointed Hope Professor of Zoology at Oxford in 1893, had both translated and promoted Weismann's work in England, sending Wallace a copy of Weismann's *Essays on Heredity*, the second volume reaching him in June 1892. Wallace had thanked Poulton graciously for the gift, but he confessed that although he was very much looking forward to reading it, he had already read the first volume and was "much disappointed with it." This was especially so of the theory of panmixia—"It seems to me the weakest and most inconclusive thing he has yet written." He went on to point out that although Weismann made claims regarding the degeneration of eyes in cave animals, as well as of anthers and filaments, "in both cases he fails to *prove* it, and apparently fails to see that his panmixia, or 'cessation of selection,' cannot possibly produce continuous degeneration resulting in the total or almost total disappearance of an organ."[24] This was not to say that Wallace denied that some degeneration might result from an alteration in the various selective pressures, but it could only "effect a reduction to the average of the total population on which selection has been previously worked" or from the "survival mean" under selection to the "birth mean" of that generation—as it would later be described by another of Huxley's former students, now a zoologist at University College, Bristol, Conwy Lloyd Morgan.[25] As we shall see, this point would later become central to Weldon's and Pearson's attempts to extend Galton's work on regression as a means to undermine the claims made about the effects of panmixia. In a subsequent letter, again to Poulton, Wallace sought to more fully explain himself. Further degeneration was obviously possible, he acknowledged, but it "must be due either to some form of selection or to 'economy of growth'—which is also, fundamentally, a form of selection."[26] This was a point he would have cause to reiterate in the discussion of the issue in *Nature*. Natural selection was not a singular force of constant intensity; the complexities of environmental change meant that although selection was always a force to be reckoned with, it necessarily varied as the environment changed with the seasons, acting now on one character and now on another. In light of this fact, he repeated his point that it made little sense to talk in terms of a "cessation of selection."[27] He would also state publicly his view that the "economy of growth" that Romanes held up as an independent factor in evolution was nothing more than an aspect of natural selection. Marshaling the views of the third director of the Royal Botanic Gardens at Kew to support his position, he wrote, "I am fully in agreement with Mr. Thiselton Dyer when he said that 'I feel more and more that natural selection is a very hard taskmaster, and that it is down very sharply on structural details that cannot give an account of themselves.'"[28]

These were exactly the points upon which Wallace felt compelled to criticize Benjamin Kidd's *Social Evolution*. Although he recognized that there were some merits to Kidd's work, he argued that there were problems with his blind adoption of panmixia. Further, Wallace was particularly keen to expose the errors in Weismann's work because Kidd applied them in an attempt to undermine socialism. It is significant that Wallace thought the book merited a notice in *Nature* as a part of the ongoing discussion of panmixia. The title of his review, "The Future of Civilisation," reflects the fact that Wallace thought the political implications of panmixia a relevant part of the scientific discussion. The editor, Norman Lockyer, clearly agreed.

Published in 1894, Kidd's *Social Evolution* was an immediate commercial success. As Bernard Lightman has pointed out, despite the fact that Kidd's book was not cheap—the first edition sold at ten shillings—it was in high demand and more than 3,300 copies sold in the first three months. The first edition was reprinted nine times, after which a second edition was published, which sold for five shillings. By the middle of 1895, nearly 11,000 copies had been sold. To put this in context, *Social Evolution* certainly rivaled *Origin*'s sales figures; including translations, between 40,000 and 50,000 copies of *Social Evolution* were sold within fifteen months of its first appearance.[29] The extent to which evolution had captured the public imagination in the second half of the nineteenth century is illustrated by recalling that the first edition of Spencer's *Social Statics*, written in 1851, sold only 250 copies. Prior to the runaway success of *Social Evolution*, Kidd had been a virtually unknown government clerk whose only previous forays into the world of science had been the occasional essay on natural history for *Longman's Magazine* or *Chambers's Journal*. However, the success of the book immediately made Kidd something of a social and literary lion. His book spoke to a broader social concern with the question of where religious belief fit in a social-evolutionary narrative, and he found himself invited to give lectures both in England and as far afield as the United States. Nonconformist ministers made it the subject of their sermons, and Henry Drummond, the author of *The Ascent of Man* (1894), another popular book that had a similar theme, rated it "an epoch-making book."[30] As a result, Kidd's views were widely sought on all manner of subjects, even though he quickly shied away from the publicity. He met with liberals, Fabians, and socialists, as well as speaking to any number of improving societies; he also met with Kropotkin at a dinner, although there is no record of the details of their conversation.[31] In *Social Evolution* Kidd advanced a theistic account of human social evolution, but—and what became a major point of contention—at the same time he also appealed to Weismann's panmixia as a means to rule out socialism. Kidd had traveled to Freiburg to meet Weismann in 1890 and had been deeply impressed.[32] Like Wells, Kidd argued that the cessation of individualistic struggle that socialism promised could only result in the degradation of society. *Social Evolution* quickly became a part of the public debate about the political consequences of evolution, and when Wallace reviewed it for *Nature* it became a part of the ongoing debate about panmixia and the nature of selection among the established scientific community as well.

Kidd's theistic account of the evolutionary progress of human society resonated with Wallace's own spiritualist convictions as well as his understanding of the limitations of natural selection. Like many who had read *Origin*, Kidd thought that because natural selection worked to preserve the fittest individuals in the struggle for life it could not therefore account for the evolution of an ethical or in any way "other-regarding" society. Thus, to the extent that natural selection accounted for human nature, man was—and could only be—a self-interested rational actor: "a creature standing with countless aeons of this competition behind him" could be nothing else.[33] Despite all that reason had achieved, reason alone could not have brought about the genuine altruism that Kidd thought not only a characteristic but also a prerequisite for modern civilized society. However, Kidd certainly believed that genuinely other-regarding sentiment was evident throughout human history, but rather than being the product of natural selection, the "fund of altruistic feeling" that had allowed society to flourish and the civilized races to prosper was "the characteristic product of the religious system associated with our civilisation."[34]

Where in social species Darwin had seen social instinct as the unifying force that had made a truly other-regarding society possible, Kidd, who accepted the Malthusian and individualistic account of selection that Darwin had given in *Origin*, saw religion as the source of the unifying altruism that held society together and drove it onward to an ever greater progress. Despite the selfish element of natural selection, Kidd argued, "it would appear that the conclusion that Darwinian science must eventually establish is that—

> *The evolution which is slowly proceeding in human society is not primarily intellectual but religious in character.*

Since man became a social creature the development of his intellectual character has become subordinate to the development of his religious character."[35]

Kidd's conclusion was that there was thus no need to fear that mankind was engaged in some "aimless Sisyphean labour," as Huxley had suggested in his "Evolution and Ethics," "breasting the long slope upwards to find when the top has been reached that our civilization must slide backwards again through a period of squalid ruin and decay." Rather, mankind could look forward to an ever greater social progress which is "always tending to secure, in an increasing degree, the subordination of the present interests of the self-assertive individual to the future interests of society, his expanding intellect notwithstanding."[36] Kidd offered an evolutionary religious salvation in an age when few others were optimistic.

Given his own tendency to see a supernatural role in human evolution, Wallace rated *Social Evolution* "thoroughly scientific in its methods" and stated that this account of the evolution of morality appeared "on the whole to be a sound one." However, he did voice significant reservations about Kidd's embrace of panmixia and his use of it to undermine the promise of a socialist future of peace and plenty.[37] Kidd's account of social evolution led from an initial state of individualistic competition to an ever more socially coherent society in which false inequalities were increasingly done away with and an equality of opportunity flourished. However, while he noted that the Factory Acts and other social legislation that stopped the capitalist exploitation of the workers were stages along the way, Kidd was clearly no socialist and he found himself nominated to the National Liberal Club and the ostensibly nonpolitical but mostly liberal Savile Club.[38]

Kidd believed that the evolution of social cohesion was the result of the growth of an altruism that was religiously inspired as a part of a wider cosmic process. In contrast, socialist schemes aimed at enforcing equality artificially through the suspension of natural processes. The difference between an artificially maintained equality and a natural and Christian equality of opportunity was telling. Echoing Spencer—at least in part—Kidd argued that the interruption of cosmic evolution could only have negative consequences for society. "The conditions of selection being suspended, such a people could not in any case avoid progressive degeneration," he warned.[39] "The evolutionist who has perceived the application of that development which the Darwinian law of Natural Selection has undergone in the hands of Weismann, is precluded at the outset from contemplating the continued success of such a society."[40]

Although Wallace could quite condone Kidd's embrace of Weismann's argument against the inheritance of acquired characters, he regretted that Kidd was "under the mistaken impression that the theory of *panmixia* leads to a continuous and unlimited degeneration," noting that "many writers have pointed out that this is an error." Restating what he had written to Poulton in private correspondence, Wallace continued, "The amount of the degeneration thus produced would be limited to that of the average of those *born* during the preceding generation in place of the average of those that had survived." Wallace thought this a particularly lamentable error "because it is used as an argument against the possibility of any form of socialism which removes the individual from the struggle for existence."[41]

Romanes was quick to respond to the points that Wallace had made in his review of *Social Evolution*, and it was here that he reminded his readers that Spencer had raised similar significant questions of the ability of either panmixia or the "economy of nutrition" to effect the amount of degeneration that Weismann described. In "The Inadequacy of Natural Selection," published in the *Contemporary Review*, Spencer had outlined his arguments against neo-Darwinism either through selection or its cessation.[42] In the process, however, Spencer demonstrated that because of his neo-Lamarckian convictions about the inheritance of acquired characters, he had failed to understand that what was really at stake in panmixia was the degenerative effects that would result from a blending of the germ plasm between individuals with divergent characters—such as good and poor eyesight, for instance. Peter Chalmers Mitchell wrote to make exactly this point, claiming that "in the matter of Panmixia, Mr. Herbert Spencer has misunderstood Weismann completely."[43] Even though Romanes knew that this was the case, having corrected Spencer on this very point in his own work, he still sided with Spencer against the neo-Darwinians, responding that Spencer had at least as firm a grasp of the issues as Wallace. Quoting Spencer, Romanes argued that in any case the significant point in Spencer's argument was that panmixia could not effect the indefinite degeneration of a population: "What is there in the state of Panmixia that determines a numerical excess of minus over plus variations, such as must be supposed if the amount of degeneration due to Panmixia alone is to proceed further than the survival-mean falling to the birth mean?"[44] Romanes's point, though, was not to show that Spencer was right, only that he had pointed out that neo-Darwinism was not up to explaining degeneration. By contrast, Romanes did believe that indefinite degeneration was possible and that this fact therefore demonstrated that there must be other evolutionary factors at work besides natural selection. Indeed, Romanes claimed to have pointed this much out as long ago as 1874. "Now this very pertinent question" of how to account for indefinite degeneration "has never been answered by Prof. Weismann," he wrote, "but what my own views have always been" are "that there are at least three very good reasons why, as soon as selection is withdrawn from an organ, the *minus* variations of that organ outnumber the plus variations, and therefore that it must dwindle in successive generations."[45]

The first of these, Romanes pointed out, was that it was now widely agreed that panmixia might account for the degeneration of the survival-mean to the level of the birth-mean—that is, that the indiscriminate interbreeding of a population that exhibited variations in a character that was not subject to selection would result in the blending out of that character to the mean of that generation, but no further. There were other factors that needed to be taken account of, however, that might account for further degeneration. Atavism, which Romanes noted, "is always at work in our domesticated varieties," was one such cause of a further degeneration, "to a very much greater amount than can be explained by the cause [i.e., panmixia] above mentioned." Even though atavism was less pronounced in the "well established species" to be found in nature, experienced naturalists recognized that "even here its occurrence is neither rare nor insignificant." Indeed, he estimated that it might effect a degeneration of a further ten to twenty percent than that obtained by the degeneration to the birth mean. This was not all. A third reason why degeneration might realistically regress beyond the average of the parental generation was that while under natural selection any failure in the transmission of hereditary characters would surely be weeded out, "as soon as natural selection ceases, . . . all imperfections will be allowed to survive, and, just as in the case of atavistic variations, will act as a dead weight on the side of degeneration."[46]

It was at this point that the zoologist and biometrician Walter F. Raphael Weldon entered the fray. Weldon, who was generally known as Raphael Weldon, had studied under Lankester in the 1870s, but he had also been inspired by the lectures of the Danish mathematician Olaus Henrici to become competent in the subject himself in addition to being a field naturalist.[47] After spending two terms at King's College, London, in 1878 Weldon was admitted to St. John's College, Cambridge, where he worked alongside William Bateson under the developmental morphologist Francis Balfour.[48] Given his training under Lankester and Balfour, it is not surprising that Weldon also chose to make marine organisms the focus of his own research: in 1881 he began his career-long association with the Zoological Station at Naples, and in 1884 was one of the founding members of the Marine Biological Station at Plymouth. Perhaps the most important turn in the development of Weldon's career occurred as a result of his reading Francis Galton's *Natural Inheritance* in 1889. Even though Galton was clearly skeptical of natural selection, Weldon was excited by the statistical methodology Galton employed in his work and looked to him as a mentor ever after. Weldon had been interested in the relationship between death rates and morphological variation in crabs and shrimp, and he immediately saw that Galton's methods could be useful in framing the kinds of questions that he was interested in. Indeed, in his early work on shrimp Weldon's findings appeared to confirm Galton's conclusions that natural selection would not change the shape of normal distribution. In 1889, the same year as this work on shrimp was published, Weldon was elected a Fellow of the Royal Society; two years later, he succeeded Lankester to the Jodrell Chair in Zoology at University College, London.[49]

It was at University College, London, that Weldon first met Karl Pearson. Pearson was particularly interested in Weldon's statistical work on crabs and the two became close friends and colleagues. Both men were convinced that it was fundamentally necessary to put the science of biology on a mathematical basis and became increasingly dogmatic in their conviction that the effects of natural selection—or its cessation—could only be understood statistically.[50] In 1894, Weldon weighed in on the debate about panmixia in the pages of *Nature* from exactly this perspective, dismissing much of the argument up to that point as little more than speculation because it was not based upon the mathematical rigor of statistical analysis. Weldon initially singled out Romanes, not because he ignored the significance of statistics, but because he invoked them without any empirical evidence to substantiate his claims. Weldon pointed out that despite all Romanes's talk of percentages of degeneracy, he had no evidence from the field to hang his numbers on and that as a result his argument was less than scientific.[51] This kind of challenge was not one that was likely to win its author many friends. While Weldon was adamant that the statistical correlation between death rate and variation was significant, the majority of his contemporaries were not convinced that the discovery of a correlation alone was proof of anything—let alone proof of something as controversial as natural selection.[52]

In his own letter to the editor of *Nature*, Romanes had previously demanded that opponents of his own reformulation of how panmixia could account for a continuous decline across generations either show the error in his reasoning or give up their objections. Weldon took up this challenge, countering that he neither accepted the principle of panmixia nor what he believed to be Romanes's flawed understanding of natural selection. Indeed, Weldon rejected all three of the statements that Romanes had made in favor of a continual degeneration, the reversion from survival-mean to birth-mean, and the role of both atavism and the imperfection of the transmission of hereditary characters in furthering degeneration. It is notable that even in 1894 Weldon was still in agreement with Galton's conclusions that selection would not alter the normal distribution in a population, and he cited evidence from Galton's work in his response to Romanes.[53]

Romanes's claims about the limits of panmictic degeneration had played the most significant part in the argument up to this point, and so it was to this that Weldon gave the most attention. "Mr. Romanes says: —*The survival mean must (on cessation of selection) fall to the birth-mean*, &c. This statement involves neglect of a way in which selection may, and often must operate," he wrote. Utilizing an example that Galton had used in his own work, he maintained that "a simple example will show this. The mean height of adult Englishmen is roughly 67½ inches; and if I offer to enrol in a regiment every adult Englishman who is more than 66 and less that 69 inches, the mean height of my regiment will, as every statistician knows, be still 67½ inches, but I shall be obliged to reject more than half the population." If the implications of this were not clear enough, Weldon drew them out: "A form of selection, involving the destruction of more than half the population, may therefore occur without affecting the mean value of the character selected." Weldon went on to communicate that he would shortly publish evidence of a study based upon "many thousands of animals of one species at many stages of growth" that demonstrated that selection can and does act is just such a fashion. "That it must so operate in many cases is obvious from the fact that many wild animals remain for several generations without sensible changes in their mean character," he concluded. If this was not indicative of selection operating in this way, it must either be the case that selection was operating but was incapable of changing the mean, or that selection was not operating at all. In any case, his point was that Romanes's conclusions hardly merited the confidence he invested in them.[54]

Having set up mathematical demonstration as the only real scientific measure of the question, Weldon dismissed Romanes's other claims almost out of hand, on the basis that "they are not demonstrated by any statistics with which I am acquainted." Certainly this was the case regarding atavism, and as for the notion that the failure in the transmission of heritable characters added significantly to the amount of the degeneration of a character, Weldon pointed out that this weakness in heredity—because present whether or not natural selection was operative—meant that it could hardly be invoked as an additional factor. Finally, he charged that neither Romanes nor Weismann—or indeed any other advocate of panmixia—had presented a single case in which they could demonstrate its occurrence with the result of a significant degeneration. By contrast, Galton, whose example of the mean stature of Englishmen he had referred to, had shown "that civilised Englishmen are themselves in a condition of Panmixia, at least in respect to several characters; especially stature and colour of the eyes"; however, far from the result being as Weismann had predicted, across generations "the mean stature of Englishmen is known to be slowly increasing." He added, "I would urge the need, which has lately been pointed out by Bateson, of a quantitative measure of the efficacy of selection."[55] The rift over whether natural selection could be demonstrated as a significant factor in the origin of new species that later opened between Pearson and Bateson over Mendelism and the continuity or discontinuity of evolutionary processes as yet lay in the future.

Predictably, Romanes attempted to deflect each of Weldon's points, although some with more success than others. Romanes pointed out that Weldon seemed to be suggesting that those, like himself, who had made arguments about the regression of the survival-mean to the birth-mean under conditions of panmixia, thought that this degeneration would occur over only one generation—this certainly seemed to be the case in the regimental analogy Weldon had used. Moreover, he argued, the same analogy was clearly a case of artificial selection and not of natural selection at all, and that "the 'cases' to which he alludes, where *natural* selection could destroy individuals nearest the mean line, while favouring those which lie at greater distances, *both above and below* this line, must be very exceptional." Further, although Weldon rejected his claims about atavism as not demonstrated by statistical analysis, Romanes thought this charge hardly fair, for, he reminded Weldon, natural selection had not been demonstrated statistically either and surely he would not rule that out of court in such a brusque manner. Indeed, Romanes was far from convinced that the building up or reduction of physical structures was even amenable to statistical analysis. In light of such radically different conceptions of the problem and how to adequately answer it, Romanes could only restate his case regarding the effects of imperfect heritability of characters upon the degeneration of that character, believing that Weldon had missed his point.[56] In echoing Bateson's call for a statistical demonstration of natural selection, it is probable that this exchange prompted Weldon to reformulate his experiments on crabs in an attempt to use statistics to demonstrate natural selection.

Debate over the efficacy of natural selection and over panmixia and the extent of the degeneration it might cause rumbled on "right wild and windy," as Karl Pearson later put it.[57] However, Pearson was no dispassionate commentator on either issue. His most telling contribution to this debate was not published in *Nature*, though, but in the much more widely read *Fortnightly Review.* He was deeply concerned at the publicity that Kidd's argument against socialism was receiving and thus he wrote for an audience that was broader than that of only his scientific peers. The title of his essay-review of Kidd's book was "Socialism and Natural Selection."

Karl Pearson was thirty-seven when Kidd's *Social Evolution* appeared in 1894; he had been only two when *Origin* was published. After studying mathematics at King's College in Cambridge, physics and metaphysics at the University of Heidelberg, and Darwinian biology under Emil Du Bois-Reymond at the University of Berlin, Pearson returned to England in 1880 to study law. He qualified but never practiced; instead, in 1891 he took up an appointment as Professor of Geometry at Gresham College in London. It was at this time that he became acquainted with Weldon.[58] Pearson was not only a student of mathematics with an interest in biology, however; he was also—and importantly—a socialist. He frequented the socialist clubs and gatherings in the capital, read and admired William Morris, and became acquainted with George Bernard Shaw.

Despite the significance of his contributions to statistics, socialism, and biology, historians of each of these fields have at best treated Pearson with ambivalence. As the historian and Pearson's biographer Theodore Porter has pointed out, historians of statistics tend to remain silent about his socialism because they would rather not acknowledge the politics of one of the founders of their field. Historians of socialism are similarly blinkered when it comes to Pearson; those who are sympathetic to the ideology they study are clearly embarrassed by Pearson's later eugenics views.[59] By contrast, Bernard Semmel went out of his way to paint Pearson as he saw him. In an unapologetic article Semmel portrayed Pearson as not only a socialist but as an aggressively imperialist Social Darwinist.[60] Here, I suggest that while this is certainly an accurate picture of where Pearson ended up, in his youth he was nothing if not a utopian visionary, and it would be remiss to underplay the significance of the very deep connection that Pearson drew between biology and politics in effecting this transformation in his politics. His own socialist beliefs developed along a similar path to those of Wells, and his eventual acceptance of panmixia also led him away from a belief that society was evolving toward the kind of socialism that Morris had portrayed in *News from Nowhere*, and toward a more overt embrace of eugenics.[61]

In addition to his formal studies in Germany, Pearson had read Marx and Lassalle as well as German medieval literature.[62] He became enamored by Germanic folk history and thus when he returned to London it was natural enough for him to sympathize with the socialism of William Morris, just as Wells had done. Not only was Morris influenced by Marx, but Marx's daughter, Eleanor Marx-Aveling, and her partner, the outspoken atheist and Darwinian socialist Edward Aveling, had sided with Morris in the split of the Social Democratic Federation; alongside Morris, they had been founder-members of the Socialist League. Morris had translated Icelandic sagas too, and drew upon his deep knowledge of English folk history in his own socialism.[63] As much influenced by Morris as by Marx, throughout the late 1880s and early 1890s Pearson sought to bring about a sea change in British politics and society by educating the wealth-owning classes to a "new morality."

Morris had frequently made the same appeal, although both men argued that in essence this was a return to an old morality, one that echoed the social obligations of preindustrial England. Pearson acknowledged not only Morris, but also Morris's own mentor, the art critic and social commentator John Ruskin. In the socialist pamphlet *Socialism in Theory and Practice* (1884) Pearson had urged upon his readers that this was neither a fantastic nor forlorn hope. "Do not think this is a visionary project," he told his readers. "At least two characteristic Englishmen, John Ruskin and William Morris, are labouring at the task; they are endeavouring to teach the capitalistic classes that the morality of a society based upon wealth is a mere immorality."[64] Thus, when in 1894 he read Kidd's *Social Evolution* he could not but concur with the criticisms that both Wallace and Weldon had raised of Kidd's particular interpretation of Weismann's theory of panmixia. However, Pearson had no sympathy with the spiritualism or supernaturalism that Wallace and Kidd shared, and he was scandalized that a man of Wallace's caliber and reputation could endorse a book in the pages of *Nature* that expressed such sentiments as a properly scientific work. No wonder the popular press embraced it as "an application of the 'most recent doctrines of science to modern society and life,'" he thought.[65]

Pearson recognized the impact that popularizers of science like Shaw and Kidd had upon the broader public, and he feared that they were misleading a public who knew no better about either socialism or biology. They were worse than the scientific speculators like Romanes, who talked of percentages of improvement or degeneration without the statistical tools to do so, or Weismann, who theorized beyond the evidence. Pearson would not say as much of Shaw, who was a good friend, but he thought that Kidd was little better than a charlatan. Ultimately, he held Weismann responsible. Pearson felt that Weismann's theory of heredity, and especially his theory of the germ plasm, had been overly speculative from the start and that where he led, others had followed—bona fide biologists as well as misguided amateurs. It was hardly surprising that certain "popular writers and the press" followed suit, "providing the public with a fluid so contaminated with the germs of muddleheadedness that it is little wonder if whole classes of the community are poisoned."[66] This much was as much a comment on the essay by the science writer and author Grant Allen, who had recently published an essay describing the debate for a public audience in W. T. Stead's *Review of Reviews*.[67] The only nonselective degeneration that Pearson could see was in the quality of Weismann's work. Weismann's *Essays upon Heredity* followed a "downward course from hard facts to complete metaphysics . . . starting with his fairly sane essay on the *Duration of Life*, and ending in the arithmetic-metaphysical muddle of his theory of amphimixis"—this last being Weismann's conception of the way in which hereditary material from each parent was combined through sexual reproduction to produce variation. Even though Pearson was sure that this was not good science, "unfortunately"—and as was only too evident from the pages of *Nature*—"a certain section of English biologists have followed him, and 'panmixia' and 'germ plasm,' ill-defined in their writings, have now reached the social platform, and are being used as absolutely unassailable arguments against the socialist movement."[68]

However, and echoing the statements that Weldon had already made in *Nature*, Pearson was adamant that "until the quantitative importance and numerical relationship of the various factors, vaguely grouped together as the theory of natural selection, are accurately ascertained, no valid argument can be based on the theory of evolution with regard to the growth of civilised human societies." Clearly a comment upon his own abilities, he added, "The great biologist of the future will be . . . a mathematician trained and bred."[69] To Pearson, this was the crux, and the reason it was so important to get the relative importance of the various factors of evolution correct, for he acknowledged that indeed, "if Mr. Kidd's theory be a correct one, then the modern socialistic movement is completely futile."[70] However, Pearson believed that Kidd was thoroughly wrong.

It was for this reason that Pearson had paid such close attention to the experiments on the effects of morphological variation on the death rates of crabs that Weldon had been busy with both at Naples and afterward on the shore below the Marine Biological Station in Plymouth. Inspired by Galton's *Natural Inheritance* to apply statistical analysis to problems in biology, Weldon had begun to develop field experiments comparing the morphologies between different marine populations of shrimp and crab in 1892. While it was Weldon's work on the statistical variation in the morphology of the common shrimp, published in the *Proceedings of the Royal Society*, that laid the foundations for the field of study in zoology that he and Pearson later coined "biometrics," it was his long-term study of crab populations both at the Zoological Station in Naples and later at the Marine Biological Association in Plymouth that he and Pearson believed provided experimental evidence of natural selection.[71] Weldon and his family had visited Malta and Naples at Easter in 1892 and had spent the following summer at the Naples Zoological Station carefully measuring and comparing the morphological variation of various characters of the crabs. Weldon had initially set out to test Galton's claim that variation in a population was subject to normal Gaussian distribution, and in the first of several papers on crabs, "On Certain Correlated Variations in *Carcinus mœnas*," published later that year, he reported that indeed this did seem to be the case, but with the significant exception of the relative frontal breadth of the carapace.[72] Initially, Weldon had attempted to break his data into two curves of normal distribution that would fit more neatly with Galton's conclusions—and incidentally might have supported a saltationary argument for the formation of new varieties, if not of new species. But he then turned to Pearson for help in interpreting the data and between them they replotted the graph as a double-humped, or bimodal, curve. In this particular character, the variation was dimorphic.[73]

This was significant, but it was also against the grain of contemporary biological thought and was certainly contrary to the work Galton had done that inspired Weldon's investigation. It is therefore likely, as Magnello suggests, that Weldon's initial attempt to divide his data into two distinct curves reflects Galton's belief that speciation must occur through discontinuous saltationary leaps. As I have indicated above, this made sense in the context of Galton's work on regression, but it also seemed to be confirmed by his experiments on pea plants that had led him to his law of "ancestral heredity." According to Galton's own experimental work, the Gaussian distribution of variation in a population suggested that any slight variations would be lost in a regression to the mean rather than accumulate to form a new species, a conclusion that seemed to be supported by what he believed to be evidence of the contribution to the hereditary material that came in diminishing amounts from parents, grandparents, and great-grandparents, etc. Based on these presumptions, Galton believed that only the occurrence of a significant novel variation—a sport—could found a new population, which would then settle around its own mean.[74] Weldon published a number of papers over the following years that made it clear that they believed that Galton was wrong in his belief that variation would always return to the mean. Instead, and with Pearson's help with the mathematics, he argued that under the right conditions natural selection would favor certain variations in such a way as to forge a divergence in morphology, shifting the mean of that character. This process would eventually warrant the classification of the organism as two distinct species. In short, what Weldon was proposing was an experimental confirmation of Darwin's theory of divergence. It was not until Weldon's 1898 Presidential Address to the 68th meeting of the British Association for the Advancement of Science, which met in Bristol that September, that he reported final confirming data from shore crabs that he and a colleague, Herbert Thompson, had collected at Plymouth Marine Biological Station in Plymouth Sound in 1893, 1895, and 1898. "There is no doubt whatever that the mean frontal breadth of crabs from this piece of shore is considerably less now than it was," he stated. He added, "I think, there can be no doubt, therefore, that the frontal breadth of these crabs is diminishing year by year at a rate which is very rapid, compared with the rate at which animal evolution is commonly supposed to progress."[75]

Initially, Weldon had performed these experiments to expose a correlation between morphological variation and death rate. He could thus arrive at evidence of evolution without entering into the sort of speculation that he thought was all too common in evolutionary biology at that time. However, in light of criticisms that correlation spoke nothing to causation and his increasing awareness of the need to demonstrate natural selection as the cause of speciation, he did go on to propose and then test the hypothesis that the morphological divergence he had recorded was indeed the outcome of natural selection. "I feel confident that this change is due to the selective destruction, caused by certain rapidly changing conditions of Plymouth Sound," he wrote. The construction of the breakwater across the mouth of the Sound, begun in 1812 and completed in 1841, not only provided a safe haven from the British fleet, but it had also radically changed the environment of the bay. Two estuaries emptied into the Sound, one on either side of the city of Plymouth, and as Weldon noted, "each of these estuaries brings down water from the high granite moorlands, where there are rich deposits of china clay. . . . One effect of the breakwater has been to increase the quantity of this fine silt which settles in the Sound itself, instead of being swept out by the scour of the tide and the waves of severe storms." Compounding this increase in sedimentation since the construction of the breakwater, not only had the towns of Plymouth, Devonport, and East Stonehouse grown significantly in size, but so too had Devonport dockyard, increasing the sewage and other refuse that was disgorged into the Sound. Weldon acknowledged that it was well known locally that many species that could be found in abundance outside of the breakwater were now unable to live in its lee and that "these considerations induced me to try the experiment of keeping crabs in water containing fine mud in suspension, in order to see whether a selective destruction occurred under these circumstances or not."[76]

The design of this experiment was really quite exquisite. Making full use of the facilities of the Marine Biological Station, Weldon collected crabs that were then placed in glass jars containing seawater "in which a considerable quantity of very fine china clay was suspended. The clay was prevented from settling by a slowly moving automatic agitator; and the crabs were kept under these conditions for various periods of time. At the end of each experiment the dead were separated from the living, and both were measured." The results of these experiments were that without exception the crabs that died were those that were distinctly broader than those that survived—"a crab's chance of survival could be measured by its frontal breadth." Not satisfied with this demonstration, Weldon went on to repeat the process using seawater carrying a courser sediment. In these experiments the death rate was much smaller—coarser silt was less selective. In one final experiment, Weldon used the fine silt directly from the beach below the Marine Biological Station and obtained the same result. Having described these experiments to his assembled audience, Weldon concluded, "I see no shadow of reason for refusing to believe that the action of mud upon the beach is the same as that in an experimental aquarium; and if we believe this, I see no escape from the conclusion that we have here a case of Natural Selection acting with great rapidity because of the rapidity with which the conditions of life are changing."[77]

In contrast to Kidd, Pearson saw no reason to invoke supernatural causes to explain the evolution of ethical and other-regarding feelings among society's members. He pointed out that Darwin had explained in *Descent of Man* exactly how this might occur through natural mechanisms via the action of the social instincts, and of the parental and filial affections in particular. "The development of social instinct and the intensification of the altruistic spirit in the higher types of gregarious animals would appear to be just as much a product of the cosmic process as the evolution of the maternal instinct in the tigress," Pearson wrote.[78] Indeed, as Darwin had pointed out, such social instincts might be exacerbated by religious belief, and Pearson had no problem in seeing the various religions in the world as evolved ideologies that aided the evolution of the groups that believed them as a result of the social coherence they conferred upon a group. He even conceded that it was doubtless true that religion had come about "not as a result of its reasonableness, but as a sanction for social conduct on the part of the unreasoning, upon whom the fear of future punishments and hope of future rewards could have an effect." However, and clearly more than half-mocking Kidd for blundering into this theological minefield, he concluded that "whether the theologian will be equally willing to see things from this standpoint is another question."[79] Kidd had meant to propose religion as God's guiding hand; as Pearson pointed out, though, it could more readily be understood as an anthropological artifact produced by natural selection.

Nevertheless, even with the evolution of altruism, the apparent tension between in-group and out-group competition remained a problem. Huxley had said as much in "Evolution and Ethics." To the extent that altruism ameliorated the struggle for existence between individuals, it weakened society. Not only would it preserve the unfit, who would thus burden society, but their survival—and thus their ability to reproduce—would weaken the average quality of the individuals who made up society, as panmixia foretold. However much the evolution of ethical sentiments between the members of society might benefit that society initially, ultimately it seemed that where altruism prevailed society would ultimately degenerate.[80]

Doubtless with Huxley's 1888 essay "The Struggle for Existence" in mind, Pearson acknowledged that it was certainly the case that "biologists of more or less authority assert that the progress of any group depends on the highest state of rivalry between individuals of the group," and that "by aid of a mysterious and novel principle termed panmixia, added by Weismann to the Darwinian theory, it is said to follow, not only that progress is impossible without natural selection, but that without natural selection degeneration must set in as certainly as death follows life." This conclusion, Pearson argued, was neither based on an adequate assessment of the facts, nor an accurate understanding of the several factors in evolution.[81]

Quoting from *Descent*, Pearson reminded his readers that Darwin had acknowledged that the struggle for existence between individuals was not the be-all and end-all of evolution. "Important as the struggle for existence has been, and still is," Darwin wrote, "yet, as far as the highest part of man's nature is concerned there are other agencies more important."[82] Darwin had listed a number of ameliorations to the individualistic struggle for existence: the effects of habit, reason, instruction, and religion.[83] Pearson certainly concurred with Darwin's views, but before reiterating them as a means to arriving at his socialist conclusions, he wanted his readers to be clear exactly what the struggle for existence might involve besides the competition between one individual and another. Indeed, also clearly aware of Kropotkin's views, which were well known throughout the socialist movement, Pearson argued that competition between individuals actually played only a very minimal part in the struggle for existence and did nothing to advance the evolution of social species.

Just as Kropotkin had argued that the predominant struggle for life that an organism encountered was likely to be with its environment rather than against other individuals of the same species, so Pearson made the case that "the struggle for existence might mean the struggle against physical nature, against disease, of group with group, or of superior race with inferior race."[84] Thus, there were many selective elements that had nothing to do with the internecine struggle of intragroup competition that Kidd had made his focus. Substantiating Kropotkin's observations with his own statistical analysis, he continued, "It is quite unproven in the case of gregarious animals of any kind, including civilised man, that the rivalry to death of individuals of the same group plays an important part in natural selection."[85] This was a point he had

FIGURE 7.2. *The Bridge of Life.* Illustration by Maria Pearson. Pearson's work included statistical data on the age distribution of death among English males between 1871 and 1880. (Frontispiece to Karl Pearson, *The Chances of Death, and Other Studies in Evolution*; courtesy History of Science Collections, University of Oklahoma Libraries)

made in his essay "The Chances of Death" (1895), a statistical study of mortality returns among 1,000 English males born in the same year that showed that death fell most frequently upon the very young and the very old. Pearson reiterated this point in "Socialism and Natural Selection." Among those who reached an age where they might really be said to be in competition with one another, competition rarely resulted in death for the vanquished, or indeed any reduction in their reproductive capacity. Among this group, he wrote, "there is no large majority which 'dies prematurely unable to reproduce itself.'" Death by chance misfortune was a much more likely cause of fatality during this period of an individual's life.[86]

Pearson maintained that these facts totally undermined not only the basis of Kidd's argument, but his credibility as someone qualified to comment on the subject. "In the first place," he wrote, "it is open to question whether Mr. Kidd has ever studied his Darwin"; certainly, it seemed that he had not read *Descent*—the very place where Darwin had stated his views on human evolution in detail. "In the second place," Pearson continued, "he can hardly have analysed the mortality tables of any civilised human community." And in the third place, and most damning in Pearson's eyes, "he has made absolutely no attempt to measure the relative importance of the various factors of natural selection in the evolution of civilised man."[87]

There "are three factors of natural selection," he argued, "intra-group struggle, physical selection, and extra-group struggle." Of the three, he noted, "it is intra-group competition that is the sole basis of the argument against socialism." However, Pearson insisted, this was not only the weakest element of the three factors of evolution, but it was the one that might be most easily ameliorated under socialism. Again in contrast to the conclusions that Kidd attempted to draw from his application of Weismann's theory to social evolution, far from the mitigation of struggle being the cause of unrestrained social and biological degeneration, Pearson argued that ending internal conflict could only benefit society. The limitation of a society's internal competition would only increase its "internal efficiency" in its encounters with other societies. Further, just as the efficiencies of socialism would make one nation more competitive in the intergroup struggle for existence, so the ongoing competition between one society and another would "more and more force the nations of Europe in the direction of socialism." The extent of one nation's sociability over another would be the measure of its collective fitness.[88] While Pearson was thus keen to refute Kidd's argument, in doing so he realized that the inefficiencies attendant upon ethical behavior might be a problem for society in the long run. If selection did not remove the weak, then perhaps panmixia was a threat after all. It was in answering the threat of panmictic degeneration, however, that Pearson became decidedly more authoritarian in his understanding of socialism than he had been.

Certainly, in the initial stages of international competition the superior military might and technological capabilities of the Western nations would see them conquer the undeveloped "savage" nations with ease. This much had been demonstrated by the imperial histories of the European powers. However, and as many of Pearson's contemporaries were now becoming aware, the imperial nations were increasingly being driven into competition with each other. Thus, regardless of the fact that panmixia might not lead to the perpetual degeneration of society that some biologists feared, Pearson became increasingly concerned that in a Darwinian struggle between advanced industrial societies, even a slight degeneration might prove fatal.[89]

Where Wells had seized upon panmixia as raising the possibility that humankind might degenerate across the generations into Morlocks and Eloi, even though Pearson thought that fears of such an extensive degeneration were unfounded, he had already admitted that in light of Galton's work there was evidence of some slight panmictic effects in the English. Initially, he had thought that these were irrelevant in the context of an overall gradual improvement, but now he thought otherwise. The careful management of the population would be necessary to maintain the internal efficiency of society in relation to its international competitors. In his arguments against Kidd's conclusions Pearson had cited the fact that in civilized society the unfit tend neither to be killed off nor to suffer any detriment to their fertility. At the time, Pearson had marshaled these facts as evidence that the struggle between individuals was not a significant factor in evolution; now, however, and as Huxley had pointed out in "Evolution and Ethics," he realized that the fact that they were not killed off had become the significant point: compounded across the generations their weakness would do nothing to curb their instinctive desire to reproduce. Pearson realized that the research that was coming out of his own lab had done enough to show that Kidd's conclusions were on a sound basis even if his understanding of Darwin was wide of the mark and his attempt to hitch them to a continuous degeneration through panmixia was wrong. As the weight of incompetents made society ever more inefficient, it would surely suffer one of two fates: in its compromised state it would either be bested by other and competing societies or it would succumb to the long-term contradictions of ethical actions in a Malthusian world.[90] As Huxley had foreseen, even if society did not fall to foreign rivals, internal conflict would ultimately reassert itself: "The theory of evolution encourages no millennial anticipations. If, for millions of years, our globe has taken the upward road, yet, some time, the summit will be reached and the downward road will be commenced. The most daring imagination will hardly venture upon the suggestion that the power and the intelligence of man can ever arrest the procession of that great year."[91]

"Evolution and Ethics" was a thoroughly anti-eugenical tract. Pearson, however, was of a different mind, and he dared to imagine a different future. This was a significant turnaround, and although he was not quite willing to admit that he might have been wrong about Weismann, by the time that the second edition of the *Grammar of Science* was published in 1900, he was at least tentatively willing to admit that Weismann might have been right, not only about panmixia but about the germ plasm as well—although he was quick to add that this was still far from proven, an example of the kind of thinking that "lies on the borderland of scientific knowledge," "outside the field of actual knowledge," and among "a range of the vaguest opinion and imagination."[92] The extent to which Pearson found it galling to admit that Weismann might been right all along is evident. Passing over his earlier jest about the degeneracy of Weismann's work, he conceded, "We will assume for the time being Weismann's main conclusion to be correct": the hereditary material was immune to changes in the environment, and panmixia might well be a real threat.[93] Accepting as much pressed home the conclusion that controlling the reproduction of a population was the only effective way to manage its quality. Pearson now thought those socialist schemes that had been aimed at raising the living conditions of the poor as a means to their improvement misguided. "We have placed our money on environment, when heredity wins at a canter," he wrote.[94] "No degenerate and feeble stock will ever be converted into healthy and sound stock by the accumulated efforts of education, good laws, and sanitary surroundings. . . . The suspension of that process of natural selection which in an earlier struggle for existence crushed out feeble and degenerate stocks, may be a real danger to society" after all. And, he concluded, this would certainly be the case "if society relies solely on changed environment for converting its inherited bad into an inheritable good."[95]

Kidd, it seemed, had been right about one thing: socialists could not afford to ignore Malthus. It now seemed that Wells had hit the nail on the head when he had said that "probably no more shattering book than the *Essay on Population* has ever been, or ever will be, written. . . . It made as clear as daylight that all forms of social reconstruction, all dreams of earthly golden ages must be either futile or insincere, or both, until the problems of human increase were manfully faced."[96] Subsequent to this realization Pearson became ever more outspoken in his support of negative eugenics strategies in hope of restricting the fertility of what was coming to be referred to as the "social problem group." This might be effected by incarceration or even forced sterilization if need be. In light of the growing economic competition that England faced from Germany and the United States, he concluded that if "we leave the fertile, but unfit, one-sixth to reproduce one-half the next generation our nation will soon cease to be a world power."[97] By the turn of the century, Pearson was arguing that "history shows me one way and one way only in which the state of civilisation has been produced, namely [through] the struggle of race with race, and the survival of the physically and mentally fitter race."[98] It was exactly these sentiments that underpinned his growing advocacy of British imperialism and that led to his call for the state, in the interest of national efficiency, to put the interests of the Commonwealth before the rights of any one individual.

As Kevles has noted, Pearson was interested in pressing forward with the eugenics agenda that was close to Galton's heart. Between 1903 and 1908, Pearson and his staff at University College published some three hundred works, including an entire series of "Studies in National Degeneration."[99] His study of the differential birthrate later became one of the most-cited studies in the field of eugenics, and it was further popularized across the socialist movement and beyond by Sidney Webb, who based his own *Decline in the Birth Rate* (1907) on research that came out of Pearson's lab.[100] The study in question had focused on several districts in London and sought to uncover the correlation between birthrate and social status. Pearson was deeply concerned at the findings. In the area covered by the study, more than half of the population had been sired by less than one-quarter of the preceding generation, but what he found really disturbing was the fact that the vast majority of those responsible for this issue came from what he believed to be the worst section of society. National degeneration was no longer just a theoretical proposition, but a stark reality that was already undermining the nation. Pearson's work thus did a lot to raise awareness of eugenic solutions to social problems. The enthusiasm that resulted found its expression in the popular Eugenics Education Society, which was founded in 1907. Like many scientists who saw the promise of eugenics, however, Pearson remained ambivalent to the Society. Many of its members were outspoken and made claims that were not substantiated by any scientific data, something that Pearson was concerned would undermine his efforts to make eugenics a respectable scientific discipline. It was only after Galton's death in 1911 that Pearson's studies were institutionally secure; Galton had left 45,000 pounds to endow the first Galton Chair of Eugenics at University College, with the stipulation that Pearson was to be the first recipient. Thereafter, Pearson was able to dedicate a significant proportion of his statistical work to eugenics questions.[101]

Pearson was aware that eugenics was a controversial subject that went against many people's conception of ethics and morality; it also raised significant questions about the extent to which the state might intervene into the lives of individual citizens. However, in light of the seriousness of the threat that society's weakest members posed to England's future, it only made sense to employ the insights of evolutionary biology to the management of the "social problem group." Contrary to his earlier position that society needed to return to the old morality of noblesse oblige—which he had recognized and admired in Morris's socialist vision—Pearson argued that in light of the evidence of the degeneration that resulted from panmixia the time had come to set old ethical standards aside and embrace a new morality that was based upon the latest science. Eugenic strategies were more appropriate for the present stage of England's social evolution. Again, it was Pearson's conversion to Weismannism that had given him the lead in this direction; in the *Grammar of Science* he stated, "Now this conclusion of Weismann's—if it be valid, and all we can say at present is that the arguments in favour of it are remarkably strong—radically affects our judgment on the moral conduct of the individual, and on the duties of the state and society towards their degenerate members."[102] As Darwin had made clear in *Descent*, humanity had acquired its standard of ethical judgment by natural selection working in such a way as to fashion the appreciation of those actions that best facilitated the survival of the community. However, while the development of the social instincts had initially benefited society, it now seemed that the preservation of the unfit at the expense of the commonweal was harmful. Under capitalism people chose to serve their own ends rather than those of society, Pearson thought. "In all problems of this kind the blind social instinct" of selfless charity "and the individual bias" of capitalist individualism "at present form extremely strong factors of our judgment," he wrote. "Yet these very problems are just those which, affecting the whole future of our society, its stability and its efficiency, require us, as good citizens, above all to understand and obey the laws of healthy social development."[103]

Pearson believed that what this meant in practice was the replacement of blind social instinct with a conscious social ethic that would be more discerning in its goal. Darwin had recognized that the apparently brutal infanticide and heartless treatment of the old among savage nations were necessary and even ethical—the old of some Native American tribes, for instance, felt it a keen social duty to die when they became a burden upon their fellows. Pearson was not quite advocating a return to such practices, for there were more humane ways of achieving the same effect. He wrote: "If society is to shape its own future—if we are to replace the stern processes of natural law, which have raised us to our present high standard of civilisation, by milder methods of eliminating the unfit—then we must be peculiarly cautious that in following our strong social instincts we do not at the same time weaken society by rendering the propagation of bad stock more and more easy."[104]

What was needed was "a check to the fertility of the inferior stocks." Panmixia, the interbreeding of good with bad stocks such as Shaw proposed as a means of social improvement would only drag everybody down. What was called for were "new social habits and new conceptions of the social and the anti-social in conduct."[105] At his most excessive, Pearson could outdo even Shaw for making statements that were at once chilling and prescient. In *The Moral Basis of Socialism* (1887), for instance, he unflinchingly embraced the conclusion that "socialists have to inculcate that spirit which would give offenders against the State short shrift and the nearest lamp-post."[106] Some, it seemed, offended simply by being born. For Pearson, it was science that would lead the way in the revision of ethics and politics. As he concluded: "The 'philosophical' method can never lead the way to a real theory of morals. Strange as it may seem, the laboratory experiments of a biologist may have greater weight than all the theories of the state from Plato to Hegel!"[107]

The debate about panmixia was thus more than just an offshoot of the broader argument over whether neo-Darwinism or neo-Lamarckism gave a better account of the origin of new species. Certainly, panmixia was a part of this debate, but there was clearly much more at stake too. There were professional rivalries and competing priority claims as well as pressing questions about the causes as well as the extent of any evolutionary degeneration. While the big names in Darwinian biology were divided on these issues, the young naturalist Raphael Weldon intervened, arguing that these were questions that could only be objectively answered using the new tools of statistical analysis—a suggestion that was controversial, as Weldon clearly intended to make mathematics the measure of all things in the science of biology.

It was Alfred Russel Wallace's review of Benjamin Kidd's popular work *Social Evolution* that alerted all comers to the public appeal of panmixia and the social implications of Weismann's theories. The Malthusian political economy that had been debated since 1798 had become central to the presumptions of English industrial society; Martineau, Mill, and the success of Darwin's *Origin* had seen to that. With competition the presumed motor of social and economic progress, to many it only made sense that a lack of competition would lead to a decline. While Kidd saw this as a solid argument against a socialist future of ease and abundance, he did not believe that humankind was condemned to a life that was dominated by a Darwinian struggle of each against all. Rather, he believed the spread of

Christianity would ensure the extension of justice and compassion to all the peoples of the world. It was the immense popularity of Kidd's book that prompted Karl Pearson to respond. He countered Kidd's claim by arguing that in terms of their relative strengths as evolutionary factors, the competition between members of the same group was much weaker than the competition between different groups. Thus, on balance it was intergroup competition that drove social evolution, leading not to an ever greater spread of Christianity but to the spread of socialism. Nevertheless, Pearson's confidence in his own conclusions were weakened by the research that came out of his own lab that indicated a differential birthrate by social class; those he considered the weakest members of society had been responsible for the greater part of the paternity of each generation across the previous fifty years. As a result, he came to believe that panmixia was indeed a significant factor in English fin-de-siècle social evolution. In response to these findings, Pearson argued that socialism could manage this threat to England's national efficiency by more effectively subordinating the interests of the individual to the good of the community as a whole—and by adopting eugenic strategies to this end. The forces of natural selection may well have forged ethical sentiments from the social instincts, as Darwin had described, but the time had come to apply reason to evolution and make it a conscious process that was more discriminating. Pearson increasingly came to believe that a combination of imperial conquest abroad and eugenics at home was the best way to ensure that England's evolution continued in the right direction.

Pearson had written all this in the years before Europe was plunged into war. The realities of trench warfare undermined the faith of all but the most dogmatic that war between modern industrial nations could possibly be a means to social progress. The Great War of 1914 cast a shadow across Europe that signaled the end of innocence in the politics of evolution.

CONCLUSION

Political Descent: Anticipations of the Twentieth Century and Beyond

What should we make of this history of the politics of evolution in which the moral meaning of Malthus played such an important role? First, I believe this book has implications for how we think about the history of evolution in the Victorian period. Adrian Desmond has drawn our attention to the fact that the politics of evolution predated the publication of Darwin's *Origin of Species* by half a century and that in Britain evolution was a part of a deeply radical political culture. He has demonstrated that Lamarckian ideas of a progressive evolution from below were used to legitimize arguments for political change that would alter politics and society from the bottom to the top. In *Political Descent* I have taken up this radical conception of evolution as my starting point and followed the ways in which different evolutionary ideas were utilized in support of one politics or another into the twentieth century. Evolutionary politics were popular among English radicals across the nation—who embraced the ideas of Erasmus Darwin as well as those of Lamarck. Those who were first called "Darwinians" were born of this tradition. Significantly, while Charles Darwin grew up in a comfortable Whig household, he was clearly familiar with his grandfather's views. If he had not given them much thought before he attended Edinburgh University to study medicine, he certainly did so once he was there, and the fact that his Edinburgh tutor, Robert Grant, was so appreciative of both Erasmus Darwin and Lamarck opened the door to what may have been Darwin's first evolutionary studies in natural history.

Herbert Spencer was more deeply embedded in this radical Lamarckian tradition, despite a childhood in which he had been set to study Harriet Martineau's Whiggish Malthusian primers. Adaptation to circumstance and the inheritance of acquired characters informed his politics throughout his career. Even after the publication of *Origin*, natural selection was only ever a secondary mechanism in his understanding of evolution. Martineau had a different effect upon Darwin. As Ashley Nelson has most recently pointed out, it was under the patronage of Lord Brougham that she had turned her considerable literary skills to the popularization of Malthusian political economy, and her success in naturalizing a Whig moral and political economy had been unequivocal.[1] This framed Darwin's 1838 reading of Malthus. Darwin's emphasis upon Malthus in *Origin* was a significant factor in both the eventual acceptance and subsequent interpretation of evolutionary ideas in mid-Victorian England. Thus, I contend that from 1859 there existed two rival traditions of evolutionary politics in Victorian England—the one Malthusian, focusing upon the individual in the evolution of society through competition; the other radical, predominantly Lamarckian, and deeply anti-Malthusian, inclined to emphasize the role of social cohesion as a means to social progress, in which individual interests tended to be subordinated to the welfare of the group. This set the stage for the subsequent development of evolutionary politics and ideas, since everyone who dealt with evolution had to take a position on the moral meaning of Malthus. While it is clear that some advocates of free trade, notably William Rathbone Greg, utilized aspects of Darwin's work to naturalize their politics, the most-well-known defender of laissez-faire, Herbert Spencer, found Malthus largely irrelevant. Indeed, beyond the 1860s it was the new liberals, Thomas Henry Huxley among them, who wrestled with how society might best surmount the social problems that Malthus had theorized. Those who took up the Malthusian reading of man's place in nature included notable figures among the Fabians—D. G. Ritchie, the Webbs, H. G. Wells, and Shaw. On the other side of the debate, many of the more-revolutionary socialists rejected Malthus and emphasized the role of mutualism and cooperation in evolution. They appealed to Lamarckian mechanisms of adaptation and inheritance to support their theories of social change. Kropotkin and Morris were particularly influential in this regard.

There were individuals who crossed the aisle, of course. For instance, while there were very few naturalists who rejected Lamarck altogether, there were some; and there were also Lamarckians who acknowledged Malthusian aspects in nature. That said, however, for the majority, it became much more a question of emphasis regarding how they framed the Malthusian or anti-Malthusian aspects of nature in relation to evolutionary understandings of progress or degeneration. As I have shown in my concluding chapter, this concern with progress and degeneration increasingly took center stage from the 1880s through to the First World War.

A second point I want to make in light of *Political Descent* is that this history has implications for how we think about certain controversial aspects of natural selection in biology today, and while I shall comment only briefly on this here, the present-day levels-of-selection debate is rife with political presumptions, even if they are less overtly stated than was the case in the period I have written about here. A brief recapitulation of the main thread of my argument will be useful in further drawing out these conclusions.

I opened *Political Descent* with Darwin's *Beagle* voyage and his first encounter with the natives of Tierra del Fuego, and even though Darwin did not read Malthus until years later, it is clear that he was already making connections between the natural economy and political and moral economy. As I have indicated, he was provoked to think deeply about the natural-historical as well as the social and political differences between the Fuegians and English as he attempted to explain the gap "between savage and civilised man"—Darwin could not help but think that it was the Fuegians lack of private property that held them back. When Darwin did read Malthus, he did so through the filter of Harriet Martineau's political tracts. Martineau had played a crucial role in promoting the program set forth by Henry, Lord Brougham, for political reform.[2] Darwin was well aware of the political associations of transmutation when he wrote *Origin*, but by emphasizing Malthus and distancing himself from Lamarck and the author of the controversial *Vestiges of the Natural History of Creation*, he successfully managed to reframe the politics of evolution around Whig interests.

This had been Robert Chambers's intention in publishing *Vestiges*, but while he succeeded in making transmutation a sensational subject of society conversation, he failed to convince the Oxbridge dons who dominated the scientific community that transmutation was a fit subject for scientific inquiry. Darwin's success had as much to do with context as it did with content, although both were important. It was in the mid-1840s that Chambers had published his attempt to draw evolutionary ideas into the service of middle-class ideology, but even among young radical Whigs like Huxley, who might otherwise have been sympathetic to Chambers's aims, *Vestiges* was too much the work of an amateur and speculator to pass muster. Those, like William Carpenter, who were impressed by what Chambers was trying to do, gave what support they could in secret, and if they wrote favorably on the subject, they availed themselves of the cloak of the reviewer's anonymity. Like *Vestiges* itself, transmutation remained an idea that was simply too radical and too controversial for anyone to put their name to and retain their respectability. As Desmond has made clear, support for transmutation implied support for atheism and the very worst excesses of Francophile revolutionism.[3]

It was with the publication of *Origin of Species*, in which Darwin articulated a conception of evolution that was thoroughly compatible with the neo-Malthusian Whig political economy, that evolution finally became acceptable among the scientific community. *Origin* did little to persuade the established Oxbridge community of scientific men—men like Sedgwick or Whewell—but the fact that it was the product of one of their own students and a man they had come to respect as a serious scientist gave them pause. Their response was more measured, if not more positive than the reception they had given *Vestiges.* Over the course of the 1860s, however, the young and rising naturalists to whom Darwin appealed were busy cementing their influence in the scientific societies, so that by the end of the decade the vast majority of practicing men of science were on board. Even if there remained questions about the efficacy of natural selection to create new species, as Huxley had made clear, Darwin had made evolution a "Whitworth gun in the armoury of liberalism," and despite persistent ambivalence among men of science about natural selection, to a broader public the significance of the struggle and selection was paramount. *Origin* served to naturalize Whig political economy and thus prompted the birth of an alternative tradition in evolutionary politics that developed in opposition to the earlier, deeply Lamarckian tradition that had been associated with revolution and a more-revolutionary radicalism.

Having begun with Darwin's voyage and his embrace of Malthus, in chapter 2 I returned to give a more-detailed consideration of the earlier politics of evolution. Darwin could not have been unaware of the radical politics of evolution. The radicalism of transmutation before Darwin is a subject that has been tackled in detail and well by Adrian Desmond and, more recently, by Paul Elliott.[4] Under the influence of Erasmus Darwin and Jean Baptiste Lamarck, as well as William Godwin, those who argued for a radical transformation of society refuted Malthus's conception of natural limits by invoking the political promise of transmutation. In the years before Darwin published *Origin*, evolution was a radical Francophile doctrine. Based upon Jean-Baptiste Lamarck's *Zoological Philosophy*, it was more closely associated with revolutionary politics than with English natural history. Indeed, adherence to transmutationist ideas was widespread across the radical community, in London as well as in the industrial North and Midlands. As Desmond and Elliott have shown, religious nonconformity, political reformism, and transmutation went hand in hand among both radical industrialists and their employees, as well as among the medical students in London and Edinburgh. It was only in the wake of 1832 and the formation of modern class society that this changed. E. P. Thompson has highlighted the 1832 settlement as the signal moment in the formation of English class identities,[5] and it is clear that while those with property and an independent living were brought into the fold of the establishment, a great many were left out in the cold. Radicalism was split, and it is not too egregious a simplification of the complexities of the situation to say that those who benefited from 1832 went on to form the backbone of English political liberalism, while many of those who were excluded sought redress through Chartism and, later, socialism.

Herbert Spencer was a radical who was deeply sympathetic to political reform and to the progress of the People's Charter. Spencer had first turned to journalism as a Chartist sympathizer. Well schooled in nonconformity, radicalism, and reform, Spencer was also familiar with transmutationist ideas garnered from Lamarck, Erasmus Darwin, and the contrary arguments of Charles Lyell. What later critics have identified as heartless individualism on his part was actually the product of the nonconformist moral emphasis upon independence. Victorian notions of masculinity, particularly among the rapidly expanding middle class who made their money through hard work and long hours, were grounded in the presumption of a man's ability to provide not only for himself but for his family as well. Further, nonconformists were deeply skeptical of the Anglican governance of "Old Corruption." As was the case for most radicals, Spencer's interest in transmutation began with what he believed it might say to humanity—and to the political possibilities for the future development of our species in particular. Even though he saw evolution as a universal cosmic process, Spencer believed that the evolutionary history of mankind revealed telling political lessons for mankind that his countrymen would need to take notice of if England was to remain in the vanguard of history. In light of what he perceived to be the evolutionary implications of an overreaching state, Spencer was skeptical of any state intervention. To Spencer's mind, state-mandated charity quite literally bred dependence.

History—and historians—have been unkind to Spencer. Given his unrelenting opposition to the state, to imperialism, and to the oppression of one man by another, in a world in which the politics of Darwinism have been tainted by Nazism, the Holocaust, and the worst excesses of the eugenics movement, it is an injustice as well as an irony that Spencer has become the scapegoat for all the ills of "Social Darwinism." Although Spencer did eventually embrace natural selection, Bob Richards is correct to point out that it only ever amounted to a secondary element in what was otherwise a thoroughly Lamarckian worldview.[6] An individualist, Spencer viewed liberty as the means by which mankind might fulfill their natural social instincts, and social life as the means by which they might develop a truly ethical regard for one another. Both were the basis of his utopian-socialist hopes for the future.

In his early work, and in *Social Statics* in particular, Spencer gave particular consideration to the natural conditions of humanity in the context of what they might say to normative action. Extending utilitarian ideas into the field of natural history, Spencer argued that it was the result of an evolved social instinct that humanity had advanced to the extent that it had. There was, he believed, a natural basis to ethics to be found in instinct, rather than one grounded in the reason familiar to traditional Enlightenment philosophies.

It was only in light of the critical acclaim given to *Origin* that Spencer's views became popular. Spencer acknowledged Darwin's work, notoriously describing it in terms of the "survival of the fittest," and this, combined with the fact that Spencer had spoken about the human implications of evolution whereas Darwin had remained silent, allowed his views to be taken to be orthodox Darwinism. Spencer's close friendship with Huxley and later membership of the X Club did nothing to expose the differences between the two men.

In the following years, many people appropriated Darwin's name, the Malthusian idea of competition that he had made central to natural selection, and Spencer's radical critique of the state to endorse their own opinions. This was especially so in the case of the Manchester political economists. Their mouthpiece, the *Manchester Guardian*, argued that Darwin had proven their doctrine as thoroughly grounded in nature. Darwin's concern to correct this view was one of his leading motivations for writing *Descent of Man*.

In *Descent*, Darwin was keen to have his own voice heard and to correct misconceptions in the now very public discussion of the implications of evolution for mankind. Spencer had commented on the significance of a natural-historical account of the development of human morals and ethics—a question that Darwin had discussed in his notebooks in the later 1830s and early 1840s. One of the pressing questions of the day was the extent to which evolution could account for the ethical nature of humanity. If natural selection favored those individuals who looked out for themselves, then how could other-regarding sentiments have arisen? Among Darwin's contemporaries there were two responses to this question. The first, offered by Alfred Russel Wallace, Darwin thought a disturbing turnaround. Wallace had suggested that since natural selection could not account for this aspect of humanity, then this must be evidence of a divine intervention. A second explanation, which at least kept the argument within the bounds of material explanation, was the Benthamite suggestion that reasoned self-interest could quite readily account for the evolution of reciprocal acts of kindness. Darwin was not satisfied with either one. The first he could not fathom; Wallace, who had been his biggest ally in advancing natural selection, had strayed well wide of the path. The second, while doubtless true enough, was not the whole story. Reasoned self-interest might certainly act in such a manner, and might help develop actions to help others, but Darwin was reluctant to conclude that such a low motive as self-interest was either the sum or the source of human moral action.

In spite of the tendency of those who read *Origin* to emphasize the competitive individualism inherent to natural selection, Darwin had been clear to point out that self-interest was not the only element of his theory. Even in *Origin* he had included an account of the evolution of what appeared to be genuinely other-regarding sentiments among the social insects, in which he described individuals acting for the good of the "social community."[7] In *Descent* he sought to extend this analysis to explain how other-regarding sentiments could quite easily be the outcome of evolution in mankind—another eminently social species. Just as he had done in his account of the evolution of the social insects in *Origin*, Darwin, like Spencer, also grounded his explanation in instinct rather than reason.

Darwin was clear to point out that genuinely other-regarding human moral sentiments could be quite adequately accounted for without having to appeal to divine intervention. Sociability, intelligence, a level of language sufficient to establish public opinion, and habit were all that were necessary. Of these elements, perhaps the most significant was public opinion. Darwin had read Adam Smith's *Theory of Moral Sentiments* shortly after his return from the *Beagle* voyage and he was aware that Smith had devoted significant attention to the importance of public opinion in setting prevailing moral standards of behavior. However, Darwin had also taken note of David Hume's comparison of human reason to animal instinct. The language of public opinion was therefore not that of the self-interested rational actor—something that Smith had presumed—but was a vocalization of the animal instinct to preserve one's group by praising actions that aided the group and penalizing actions that were harmful to it.

Darwin had recognized that there were tensions between the instinct for self-preservation and any instinct that might work for the good of the group, and he admitted that, all things being equal, one would expect that self-interest and individualism would prevail. However, as he was keen to point out, in nature not all things were always equal. Public opinion, in particular, could be quite discriminating in this regard. Although he did not refer to it explicitly in his chapter on the moral sense, in the latter half of the book Darwin went on to explore the outcomes of sexual selection as one significant aspect of such discrimination. Sexual selection was as much about female choice as it was about competition between males, and while both were significant when it came to the evolution of other-regarding sentiments, it seems that Darwin thought that female choice was the more so. Competition between males to win females was clearly important, but in many of the cases that Darwin discussed, it was competition to win the attention of the females rather than to win ownership of them per se that was important, the women having the freedom to choose from among those who were successful in war. As Darwin went on to explain, anthropologists had recognized that these choices reflected culturally and historically determined standards for the qualities that women looked for in a husband: a willingness to defend the tribe as well as to defend one's own family was significant.

In their own recent work, *Darwin's Sacred Cause*, the historians Adrian Desmond and James Moore have argued that Darwin was motivated to formulate his theory of common ancestry as a part of a broader political antislavery agenda. While they have failed to convince the majority of their readers that this was Darwin's motivation, what they have succeeded in doing is exposing the politics at stake in the question of evolution—politics of race as well as class and sex. It has long been observed that Darwin's views on evolution reflected the political-economic ideas of his time and of his class, and as I have argued here, by claiming Malthus in the way he did, Darwin intentionally signaled that transmutation was no longer a dangerous Francophile theory but was quite amenable to a liberal Whig science. While the evidence for Darwin's motivation is largely circumstantial—he was ever keen to present himself as being an objective scientist—what is beyond dispute is that when Darwin wrote *Descent* he was well aware that he was entering an arena that was as much about politics as it was about natural history, and as a part of his own agenda he was thoroughly determined to counter the views of humankind that were being put forward by the Manchester political economists.

It was not only laissez-faire liberals who had appropriated Darwin's views. In the second half of the century, many other English liberals found laissez-faire wanting too. These "new liberals" argued that the social dislocation of the Industrial Revolution required a social response. John Stuart Mill had begun to theorize the socialization of liberalism in the 1860s and others were quick to follow his lead. By the mid-1880s, in the midst of what later became known as the "socialist revival," an increasing number of people argued for thoroughly socialist remedies to England's industrial growing pains, and they too invoked evolution in justification of their politics. Indeed, as I have shown in chapter 4, there was little in post-*Origin* politics that was not colored by evolutionary views of one kind or another. Significantly, and as I hope to have made clear here, this was not merely a case of politicians appropriating Darwin's name to bolster their preexisting views, for there were a number of instances in which individuals altered their political strategies in light of what they learned about evolution. More to the point, however, as Margaret Schabas and Donald Winch have shown, there was a long and established history of moral, political, and natural economies being seen as intimately connected.[8]

Following the emphasis upon the social nature of humanity, new liberals were keen to stress mutual aid and cooperation as a fundamental aspect of evolution. Certainly, there were those, like Walter Bagehot, who pointed out that cooperation had developed in the context of warfare against other nations, but it was a departure from the laissez-faire individualism that had become associated with Darwinism in *Origin*'s wake nonetheless. This was also a reinterpretation of the moral meaning of Malthusian struggle; in this instance, it was the competition between nations that drove social evolution. As I have shown in chapter 4 though, this is not to say that all Malthusians rejected Lamarckism or that all anti-Malthusian's rejected natural selection. There were people who embodied these extremes of course, but for the most part the distinction was a matter of emphasis. Also, while I have sought to draw out the significance of two traditions of evolutionary politics—the one Malthusian, the other anti-Malthusian and Lamarckian—in practice these differences sometimes took a back seat to more-pressing concerns. This was the case in the context of the identity crisis of English liberalism, for example, in which lines had been drawn between those who advocated for laissez-faire and those who argued that some level of state intervention was a necessity in responding to the social consequences of industrialism and urbanization. However one understood the processes of evolutionary change colored the views of the various participants in this debate. This is most clearly illustrated in the different positions that Herbert Spencer and Thomas Huxley took on this subject.

While historians have paid significant attention to the likes of Huxley and Spencer in this regard, what I have tried to emphasize here is that this was very much only half of the story. Despite the efforts of a number of historians, the history of the relationship between socialism and evolution still does not receive the attention it merits. From the 1880s, England witnessed the emergence of a large and diverse socialist movement. English socialism had its roots in the London radical clubs, and as a result of radical ambivalence to Malthus, many socialists also struggled to come to terms with the Malthusian elements of Darwin's work. Most formulated a socialism that was informed by a reading of Darwin without Malthus. This was true of the founder and leading light of the Democratic Federation, Henry Hyndman. Hyndman embraced Marxism and popularized an evolutionary historical materialism in England. Rejecting Malthus, he argued that struggle and selection were the driving dynamic of human history, but he insisted that this was by no means a natural or inevitable circumstance. Rather, Hyndman argued that competition was a product of unjust social relations, and instead of acting upon physiology as it had in humanity's early history, now, natural selection had come to act on the prevailing mode of production of each society.

In contrast to this anti-Malthusian tradition in socialism, there were some socialists who embraced Malthus as a fundamental truism. Those who did so were predominantly in the liberal-socialist Fabian Society, although again, the Fabians were such a diverse grouping that it is difficult to generalize without noting the few but notable exceptions.

The division between liberals and socialists on the place and significance of Malthus in evolution is perhaps best exemplified by the difference of opinion on this matter between Thomas Huxley and Peter Kropotkin. This has been noted before, of course, but here I have taken my analysis further than that of other scholars. First, what I have suggested here is that the differences between the two over the political implications of evolution did not represent the extremes on this issue. Instead, and in the context of the broader political arena, it was a battle for the middle ground that lay between out-and-out individualism on the one hand and state socialism on the other. As a result, Huxley found himself fighting a war on two fronts: he rejected the out-and-out laissez-faire ideals of classical liberals like Herbert Spencer, but at the same time he remained wary of the collectivism that threatened socialism. This much is illustrated by the evident tensions between his two essays "Administrative Nihilism" and "The Struggle for Existence." Kropotkin had similar reservations. He admired a lot of what Spencer had had to say in his *Synthetic Philosophy* about the gradual evolution of humanity toward a socialist future, and like Spencer he was ardently opposed to state power. Unlike Spencer, though, he thought the private ownership that had led to capitalism was a wrong turn in an otherwise progressive social evolution. Further, and what I want to emphasize here, Kropotkin's argument with Huxley was not just that he believed that Huxley had overemphasized Malthusian competition at the expense of mutual aid, but that Huxley had misrepresented Darwin's views on human evolution in the process. While in "The Struggle for Existence" Huxley had emphasized the Malthusian and individualistic aspects of Darwinism to be found in *Origin*, Kropotkin quoted *Descent of Man* to contradict him. Kropotkin maintained that when it came to understanding human evolution, it was he and not Huxley who was actually the more "Darwinian" of the two. Finally, although Kropotkin was keen to correct Huxley's reading of Darwin, he also pointed out that Darwin had got several things wrong too. Not only had Darwin taken his own metaphors too literally on occasion, but he had also given too exclusive an emphasis to the parental and filial affections when it came to accounting for the origins of other-regarding sentiments. Kropotkin argued that brotherhood and sisterhood were equally important and that in the context of the historical development of human societies, selflessness might easily extend beyond the bounds of biological kin relationships.

After 1888, Kropotkin dedicated much of the rest of his life to substantiating his argument that mutual aid was a more significant factor in the evolution of social species than was competition. Echoing Spencerian and Lamarckian ideas of adaptation to environment and the inheritance of acquired characters, Kropotkin theorized a much deeper historical account of the evolution of ethics than Huxley had acknowledged in either his 1888 essay "Administrative Nihilism" or in his 1893 Romanes lecture, "Evolution and Ethics." Further, he made it clear that he believed that he was much closer to Darwin's thoughts on the subject than the man who had made his name as "Darwin's Bulldog."

By the 1880s, the mechanisms of heredity had become as politically controversial as the issues that Malthusianism had raised. The work of the German naturalist Friedrich Leopold August Weismann appeared to rule out the Lamarckian mechanisms that had become so central to the strategies for change articulated by prominent socialists like William Morris. In chapter 6, I showed how a commitment to Malthusian ideas and a newfound appreciation for developments in the study of heredity influenced the young socialist and science-fiction writer H. G. Wells. Initially impressed with the socialist-utopia Morris had portrayed in *News from Nowhere*, Wells had begun to doubt Morris's conclusions in light of the anti-Malthusian tenor of the book. Wells had studied under Huxley, and from the 1890s began to reconsider his socialism in light of his substantial knowledge of biology. Further, after having read Weismann in 1894, Wells rejected the Lamarckian belief in the inheritance of acquired characters. As a result, he concluded that Morris's presumptions about social change, which were based upon Lamarckian understandings of adaptation and inheritance, were wrong. Wells was also troubled by Morris's conception of what life might be like under socialism. Morris had proposed that having won the struggle for existence, humanity might settle into an "epoch of rest." In light of Weismann's theory of panmixia, however, it seemed to Wells that Morris was unknowingly advocating for social conditions that would cause the social and biological degeneration of the human species.

Weismann may not have convinced everyone, but he convinced Wells, and Wells was well on the way to becoming one of the most influential writers in the country. George Bernard Shaw took issue with Wells over his friend's acceptance of Weismann's conclusions. Losing faith in Darwinism precisely because of the ways it had been used to political ends, Shaw argued instead that Lamarckism was far from disproven, and he pinned his own faltering faith in the future evolution of humanity upon it accordingly. It was the horror of the First World War that ultimately convinced him that Darwinism held no promise of a brighter tomorrow.

In Britain, Weismann's work did nothing to relieve pressing concerns about evolution and what the future might hold in store for humanity. Weismann's germ-plasm theory undermined the inheritance of acquired characters, while his theory of panmixia threatened evolutionary degeneration for any organisms that were not kept up to the mark by a vigorous natural selection. While the merits of panmixia were debated in the pages of *Nature*, Grant Allen's accessible essay on the ongoing debate on heredity for W. T. Stead's *Review of Reviews* brought it to public attention.[9] It was the publication of Benjamin Kidd's book *Social Evolution* in 1894, however, that did the most to make discussion of panmixia a public concern. Kidd had been deeply impressed by Weismann's work, especially his work on retrogression, and had visited him in Freiburg in 1890.[10] Although he largely agreed with the socialist critique of capitalism, Kidd suggested that Weismann's theory of panmixia undermined the aspirations of many in the socialist movement; he shared Wells's conclusions that the epoch of peace and plenty that Morris had written about, for example, would indeed only lead to social degeneration. Kidd was not a harbinger of doom, however, for while he argued that as a material theory natural selection could only make for competition and its cessation could only bring degeneration, he believed that the spread of religious belief had established a mutual ethical concern among mankind that ameliorated both tendencies. Kidd was as reluctant as Darwin had been to ground the origin of mankind's ethical capacities in self-interested reason; rather, he contended that the truly progressive forces that had driven mankind's advance were those that operated in spite of and in opposition to the individual man's rational self-interest. Kidd admitted that socialists were morally right to challenge the individualism of capitalist social relations, but argued that their alternative could only lead to a panmictic degeneration. Religion, which to Kidd's mind was a non-rational aspect of mankind's existence, would prevent the otherwise inevitable collapse of Western civilization. As Kidd's biographer, Crook, has pointed out, it was this attempt to provide an evolutionary justification for religion that made Kidd's book so popular—it was one of many on the subject that sold well.[11]

The popularity of Kidd's book brought Karl Pearson into the fray. He had been critical of the ways in which different advocates of biological degeneration had gone about trying to prove their case, and he was initially skeptical of Weismann's claims as well. More than this, though, he was outraged to see Kidd use Weismann's biology to attempt to undermine socialism in such a fashion. Pearson attacked Kidd in the *Fortnightly Review*, laying out his case in "Socialism and Natural Selection." Pearson argued that of the various factors in natural selection, the in-group competition that Kidd emphasized was in fact much weaker than the competition between groups—mutual aid and sociality did not require any supernatural intervention. He acknowledged Kidd's point that religious belief might have a role to play in enhancing social solidarity, but this did nothing to establish religion as anything more than an anthropological artifact. Solidarity could be achieved much more rationally through socialism, he argued. However, and in light of the researches that his students and colleagues undertook into the relationship between fertility and social class, Pearson ultimately became convinced that Weismann might be right about panmictic degeneration. Evidence of a differential birthrate convinced him that the average quality of the English was in decline. In the context of international competition, Pearson became concerned that even a slight decline might still be enough to see England's defeat by a foreign aggressor. Only the careful management of the population and the subordination of the individual will to the good of the Commonwealth could ensure that England's future was secure.

Although Pearson was by no means alone in advocating what G. R. Searle has called "the quest for national efficiency"[12] as a part of a social-imperial politics that helped set the country on a war footing in the years leading up to 1914, it would be a mistake to see this as the logical outcome of New Liberal accounts of evolution. As much of this study has shown, there were many who followed Darwin's efforts to theorize an expansion of moral considerability to encompass the peoples of all races and all nations. However, by 1914, it was the imperialist view that came to the fore, and in the process made Darwinism an integral part of the nationalist ideology that helped lead the Western world into a conflict of unprecedented scale. Even though Kropotkin continued to argue for pacifism and peace in the name of our evolved capacity for mutual aid, once the fields of Europe had become mired in trench warfare the language of the struggle for existence and the survival of the fittest appeared to describe the realities of the Great War only too well.

This was the beginning of the end of an era, and it did not take long for Darwinism to lose its appeal as a means to social progress. Both Shaw and Wells had initially portrayed the war as the necessary salve that would allow a jaded Western society to arise anew like a phoenix from the flames, but with millions of young men slaughtered to gain a few yards of soil this argument became increasingly hard to maintain. Even as late as 1916, Wells argued in the pamphlet "What Is Coming" that the total destruction of Europe was necessary to clear the way for socialism, a diagnosis that Shaw repeated in *Heartbreak House* the same year. By 1918 though, Shaw had changed his tune, and in "What I Really Said about the War" he confessed that for all the devastation and death the world was further away from any utopian society than it had been before hostilities began. It was certainly not the restorative "War to End Wars" that Wells had initially perceived it to be.[13] This set the stage for Shaw's dispirited and dispiriting preface to *Methuselah*. He had come to doubt whether mankind would ever be capable of solving the problems raised by modern civilization. "Neo-Darwinism in politics had produced a European catastrophe of a magnitude so appalling, and a scope so unpredictable, that as I write these lines in 1920, it is still far from certain whether our civilisation will survive it," he wrote.[14]

The pan-European psychological impact of the war cannot be overstated. Whereas from the 1880s there was a growing concern that British society might be in decline and a few evolutionary biologists debated the extent to which evolution might lead to degeneration in the event that struggle was suspended, the heavy toll of the Western Front made it impossible for anyone to hang on to the optimism and sense of progress that had defined the Victorian era. The end of faith in nature's progress did not end the connections between evolutionary biology and politics, however. Indeed, although Shaw's preface to *Methuselah* was a harsh indictment of neo-Darwinism, it was as much an attempt to rally around Lamarckism and resuscitate both purpose and progress in a social-evolutionary context. He did so with only limited success. Others, as we have seen, turned to eugenic solutions: where nature might fail, science applied to humanity might yet succeed.

One man who was perhaps unique in having thoroughly enjoyed his experiences in the war was the young geneticist J. B. S. Haldane. Although here is not the place for a full discussion of Haldane, it is relevant to point out that despite witnessing the war firsthand—or perhaps in light of experiencing the comradeship among the ranks—he did not rule out the possibility that humans might have evolved to be capable of genuinely other-regarding altruism despite their all-too-evident abilities to maim and kill one another.[15] In what has become his most-famous statement, Haldane addressed exactly this question. In the years after the war, especially given the persistence in the public mind of the association between a "Darwinian ethic" and a self-interested struggle for existence—and despite a century and a half of socialist attempts to show that a Darwinian ethic can be so much more than this—at one of his many public lectures Haldane was pressed by a Christian in the audience on the question of whether as an atheist, socialist, and advocate of Darwinism he would be willing to give his life for his brother. The questioner presumably hoped to draw Haldane into an affirmative statement in order to show that his own altruism was inconsistent with his other commitments. However, Haldane's response showed that he had already given some thought to the evolution of ethics from a genetic perspective, and this doubtless surprised his inquisitor. "No," he answered. "But I would give my life for two brothers or eight cousins"—showing that he was already thinking in terms of what John Maynard Smith would later term kin-selection.[16]

This is provocative, but Haldane's contribution to what has been termed the problem of altruism in evolutionary biology is one part of another story. Haldane, R. A Fisher, George Price, W. D. Hamilton, and John Maynard Smith are but a few of the many who played a significant part in the study of the evolution of altruism in the context of population genetics and theoretical biology in order to describe how altruistic behaviors might not only evolve, but persist, to become what the evolutionary biologist John Maynard Smith and the mathematician George Price termed an "evolutionary stable strategy."[17] In the last pages of *Political Descent* I would like to briefly—and selectively—map that terrain because this illustrates that the same political dynamic that existed in the nineteenth century continued into the twentieth century—and is still being played out today. The biology of the story is well known, but the political side of this debate has still largely yet to be written.

This more-recent history, at least insofar as it has been written, tends to begin with the work of the ornithologist Vero Copner Wynne-Edwards in the 1960s and his book *Animal Dispersion in Relation to Social Behaviour* (1962). In it, Wynne-Edwards had made the case that natural selection might act for the good of the species on the basis that his work revealed that individuals in the bird populations he studied limited their reproductive rates in relation to the size of their flock and the availability of resources.[18] His conclusions were to prove controversial.

In essence, what Wynne-Edwards was proposing was that individuals altered their own behavior, limiting their own reproductive interests for the good of the species as a whole. While Wynne-Edwards was certainly a significant representative of the emphasis that was being placed upon group selection in the 1960s, it is important to recognize that he was but one of a number of scientists and social scientists who popularized group selection and explanations based on the presumption that natural selection favored behaviors that worked "for the good of the species." The work on social behavior of the Austrian ethologists Irenäus Eibl-Eibesfeldt and Konrad Lorenz, the American popular-science writer Robert Ardrey, and the English zoologist and ethologist Desmond Morris had gained a wide audience by the end of the 1960s. Their work, like that of Wynne-Edwards, appealed to the social and environmental politics of the day. However, the fact that Lorenz had aligned himself with the National Socialists in the 1930s and 1940s became a target for critics of group selection in their argument that there was a nefarious and totalitarian political agenda behind the theory.[19] Thus, while opponents of group selection were unclear whether it was a Trojan horse for fascism or for communism, they were united in the belief that it was not only bad politics but bad science. Notable critics included David Lack, William Hamilton, John Maynard Smith, George C. Williams, and Robert Trivers, among others. Significantly, their criticisms were summarized and reiterated by the young evolutionary biologist and popularizer of science Richard Dawkins in what was quickly to become the best-seller *The Selfish Gene* (1976). The general tenor of the criticism was that the kind of explanation that Wynne-Edwards offered was premised upon a number of substantial misunderstandings that were indicative of what Dawkins disparaged as woolly thinking. There was no need to invoke the notion that organisms acted or that selection favored those that acted "for the good of the group." Rather, and here drawing from the work Hamilton, Williams, and Maynard Smith in particular, Dawkins noted that there was a much simpler explanation. All things being equal, selection favored self-interest. However, it acted not upon the self-interest of individuals, as Darwin had largely supposed, but upon genes, as Hamilton had pointed out in 1963.[20] While Hamilton and the political scientist Robert Axelrod continued to argue that a gene's-eye view of evolution clearly demonstrated that altruism could at best only persist in the context of a reciprocally advantageous iterated relationship, in practice it was not until the end of the 1980s that group selectionist theories can truly be said to have fallen out of favor. The popularity of Dawkins's *Selfish Gene* reflected not only his ability to convey complicated scientific ideas and why they should matter to a public audience, but, as Swenson points out, its success also reflected an increased focus upon genetics in biology and echoed the neoconservative politics of the Thatcher-Reagan years.[21]

Since the 1980s, group selection has been out of favor in biology, the scientific community accepting the claims of Dawkins, Williams and Company that it was grounded upon a naive misunderstanding of evolutionary biology. However, group-selectionist explanations in biology have enjoyed something of a renaissance in the last decade and a half. The philosopher Elliott Sober and the biologist David Sloan Wilson, and more recently the myrmecologist and sociobiologist E. O. Wilson, have led a revival of the debate about how we define fitness and thus how we think about evolutionary pressures and processes. Rather than championing selection at the level of the group, the individual, or the gene, they are among a growing number of theorists who believe that selection functions at each of these levels. They argue that if we ignore the interactions between individuals within groups, between individuals and their environment, and between different groups, in favor of focusing upon the change in gene frequency within a population over time, then we see only the outcome of evolution and not its process.

What I find most interesting in this ongoing discussion has been the repeated return to the public arena that has marked this debate. Alongside scholarly publication in scientific journals, key protagonists continue to publish for a popular audience. Just a few of the most-recent book-length studies include David Sloan Wilson's *The Neighborhood Project* (2011), Robert Trivers's *The Folly of Fools* (2011), and E. O. Wilson's *The Social Conquest of Earth* (2012). Moreover, Richard Dawkins, David Sloan Wilson, and the linguist Stephen Pinker have all published articles in the popular press.[22] In each of these works the authors engage explicitly with the political implications of our evolution. They ask again, "What kind of creature is a human being?" and in consequence, "How might we live?" Are we organisms that are selfish even to our very genes? Or are we beings that are capable of real and genuinely other-regarding actions? I hope that the story I have told in *Political Descent* will show us that these authors are only the latest to wrestle with what evolutionary biology has to say to these questions—and that the questions, as well as the answers, are as dependent upon politics as they are upon biology.

AFTERWORD

Engaging the Present

Just as scientific life and political life develop in conversation with one another, so the history of science is undertaken in conversation with present-day politics and the politics of the past. In light of this, I can foresee two questions that this book might prompt. Indeed, I have been asked each of them on a number of occasions when I have talked to people about my work. The first is centered upon the present attempts in America by some Christians to marginalize the study of evolution in public schools. Alongside ongoing attempts to find a way around the First Amendment and so get a religious account of Creation into schools, a second and more-recent strategy has been to argue that Darwin and Darwinism are a dark and dirty politics—to suggest that Darwin was a racist and a sexist and that in consequence evolution is not the sort of thing that we would want to teach our children.

What does *Political Descent* say to this? If anything, in the chapters of this book that deal explicitly with Darwin, I certainly argue that he was a deeply political character and that he consciously constructed his theory of evolution by means of natural selection to vindicate and naturalize the political views that he hoped to see widely adopted. Also, Darwin was certainly a man of his time, and thus his views on race and sex are of his time as well. But—and as other scholars have also made clear—Darwin's views were generally at the most progressive end of Victorian liberal politics. Further, here I make it clear that while Darwin certainly advanced liberal politics, the version of liberalism he favored emphasized collective rather than individual responses to social problems. Far from wanting to see the individualistic struggle for existence or the "survival of the fittest" made the watchword of society, Darwin was quite explicit that he sought to naturalize a liberal politics based upon genuinely altruistic and other-regarding moral sentiments. He could not believe that our highest motives were based in mere self-interest, no matter how enlightened that self-interest might be. Thus, if the nineteenth-century politics of evolution is at all relevant to whether present-day evolutionary biology is a fit subject to be taught in schools—and I am not sure that it is—it seems to me that Darwin's own ambitions of promoting an increasingly tolerant and humanitarian society speak highly in its favor.

The second question that I have been asked regarding this study goes beyond discussion of Darwin to focus on what I might have to say on the relationship between biology and politics in the present. In the context of the expanding popularity of evolutionary psychology, many who care deeply about social justice are concerned about what the political implications of a biology of mind might be. I have commented at least briefly on my views on the relationship between biology and politics in the introduction to this volume, but here I will add that on this point I can say that while I share the opinion of those who believe that a great many of the claims of evolutionary psychologists are overly speculative, I do not think that it has been an uninteresting or fruitless speculation, or indeed that these are questions that we should be afraid of asking.

In *Origin*, Darwin famously wrote: "In the distant future I see open fields for far more important researches. Psychology will be based on a new foundation, that of the necessary acquirement of each mental power and capacity by gradation. Light will be thrown on the origin of man and his history."[1] It is clear to me that the fact of our evolution will certainly have a lot to say about the origins and history of our mental faculties, just as it has had much to say about our morphology. What is significant, though, is that whatever those findings might be, as Darwin himself made clear in *Descent of Man* and as I reiterate in this book, we have evolved to have the capacity for genuinely other-regarding moral sentiments. What I hope is clear from my consideration of Darwin's thoughts on the implications of sexual selection for the perpetuation of altruistic sentiments—even if we were to jettison the Lamarckian aspects of his theory—is that I side with those who are convinced that there are still good grounds for an evolutionary account of genuine altruism. Do I think that biology speaks to the sorts of creatures we are? I do. But I am also aware, as I hope to have shown here, that the science of biology is not an objective endeavor, and thus we need to approach any science of humanity with a deeply skeptical eye, considering that there is so much at stake.

Acknowledgments

I am grateful to a number of people for the contributions they have made to the completion of this book. First and foremost, I am indebted to Paolo Palladino, who was my PhD advisor at Lancaster University between 1997 and 2003. It was through long discussions with Paolo about evolution, socialism, and the meaning of utopia, standing that I was first challenged to think seriously about the historical connections between biology and politics. It was Paolo too who first introduced me to his own doctoral advisor, John Beatty, who was then at the University of Minnesota. At Paolo's behest, and with the financial support of the British Academy, I spent the first of two six-month periods studying with John in the Program in History of Science and Technology at the University of Minnesota, where I met faculty and graduate students whom I have subsequently come to list among my most-valued friends and colleagues. Richard Bellon, Paul Brinkman, Juliet Burba, Chris Eliot, Kevin Francis, Margot Iverson, Mark Largent, Don Optiz, Susan Rensing, and David Sepkoski stand out as having challenged me to think about the politics of biology and for offering warm friendship over the course of two Minnesota winters. It was at John's prompting too that I first encountered the rich intellectual community of scholars who make up the Columbia River History of Science Group. The discussions that I have had and the friendships I have made at the Friday Harbor Laboratories have helped to make me the historian I am today. It was John too, who first encouraged me to participate in the excellent and intellectually stimulating Dibner workshops in the history of biology at Wood's Hole.

I continued to amass intellectual debts in the course of a postdoctoral fellowship at the University of British Columbia in Vancouver. John again introduced me to his new colleagues, and again presented me with opportunities to discuss my work with others, notably with Margaret Schabas, Keith Benson, and Joy Dixon as well as faculty and student members of the Interdisciplinary Nineteenth-Century Studies Group. It was as a result of a chance discussion with Rana Ahmad that I first gave serious attention to the problem of the origin and evolution of altruism. It was during this time too that I met Bernie Lightman and his former students, Erin Jenkins and Jamie Elwick. Bernie has done more to shape my conception of nineteenth-century science than anyone.

Despite this long period of gestation, it was only following my arrival at the University of Oklahoma that I first seriously put pen to paper on *Political Descent.* Colleagues to whom I am particularly indebted are Hunter Heyck, and Stephen Weldon, who read and commented on early drafts of a number of chapters, and Steven Livesey, who offered encouragement and guidance through the practicalities and pitfalls of the timetable of writing a book. Kerry Magruder and JoAnn Palmeri at the University of Oklahoma History of Science Collections have been of real assistance in tracking down sources and providing the majority of the images that illustrate this book. Carilyn Livesey made time in her schedule to scan printed material and images. Graduate and undergraduate students at OU have also forced me to think through my ideas and how best to articulate them and I have had particularly rewarding discussions with the "recently doctored" John Stewart, and with T. Russell Hunter, Ashley Nelson, and Sarah Swenson.

In 2009, the University gave me the funding to invite a number of speakers to take part in a University-wide commemoration of Darwin's life and works. This gave me the opportunity to discuss my thoughts on the politics of evolution with some of the scholars I hold in high regard. Garland Allen, John Beatty, Joe Cain, John Lynch, Michael Ruse, Ken Taylor, Paul White, and John van Wyhe each gave their time to talk to me about their work, and to listen to me talk about mine. I am grateful too to Dean Paul Bell for his unwavering support.

I also deeply appreciate the commitment and patience of Karen Darling, Abby Collier, and Amy Krynak at the University of Chicago Press; of Ruth Steinberg, my copy editor; and of Mark Borrello and two anonymous reviewers, whose comments on the manuscript I found both insightful and challenging. Diane Paul and Erika Milam each also took time to read and comment on some very rough early drafts, while Mike Carter and Lisa Downs read through the final manuscript copy.

As anyone who has written a book will know, it is a process that places practical, intellectual, and emotional demands on others. In this regard, I am grateful to have had the support of those listed above, but I would also like here to acknowledge the patience of my daughters Sativa and Raven, who did without my company for the long hours I spent shut in my office.

Notes

Introduction

1. Hume, *A Treatise of Human Nature*, 353–67, esp. 363; G. Moore, *Principia Ethica.*

2. Trivers, "The Evolution of Reciprocal Altruism."

3. Borrello, *Evolutionary Restraints*, 69.

4. Williams, *Adaptation and Natural Selection*, 92.

5. Ibid., vii; see also 193–95.

6. Trivers, "The Evolution of Reciprocal Altruism"; Smith and Price, "The Logic of Animal Conflict."

7. See Wilson and Sober, "Reintroducing Group Selection to the Human Behavioral Sciences"; Sober and Wilson, *Unto Others*; and Wilson and Wilson, "Rethinking the Theoretical Boundaries of Sociobiology."

8. Elliot Sober and Robert Trivers, personal communications. My conversations with Sober and Trivers confirm this from both sides of the fence.

9. Ridley, *The Origins of Virtue*, 262.

10. See Weikhart, *Hitler's Ethic*; and Weikhart, *From Darwin to Hitler.* See also Arnhart, *Darwinian Conservatism.*

11. Reeves, *John Stuart Mill*, 87.

12. Reid, *Charles James Fox*, 266.

13. It was not until 1855 that duty was removed from newspapers entirely. It should be noted, of course, that many radical papers published unstamped throughout this period, their publishers often suffering fines and imprisonment as a result.

14. Habermas, *The Structural Transformation of the Public Sphere.* Richard Yeo applies the Harbermasian notion of the bourgeois public to the nineteenth century (see *Defining Science*, 28–48); for an endorsement of this approach, see also Cantor et al., *Science in the Nineteenth-Century Periodical*, 4. On the importance of and figures for expanding newspaper publication, see G. Jones, "Spencer and His Circle," 4; and Francis, *Herbert Spencer and the Invention of Modern Life*, 132.

15. Hilton, *The Age of Atonement.*

16. Reeves, *John Stuart Mill*, 87.

17. Mill is quoted in ibid.

18. E. P. Thompson, *The Making of the English Working Class*, 807.

19. The debate over the timing of the development of class and what that meant is a major one in British social history. Just a few of the major protagonists in this debate over the years have been E. P. Thompson in *The Making of the English Working Class* and "The Peculiarities of the English"; Perry Anderson in "The Origin of the Present Crisis" (see p. 23); Tom Nairn in "The English Working Class" (see p. 24); Craig Calhoun in *The Question of Class Struggle*; and Patrick Joyce in *Visions of the People*.

20. Bourdieu, *Distinction*, 53–54, 114–15.

21. E. P. Thompson, "The Peculiarities of the English," 245.

22. Nelson, "Harriet Martineau's Political Economy."

23. Evans, *The Forging of the Modern State*; Evans, *Britain before the Reform Act*; Perkin, *The Origin of Modern English Society*; Briggs, *The Age of Improvement*.

24. Desmond and Moore, *Darwin's Sacred Cause*, 5–7.

25. Bentham's body, fully preserved and dressed in his finest, is on display in a cabinet in the hall of the college he was instrumental in founding.

26. Bentham, *An Introduction to the Principles of Morals and Legislation*, 311n.

27. Malthus, *An Essay on the Principle of Population*.

28. Darwin, *On the Origin of Species by Means of Natural Selection*, 1st ed. (1859), 5. Unless otherwise noted, all subsequent citations to Darwin's *Origin of Species* are to this first edition.

29. Desmond, *The Politics of Evolution*, 6.

30. Ibid.

31. Elliott, "'The Derbyshire Darwinians.'"

32. Desmond, *The Politics of Evolution*, 3.

33. Bowler, *The Non-Darwinian Revolution*; G. Jones, *Social Darwinism in English Thought*; Lightman, "Darwin and the Popularization of Evolution."

34. Darwin, *On the Origin of Species by Means of Natural Selection*, 1–6.

35. Laura J. Snyder has argued that Whewell is often mischaracterized as a conservative by many of his contemporaries as well as by subsequent historians. He supported the 1832 extension of the franchise but thought that further reform should come only slowly. This caution is what led those like John Stuart Mill, who held more-radical views, to think Whewell more conservative than he was (see Snyder, *Reforming Philosophy*, 28).

36. An evolutionary argument could also be marshaled to support a distinct racial hierarchy, of course, and Darwin himself thought the races of man were varieties that were incipient species.

37. Secord, *Victorian Sensation*.

38. See Snyder, *The Philosophical Breakfast Club*.

39. Hale and Smith, *Negotiating Boundaries*, xxxv.

40. Darwin, *The Voyage of the "Beagle,"* 203, 191.

41. Winch, *Riches and Poverty*; Schabas, *The Natural Origins of Economics*.

42. Darwin, *On the Origin of Species by Means of Natural Selection*, 488.

43. Darwin, *The Descent of Man, and Selection in Relation to Sex*, 1st ed. (1871), 1:71. Unless otherwise noted, all subsequent citations to Darwin's *Descent of Man* are to this edition.

44. *Karl Marx, Friedrich Engels, Collected Works*, 41:232.

45. Shaw, preface to *Back to Methuselah*, lix. See also Oldroyd, *Darwinian Impacts*, 225–43.

46. Collini, *Liberalism and Sociology*, 9.

Chapter One

1. Darwin recorded in his autobiography: "My Descent of Man was published in Feb. 1871. As soon as I had become, in the year 1837 or 1838, convinced that species were mutable productions, I could not avoid the belief that man must come under the same law" (Darwin, *Autobiography*, 130). John C. Greene has noted that annotations in Darwin's copy of Robert G. Latham's *Man and His Migrations* show that he intended to quote some of Latham's "excellent remarks" on the geographic dispersal of man and extermination of intermediate races as a result of ensuing competition. Greene also notes that at the time Darwin read Latham's work he was intending to cover this material in chapter 6 of his big book on natural selection, which he had provisionally entitled "Theory Applied to the Races of Man" (Greene, "Darwin as a Social Evolutionist," 5). See also R. J. Richards, "Darwin on Mind, Morals, and Emotions," in which the author notes that Darwin began his speculations on man "in the initial pages of his first transmutation notebook" (Notebook B, 92). Transcriptions and commentary on Darwin's "M" and "N" notebooks, together with what he termed his "Old and Useless Notes" have been published in Gruber, *Darwin on Man*. Full transcriptions of all of Darwin's post-voyage notebooks are available in *Charles Darwin's Notebooks*, ed. Barrett et al.

2. Darwin, Notebook M, 84, in Gruber, *Darwin on Man*, 281. See also Herbert, "The Place of Man in the Development of Darwin's Theory of Transmutation"; and J. Hodge, "The Notebook Programme and Projects of Darwin's London Years."

3. J. Hodge, "The Notebook Programme and Projects of Darwin's London Years." This point is also made by Sandra Herbert and David Kohn in their introductory essay to *Charles Darwin's Notebooks*, ed. Barrett et al., 8.

4. Van Wyhe, "'My Appointment Received the Sanction of the Admiralty,'" 316–26, esp. 317–18 and 321. Van Wyhe dates the origin of the "companion" myth to J. W. Gruber, "Who Was the *Beagle*'s Naturalist?" It has been most prominently reiterated in S. J. Gould, "Darwin's Sea Change, or Five Years at the Captain's Table"; Desmond and Moore, *Darwin*, 101–4; and Browne, *Charles Darwin: Voyaging*, 144–49.

5. Various figures have been given for the full complement of those who sailed on *Beagle*. Seventy-three is the most frequent figure given, which is based on Darwin's stated recollection, although he also gave slightly different numbers in different accounts. John Woram has reconstructed a detailed list of the ship's company and passenger list, in which he comments on the various numbers that have been suggested. Interestingly, he has identified an additional five men whom both Darwin and FitRoy seem to have overlooked in their various written accounts. This would put the figure at seventy-eight, although the numbers fluctuated in the course of the voyage (see Woram, *Ship's Company, Passenger List, and Fate of HMS "Beagle*," http://www.rockvillepress.com/tierra/texts/CREWLIST.HTM).

6. A. Secord, "Corresponding Interests"; Barton, "'Men of Science'."

7. Van Wyhe, "'My Appointment Received the Sanction of the Admiralty'", 325.

8. See S. J. Gould, "Darwin's Sea Change, or Five Years at the Captain's Table."

9. Schabas, *The Natural Origins of Economics*.

10. Desmond and Moore, *Darwin*, 267.

11. Darwin, *Journal of Researches into the Natural History and Geology of the Countries Visited during the Voyage of H.M.S. "Beagle" Round the World*, 2:305.

12. Desmond and Moore, *Darwin*, 33–40.

13. Ibid., 70–71.

14. Young, "Malthus and the Evolutionists," 24.

15. Darwin, *Life and Letters*, 27.

16. Browne, *Charles Darwin: Voyaging*, 234.

17. Darwin, *Narrative of the Surveying Voyages of His Majesty's Ships "Adventure" and "Beagle" between the Years 1826 and 1836*, 3:228.

18. Gayon, *Darwin's Struggle for Survival*, 36–77.

19. *Charles Darwin's "Beagle" Diary*, 223.

20. Darwin, *Narrative of the Surveying Voyages of His Majesty's Ships "Adventure" and "Beagle" between the Years 1826 and 1836*, 3:234.

21. *Charles Darwin's "Beagle" Diary*, 223.

22. Ibid., 220.

23. Darwin, *Narrative of the Surveying Voyages of His Majesty's Ships "Adventure" and "Beagle" between the Years 1826 and 1836*, 3:222, 234.

24. Ibid., 3:235–36.

25. Fitzroy, *Narrative of the Surveying Voyages of His Majesty's Ships "Adventure" and "Beagle" between the Years 1826 and 1836*, 2:232.

26. Keynes notes that the story of Fuegian cannibalism is absent from Darwin's *Beagle* diary, the first recorded account coming only in the 1845 edition of the *Journal of Researches into the Natural History and Geology of the Countries Visited during the Voyage of H.M.S. "Beagle" Round the World* (see *Charles Darwin's "Beagle" Diary*, 224). For a thorough publication history of the *Journal of Researches*, also published under the title *The Voyage of the "Beagle,"* see R. B. Freeman, bibliographical introduction to *Journal of Researches*, http://darwin-online.org.uk/EditorialIntroductions/Freeman_JournalofResearches.html.

27. Darwin, *Narrative of the Surveying Voyages of His Majesty's Ships "Adventure" and "Beagle" between the Years 1826 and 1836*, 3:236.

28. Darwin, *Journal of Researches into the Natural History and Geology of the Countries Visited during the Voyage of H.M.S. "Beagle" Round the World*, 2:237.

29. *The Autobiography of Charles Darwin*, 49. Both in *Origin of Species* and afterward, Darwin repeatedly sought to distance his own ideas from those of Lamarck and his grandfather. This was certainly in order to underline his originality, but also to distance himself from the radical associations of the former's transmutationist theories.

30. McNeil, *Under the Banner of Science*.

31. Elliott, *The Derby Philosophers*. See also Elliott, "Erasmus Darwin, Herbert Spencer, and the Origins of the Evolutionary Worldview in British Provincial Scientific Culture."

32. McNeil, *Under the Banner of Science*, 1–7.

33. G. Jones, "Spencer and His Circle"; R. J. Richards, *Darwin and the Emergence of Evolutionary Theories of Mind and Behavior*, esp. 31–39.

34. Erasmus Darwin, *The Temple of Nature*, 14.

35. Godwin, *Enquiry concerning Political Justice*, in *The Anarchist Writings of William Godwin*, 61.

36. Erasmus Darwin, *The Temple of Nature*, 9, 77.

37. Schabas, *The Natural Origins of Economics*; Winch, *Riches and Poverty*.

38. Winch, *Riches and Poverty*, 199. Winch draws attention to E. P. Thompson's contrast between what he termed the "moral economy" of eighteenth-century England and what he perceived to be an amoral capitalist political economy that was based exclusively upon money (see Thompson, "The Moral Economy of the English Crowd in the Eighteenth Century").

39. Peel, *Herbert Spencer*, 61.

40. Schweber, "Darwin and the Political Economists," esp. 258–59.

41. Browne, *Charles Darwin: Voyaging*, 8.

42. *Charles Darwin's "Beagle" Diary*, 223.

43. Darwin, *Narrative of the Surveying Voyages of His Majesty's Ships "Adventure" and "Beagle" between the Years 1826 and 1836*, 3:242. It is important to note here too that Erasmus Darwin had grounded the origin of social feeling and sentiment in the awakening of sexual desire and the ties that united man and woman (see Canto II of *The Temple of Nature*).

44. Nelson, "Harriet Martineau's Political Economy." At the March 2013 meeting of the Columbia River History of Science Group at the Friday Harbor Laboratories, Nelson presented significant work on Martineau's role as an intellectual broker and patron in London Society throughout the 1830s and into the 1850s.

45. Darwin, *The Voyage of the "Beagle,"* 242. It is worth pointing out here that in making these comments about the total lack of the concept of private property among the Fuegians, Darwin clearly disregarded the apparently contrary observations he had made of the natives while the crew had been distributing gifts to them in their boats and bartering trinkets for the fish they had caught. Darwin had witnessed a scene that indicated not only an appreciation of the concept of trade, but of the principle of justice that made it possible. Only a few pages prior to his harsh conclusions about the Fuegians, he had noted that "if any present was designed for one canoe, and it fell near another, it was invariably given to the right owner" (ibid., 242).

46. Darwin's recollections are slightly erroneous over the date. His notebooks indicate that he read Malthus between September 28 and October 12 of that year.

47. Darwin, *Autobiography*, 120.

48. For an account of this debate, see Mayr, *One Long Argument*, 73. As Sandra Herbert and David Kohn note, a significant number of pages that Darwin had excised from his notebooks and set aside for later use were found in 1961 and 1967 (*Charles Darwin's Notebooks*, ed. Barrett et al., 7).

49. Mayr, *One Long Argument*, 72–73. See also De Beer, "The Origin of Darwin's Ideas on Evolution and Natural Selection"; S. Smith, "The Origin of 'The Origin'"; Gruber, *Darwin on Man*; Kohn, "Theories to Work By"; and Limoges, *La selection naturelle*.

50. *Charles Darwin's Notebooks*, ed. Barrett et al., 83, 167.

51. Ibid., 83.

52. Herbert, "The Place of Man in the Development of Darwin's Theory of Transmutation. Part II," 216.

53. Schweber, "The Origin of the 'Origin' Revisited," 231–32.

54. *Charles Darwin's Notebooks*, ed. Barrett et al., 375.

55. Ibid., 375n.

56. Darwin, *On the Origin of Species by Means of Natural Selection*, 62; Lyell, *Principles of Geology*, 2:131.

57. *Charles Darwin's Notebooks*, ed. Barrett et al., 375.

58. Depew and Weber, *Darwinism Evolving*, chap. 5, esp. 127.

59. Sulloway, "Geographic Isolation in Darwin's Thinking," 37–38; Beatty, "Speaking of Species."

60. Sulloway, "Geographic Isolation in Darwin's Thinking," 31–32.

61. On Cuvier, see Rudwick, *Bursting the Limits of Time*, 349–415; and Wilkins, *Species*, 117. On Owen, see Amundson, "Richard Owen and Animal Form"; and Bowler, *Evolution*, 124–29.

62. Georges Cuvier, *Le règne animal . . .*, quoted in Ruse, *Darwin and Design*, 61.

63. Rudwick, *Bursting the Limits of Time*, 359, 368.

64. Ruse, *The Darwinian Revolution*, 12–15.

65. Rudwick, *Bursting the Limits of Time*, 392–99.

66. Jameson, introduction to *An Essay on the Theory of the Earth*, by George Cuvier, esp. ix.

67. See Snyder, *The Philosophical Breakfast Club*.

68. See the introductory essay in Hale and Smith, *Negotiating Boundaries*, xxxi–xlv, esp. xxxv–xxxvi, and also pp. 13–15.

69. Ibid., xxxv.

70. William Whewell (1845), quoted in Ruse, *Darwin and Design*, 80.

71. Jody Hey has recently argued that there is a significant difference between what Mayr terms "population thinking" in this instance—i.e., the emphasis upon variation among individuals in a population—and the concept of a population as a biological entity. Hey points out that, paradoxically, population thinking is therefore not about populations at all (see Hey, "Regarding the Confusion between the Population Concept and Mayr's 'Population Thinking'").

72. *Charles Darwin's Notebooks*, ed. Barrett et al., 375–76. Comparing the transcription in Barrett with that on the Darwin Online website, which also reproduces a scanned copy of the original, it seems that Darwin wrote, "The final cause of all this wedging, must be to sort out proper structure and adapt it to change," and not, as Barrett et al. contend, "The final cause of all this wedgings, must be to sort out proper structure and adapt it to changes" (see Van Wyhe, *The Complete Work of Charles Darwin Online*, http://darwin-online.org.uk/content/frameset?viewtype=side&itemID=CUL-DAR123.-&pageseq=1).

73. Mayr, *One Long Argument*, 46.

74. Schweber, "Darwin and the Political Economists," 196. Young makes this point too, citing the work of Sydney Smith in doing so (see Young "Malthus and the Evolutionists," 26).

75. Desmond, *The Politics of Evolution*; Elliott, *The Derby Philosophers*.

76. See Peterson, "The Malthus-Godwin Debate, Then and Now."

77. Young, "Malthus and the Evolutionists," 26.

78. See Peterson, "Malthus and the Intellectuals"; and Schweber, "Darwin and the Political Economists," 196.

79. Robert Southey, quoted in Peterson, "Malthus and the Intellectuals," 472.

80. Hazlitt, *The Spirit of the Age, or, Contemporary Portraits*, 156.

81. Hazlitt, *A Reply to the Essay on Population*. See also Albrecht, "Hazlitt and Malthus."

82. Cobbett, *Advice to Young Men and (Incidentally) to Young Women*, 84.

83. Desmond, *The Politics of Evolution*, 14.

84. *Autobiography of John Stuart Mill*, 89.

85. Frances Darwin, *The Life of Charles Darwin*, 30; Darwin, *Autobiography*, 82.

86. Darwin, *The Zoology of the Voyage of HMS "Beagle,"* http://darwin-online.org.uk/EditorialIntroductions/Freeman_ZoologyOfBeagle.html.

87. *Charles Darwin's Notebooks*, ed. Barrett et al., 10.

88. Young, Malthus and the Evolutionists.

89. Desmond and Moore, *Darwin*, 248.

90. *The Correspondence of Charles Darwin* [hereafter cited as *CCD*], vol. 4, appendix 4, 447.

91. According to Darwin's reading notebooks, he read Godwin's 1798 *Memoir of the Author of "A Vindication of the Rights of Woman": Of Population. An Enquiry Concerning the Power of Increase in the Numbers of Mankind, Being an Answer to Mr. Malthus' Essay on That Subject* (1820)*; and Transfusion. By the Late William Godwin, Junior* (1835) (see *CCD*, vol. 4, appendix 4, 524).

92. Young, "Malthus and the Evolutionists." This point has also been appreciated by Gruber in *Darwin on Man.* On Godwin's influence on Malthus, see Petersen, "The Malthus-Godwin Debate, Then and Now." It is well to note, however, that Darwin would change his tune on this point when he did eventually come to publish on man.

93. Darwin, *Autobiography*, 120.

94. Darwin, *On the Origin of Species by Means of Natural Selection*, 5, 63.

95. Ibid., 1–3.

96. Charles Darwin to Charles Lyell, 6 June 1860, *CCD*, 8:242.

97. Desmond and Moore, *Darwin*, 267.

98. Karl Marx to Friedrich Engels, 18 June 1862, in *Karl Marx, Frederick Engels: Collected Works*, 41:381.

99. T. H. Huxley, "The Origin of Species."

100. Reports of the successful testing of the Whitworth had appeared in the recent press (see the *Times* [London], 23 April 1859).

101. Charles Darwin to Charles Lyell, 4 May 1860, *CCD*, 8:189.

102. *Manchester Guardian*, 20 April 1860, 4.

103. Briggs, *Victorian Cities*, 125.

104. Ibid., 90.

105. McNeil, *Under the Banner of Science*, 66.

106. Darwin, *On the Origin of Species by Means of Natural Selection*, 62.

107. Ibid., 62–63.

108. Ibid., 62.

109. Malthus, *An Essay on the Principle of Population, or, A View of Its Past and Present Effects on Human Happiness*, 1:6.

110. *Charles Darwin's Notebooks*, ed. Barrett et al., 375; Darwin, *On the Origin of Species by Means of Natural Selection*, 62.

111. Darwin, *On the Origin of Species by Means of Natural Selection*, 75.

112. Darwin, *Autobiography*, 120–21.

113. Schweber, "Darwin and the Political Economists," 214; Charles Darwin to Joseph Dalton Hooker, 22 August 1857, *CCD*, 6:443–44; Charles Darwin to Asa Gray, 5 September 1857, *CCD*, 6:445–50; Darwin, *On the Origin of Species by Means of Natural Selection*, 116 (quotation).

114. Darwin, *On the Origin of Species by Means of Natural Selection*, 93. Schabas points out that there are significant differences between Milne Edwards's ideas and those that Smith articulated in *Wealth of Nations.* However, I believe Schweber is correct in his assertion that the division of labor in Darwin's theory of the divergence of character owes more to Smith than to Milne Edwards, despite the fact that it is the latter he cites.

115. Schweber, "Darwin and the Political Economists," 195–289.

116. Paul and Beatty, "Discarding Dichotomies, Creating Community."

117. Ibid.

118. Schabas, "The Greyhound and the Mastiff."

119. Schweber, "Darwin and the Political Economists," 259.

120. Schweber, "Darwin and the Political Economists," 231.

121. Schweber, "Scientists as Intellectuals", 13.

122. Darwin lists Stewart under books to be read [Dar 119:3v] (*CCD*, 4:438). The bracketed information, here and elsewhere, is the notation assigned by the Darwin Archive at the Cambridge University Library. See also Stewart, "An Account of the Life and Writings of Adam Smith LL.D."

123. Gruber, *Darwin on Man*, 296.

124. For Darwin's reading lists, see *CCD*, vol. 4, appendix 4. Darwin lists Stewart under books to be read [Dar 119:3v] (*CCD*, 4:438), and he lists Smith's *Theory of Moral Sentiments* as having been read [Dar 119:12a] (*CCD*, 4:465). Darwin recorded in his Notebook M that he had read Dugald Stewart on Smith the late August of 1838 (Darwin, Notebook M, 108, in *Charles Darwin's Notebooks*, ed. Barrett et al., 546). Silvan Schweber has argued that Smith's explanation of the division of labor from his *Wealth of Nations* was central to Darwin's theory of divergence (Schweber, "Darwin and the Political Economists, 265). Schabas, however, has pointed out that even if, as is likely, Darwin imbibed Smith's theory of the division of labor from his reading and discussion of political economy at home and in society, it seems strange that he did not cite Smith's influence in this regard (Schabas, "Ricardo Naturalized," 48). This is a point which is echoed by Diane B. Paul and John Beatty ("Discarding Dichotomies, Creating Community"). I shall argue in chapter 3 that Smith's *Theory of Moral Sentiments* was important for Darwin's account of human evolution later on.

125. Darwin, *On the Origin of Species by Means of Natural Selection*, 81.

126. Dov Ospovat cites David Kohn's work to support his own contention that this was a significant break from a conventional natural-historical understanding of adaptation and of the economy of nature, which only occurred after he had read Malthus, and not, as Camille Limoges had suggested, at an earlier period of Darwin's thinking about the nature of adaptation. Prior to reading Malthus, Darwin still characterized nature in terms of harmony and balance (see Ospovat, *The Development of Darwin's Theory*, 37–38).

127. Darwin, *On the Origin of Species by Means of Natural Selection*, 112.

128. Ibid., 90–91.

129. R. J. Richards, *Darwin and the Emergence of Evolutionary Theories of Mind and Behavior*, 146–48; Van Wyhe, "Mind the Gap." Darwin also notes that he was also held up by the difficulty of explaining the divergence of character (Darwin, *Autobiography*, 120).

130. Darwin, *Autobiography*, 120.

131. Charles Darwin to Alfred Russel Wallace, 22 December 1857, in *CCD*, 6:515.

132. *The Red Notebook of Charles Darwin*, 80.

133. Darwin, Notebook M, 74, in *Charles Darwin's Notebooks*, ed. Barret et al., 536; and Darwin, Notebook M, 57, ibid., 532.

134. Charles Darwin to Charles Lyell, 11 October 1859, *CCD*, 7:345.

135. Ibid.

136. Darwin, *On the Origin of Species by Means of Natural Selection*, 488.

137. He was certainly later to dismiss the attempts made by German socialists to adopt his theory, writing to Dr. Scherzer of the "foolish idea [that] seems to prevail in Germany on the connection between Socialism and Evolution through Natural Selection" (Charles Darwin to Dr. Scherzer, 26 December 1879, in *The Life and Letters of Charles Darwin*, 2:413.).

138. Duke of Argyll, *The Reign of Law*, 157.

139. Charles Darwin to Thomas Huxley, 18 February 1863, *CCD*, 11:148.

140. Lyell, *The Geological Evidences of the Antiquity of Man*, 505.

141. Charles Darwin to Charles Lyell, 4 February 1863, *CCD*, 11:114.

142. Charles Darwin to Asa Gray, 23 February 1863, *CCD*, 11:166; Charles Darwin to J. D. Hooker, 24[–25] February 1863, *CCD*, 11:172–76.

143. Charles Darwin to J. D. Hooker, 24[–25] February 1863, *CCD*, 11:174.

144. Charles Darwin to Alfred Russel Wallace, 26 January 1870, *CCD*, 18:18; Wallace, "The Origin of Human Races and the Antiquity of Man Deduced from the Theory of 'Natural Selec-

tion'". Joel S. Schwartz has contended that Darwin was skeptical about Wallace's 1864 paper, being more interested in Wallace's recent paper on variation. However, the fact that Darwin wrote to his close friends and colleagues in praise of Wallace's paper suggests otherwise (Schwartz, "Darwin, Wallace, and the *Descent of Man*," 272).

145. J. D. Hooker to Charles Darwin, 14 May 1864, *CCD*, 12:182. See also Darwin's reply, 15 May 1864, in which he agrees that Wallace had written "a capital paper" (*CCD*, 12:183).

146. Alfred Russel Wallace to Charles Darwin, 10 May 1864, *CCD*, 12:173–74.

147. Charles Darwin to Alfred Russel Wallace, 28 May 1864, *CCD*, 12:216. See also Kottler, "Alfred Russel Wallace, the Origin of Man, and Spiritualism"; Schwartz, "Darwin, Wallace, and 'The Descent of Man'"; and Ruse, "Alfred Russel Wallace, the Discovery of Natural Selection, and the Origins of Humankind." The main sources for Wallace are *My Life: A Record of Events and Opinions*, and *Alfred Russel Wallace: Letters and Reminiscences*. See also Shermer, *In Darwin's Shadow*; Raby, *Alfred Russel Wallace*; and Martin Fichman, *An Elusive Victorian*.

148. Wallace primarily intended his paper to address the issue of the unity of mankind, and thus he was entering the lion's den to deliver such a paper to the Anthropological Society, the majority of whom were ardent polygenists. According to the full account of Wallace's paper in the *Anthropological*, his presentation made few converts, even though he placed the point of divergence so far in the past that one might still suppose significant differences in intellectual development between the races (see *CCD*, vol. 12, n. 7; and Stocking, *Victorian Anthropology*, 245–57).

149. Wallace, "The Origin of Human Races and the Antiquity of Man Deduced from the Theory of 'Natural Selection,'" clxii.

150. Ibid.

151. Ibid.

152. Wallace, "Development of the Human Races under the Law of Natural Selection," 317.

153. Stocking, *Victorian Anthropology*, 238–73. Sex as well as race was a dividing issue between the Ethnological Society and Anthropological Society (see E. Richards, "Redrawing the Boundaries," 119–42, esp. 126–27).

154. Wallace, "The Origin of Human Races and the Antiquity of Man Deduced from the Theory of 'Natural Selection,'", clxi, clxvi.

155. Ibid.

156. Ibid., clxiv.

157. The boisterous discussion that followed Wallace's presentation is reported in detail on pp. clxx–clxxxvii.

158. Charles Darwin to Joseph Dalton Hooker, 22 May 1864, *CCD*, 12:204.

159. Ibid.

160. Wallace, "Development of Human Races under the Law of Natural Selection," 313.

161. Wallace, "The Origin of Human Races and the Antiquity of Man Deduced from the Theory of 'Natural Selection,'" clxix.

162. Charles Darwin to J. D. Hooker, 22 May 1864, *CCD*, 12:204.

163. On Wallace's position on the evolution of man, see Raby, *Alfred Russel Wallace*; Kotler, "Alfred Russel Wallace, the Origin of Man, and Spiritualism"; and R. J. Richards, *Darwin and the Emergence of Evolutionary Theories of Mind and Behavior*, 161–84.

164. Alfred Russel Wallace to Charles Darwin, 24 March 1869, *CCD*, 17:155.

165. Charles Darwin to Alfred Russel Wallace, 27 March 1869, *CCD*, 17:157. See also Raby, *Alfred Russel Wallace*, 201.

166. [Wallace], "Geological Climates and the Origin of Species."

167. Raby, *Alfred Russel Wallace*, 202.

168. Wallace, "The Limits of Natural Selection as Applied to Man," 359.

169. Wallace, "Geological Climates and the Origin of Species," 394.

170. Charles Darwin to Alfred Russel Wallace, 14 April 1869, *CCD*, 17:175.

171. Charles Darwin to Alfred Russel Wallace, 26 January 1870, *CCD*, 18:17.

172. Alfred Russel Wallace to Charles Darwin, 2 July 1866, *CCD*, 14:227.

173. Darwin, *On the Origin of Species by Means of Natural Selection*, 5th ed., 91.

174. Charles Darwin to Alfred Russel Wallace, 28 May 1864, *CCD*, 12:216–19 (Darwin offers Wallace his notes on man on p. 217). In "Alfred Russel Wallace, *The Origin of Man*, and Spiritualism," Malcolm Kottler suggests that rather than a genuine offer, Darwin's communication was in fact an attempt to stake a priority claim over Wallace. As Joel S. Schwartz points out, however, Kottler does not provide any evidence to support this view (Schwartz, "Darwin, Wallace, and *The Descent of Man*.")

175. [Mivart], Review of "The Descent of Man, and Selection in Relation to Sex, by Charles Darwin FRS etc."; Mivart, *The Genesis of Species*.

176. Richards, *Darwin and the Emergence of Evolutionary Theories of Mind and Behavior*, 187.

Chapter Two

1. Darwin, *On the Origin of Species by Means of Natural Selection*, 488.

2. Charles Darwin to Charles Lyell, 4 May 1860, *CCD*, 8:189.

3. Desmond, *The Politics of Evolution*; Elliott, *The Derby Philosophers*.

4. Himmelfarb, *Darwin and the Darwinian Revolution*, quoted in R. J. Richards, *Darwin and the Emergence of Evolutionary Theories of Mind and Behavior*, 243.

5. R. J. Richards, *Darwin and the Emergence of Evolutionary Theories of Mind and Behavior*, 243–46.

6. Ruse, *Monad to Man*.

7. See M. W. Taylor, *The Philosophy of Herbert Spencer*; and Francis, *Herbert Spencer and the Invention of Modern Life*.

8. R. J. Richards, *Darwin and the Emergence of Evolutionary Theories of Mind and Behavior*, 241.

9. J. R. Moore, "Herbert Spencer's Henchmen."

10. R. J. Richards, *Darwin and the Emergence of Evolutionary Theories of Mind and Behavior*, 243–44.

11. Hofstadter, *Social Darwinism in American Thought*, 31.

12. Shapin, "A Man with a Plan."

13. Andrew Carnegie, quoted in Shapin, "A Man with a Plan."

14. See chap. 1.

15. R. J. Richards, *Darwin and the Emergence of Evolutionary Theories of Mind and Behavior*, 253–56.

16. Spencer, *The Principles of Sociology*, 3:610; R. J. Richards, *Darwin and the Emergence of Evolutionary Theories of Mind and Behavior*, 266.

17. Uglow, *The Lunar Men*; P. Jones, *Industrial Enlightenment*; Elliott, *The Derby Philosophers*.

18. Elliott, "Erasmus Darwin, Herbert Spencer, and the Origins of the Evolutionary Worldview in British Scientific Culture."

19. Spencer, *An Autobiography*, 1:94.

20. Ibid., 1:92; Peel, *Herbert Spencer*, 7–11; Francis, *Herbert Spencer and the Invention of Modern Life*, 24, 248.

21. The following biography of Spencer's early life is based upon the account given by Michael Taylor in *The Philosophy of Herbert Spencer*, 10–11.

22. George Eduard Biber's *Henry Pestalozzi and His Plan of Education* was the most influential contemporary source of Pestalozzi's ideas in England.

23. Spencer, *An Autobiography*, 1:104.

24. Taylor, *The Philosophy of Herbert Spencer*, 10–11.

25. Spencer, *An Autobiography*, 1:125–26.

26. Ibid., 1:31, 127.

27. G. Jones, "Spencer and his Circle," 2.

28. Spencer, *An Autobiography*, 1:130, 140.

29. McCalman, *Darwin's Armada.*

30. Raby, *Alfred Russel Wallace*, 194.

31. Spencer, *An Autobiography*, 1:200–201.

32. Ibid., 1:507.

33. Wilkins, *Species.* On Lamarck's nominalist views on species as contrasted to the essentialist views of Cuvier, see ibid., 104–5.

34. Ibid., 108–9.

35. Rudwick, *George Cuvier, Fossil Bones, and Geological Catastrophes*, 1. See also Rudwick, *Bursting the Limits of Time*, 353–55.

36. Desmond, *The Politics of Evolution*, 333.

37. Spencer, *An Autobiography*, 1:237.

38. R. J. Richards, *Darwin and the Emergence of Evolutionary Theories of Mind and Behavior*, 262–68.

39. In these early days Spencer's political views were closer to social anarchism than to the individualist libertarianism with which he has subsequently become most closely associated.

40. Spencer, *An Autobiography*, 1:411–12.

41. Ibid., 1:487.

42. Spencer, *Social Statics*, 26.

43. John Stuart Mill would address this point explicitly in Mill and Comte, *"Utilitarianism," "Liberty," "Representative Government,"* chap. 3.

44. R. J. Richards, *Darwin and the Emergence of Evolutionary Theories of Mind and Behavior*, 262–68.

45. Spencer, *Social Statics*, 30–31.

46. Ibid., 30–31.

47. Ibid.

48. Ibid., 24–25.

49. Ibid., 44–45.

50. Spencer, *An Autobiography*, 1:399–400.

51. Spencer, *Social Statics*, 45–46.

52. Ibid., 33–34.

53. Ibid.

54. Spencer, *An Autobiography*, 1:399–400.

55. Ibid., 1:500.

56. R. Smith, "William Carpenter (1813–1885)"; Desmond, *The Politics of Evolution*, 213–14.

57. Secord, *Victorian Sensation*, 465–66. Carpenter would later put his name to several favorable reviews of *Origin of Species* after Darwin encouraged him to do so (see W. B. Carpenter, "Darwin on the Origin of Species" and "The Theory of Development in Nature," both reprinted in Lynch, *Darwin's Theory of Natural Selection*). See also Charles Darwin to W. B. Carpenter, 18 November 1859, *CCD*, 7:378–79; Charles Darwin to W. B. Carpenter, 6 January 1860, *CCD*, 8:21; and Charles Darwin to W. B. Carpenter, 6 April 1860, *CCD*, 8:144.

58. Desmond, *The Politics of Evolution*, 213–20.

59. W. B. Carpenter to John Herschel, 29 November 1839, quoted in ibid., 219.

60. Carpenter had acknowledged in his 1839 *Principles of General and Comparative Physiology* that the suggestion that "acquired powers are sometimes propagated as hereditary instincts"—views that were central to Lamarckian and Geoffroyan biology—"seems probable" (423). On Carpenter's views, his reviews, and his contributions to *Vestiges*, see Desmond, *The Politics of Evolution*, 213–14; and Secord, *Victorian Sensation*, 465–56.

61. Chambers had chosen Churchill not only to distance *Vestiges* from his own publishing business, but because he specialized in medical and scientific works, publishing both the *British and Foreign Medical Review* and the *Lancet*. He could also be counted on to be discrete (see Secord, *Victorian Sensation*, 111–15).

62. Secord, *Victorian Sensation*, 465. The guinea had officially been replaced by the pound as the unit of currency in 1816, but for many years afterward professional fees were still frequently quoted in guineas. Whereas the guinea had been minted in gold and was thus of variable value, in this case the term refers to the amount of twenty-one shillings (the pound was worth twenty shillings).

63. Spencer, *An Autobiography*, 1:445.

64. Francis, *Herbert Spencer and the Invention of Modern Life*, 146. The quotation is from Charles Kingsley to T. H. Huxley, 31 October 1860, Huxley Papers (Imperial College, London), Scientific and General Correspondence, Series 1K, box 19, p. 195.

65. Martineau, *Autobiography*, 624. Martineau would go on to have an increasing financial stake in the *Westminster Review*, paying off creditors in 1854, one of the most aggressive of whom, William B. Hodgson, was hoping thereby to advance her brother, James Martineau.

66. Reeves, *John Stuart Mill*, 1–2. Mill had been arrested in his youth for distributing birth control information.

67. Thomas Doubleday is quoted in Spencer, "A Theory of Population, Deduced from the General Law of Animal Fertility," 470.

68. This natural mechanism for the maintenance of species anticipates the kind of "for-the-good-of-the-species" argument later articulated by Vero Copner Wynne-Edwards in the work that culminated in his *Animal Dispersal in Relation to Social Behaviour*. Wynne-Edwards, however, suggested that populations self-regulated through social conventions informed by what he called "epideictic" behavior—behavior through which populations actively assessed their own population density (see Borrello, *Evolutionary Restraints*, 69).

69. Spencer, "A Theory of Population," 470.

70. Harriet Martineau had read and begun to think seriously about translating Comte in April of 1851 (Martineau, *Autobiography*, 586–87; Spencer, *An Autobiography*, 1:515).

71. Charles Darwin to Herbert Spencer, 23 February 1860, *CCD*, 8:106.

72. Herbert Spencer to Charles Darwin, 22 February 1860, *CCD*, 8:98–99.

73. Spencer, "A Theory of Population," 476.

74. Spencer, *The Principles of Psychology*, 620.

75. Spencer, *The Principles of Sociology*, 2:664.

76. Spencer, *Social Statics*, 452–53.

77. Ibid., 455.

78. Spencer, *An Autobiography*, 1:477.

79. E. Richards, "Redrawing the Boundaries," 121. As Richards notes, this became even more so following the publication of *Descent of Man* and Darwin's views on sexual selection. Richards notes that Spencer's belief in the biological limitations of women informed a conservative view of women's education and their place in society.

80. Spencer, "A Theory of Population," 488.

81. Ibid., 492.

82. Spencer, *Education*, 280–81.

83. Spencer, *An Autobiography*, 1:578.

84. Schiebinger, *The Mind Has No Sex?*; S. J. Gould, *The Mismeasure of Man.*

85. Spencer, *An Autobiography*, 1:539.

86. Spencer, "A Theory of Population," 497–98.

87. Ibid., 498.

88. Malthus, *An Essay on the Principle of Population*, 357.

89. Spencer, "A Theory of Population," 492–93.

90. Charles Darwin to Charles Lyell, 25 February 1860, *CCD*, 8:109–10. C. U. M. Smith has suggested that Darwin's comment is indicative of "perhaps just a touch of anxiety about his own priority" (see Smith, "Evolution and the Problem of the Mind: Part 1. Herbert Spencer," 67).

91. D. Freeman, "The Evolutionary Theories of Charles Darwin and Herbert Spencer" [with comments and replies]; R. J. Richards, *Darwin and the Emergence of Evolutionary Theories of Mind and Behavior*, 243–329.

92. Darwin, *The Origin of Species by Means of Natural Selection*, 6th ed., 428.

93. Spencer, *An Autobiography*, 239.

94. From 1864, the *Leader* became the mouthpiece for a united front of writers who sought to take science to the people, as well as Spencer, who by that time had become a significant public figure. Huxley, Tyndall, Kingsley, Norman Lockyer, and many other notables lent their name—and their money—to the cause.

95. [Herbert Spencer], "The Development Hypothesis," *The Leader*, 20 March 1852. Whewell's argument is made in his *Astronomy and General Physics Considered with Reference to Natural Theology*, which is also quoted as a preface to each edition of Darwin's *Origin of Species.* The quotation reads, "But with regard to the material world, we can at least go so far as this—we can perceive that events are brought about not by insulated interpositions of Divine power, exerted in each particular case, but by the establishment of general laws."

96. Spencer, *An Autobiography*, 1:201.

97. Ibid., 1:438.

98. Elliott, *The Derby Philosophers*, 190–213.

99. Spencer, *An Autobiography*, 1:564–65.

100. Ibid., 2:75.

101. Ibid., 1:538–39.

102. Spencer, *Social Statics*, 23.

103. Snyder, *Reforming Philosophy*, 8.

104. Peel, *Herbert Spencer*, 113–19.

105. Hale and Smith, "William Whewell."

106. This much would become clearer in Mill's 1869 essay *On the Subjection of Women* (see pp. 118–21).

107. A. Smith, *An Inquiry into the Nature and Causes of the Wealth of Nations*, 120.

108. Hawkins, *Social Darwinism in European and American Thought*, 82–103.

109. Spencer, *The Principles of Psychology*, 612.

110. Spencer, *An Autobiography*, 1:546.

111. Spencer responded to charges that *The Principles of Psychology* advanced atheism in a letter to the *Nonconformist*, 23 January 1856 (quoted in Francis, *Herbert Spencer and the Invention of Modern Life*, 112).

112. Spencer would be drawn into debate over these issues repeatedly in the 1880s, one such notable exchange between Spencer and Frederic Harrison occurring in the pages of both *Nineteenth Century* and the *Pall Mall Gazette* (see Spencer and Harrison, *The Insuppressible Book*).

113. Brewster, Review of *Vestiges of the Natural History of Creation*, by Robert Chambers.

114. Lightman, "Robert Elsmere and the Agnostic Crises of Faith," 301–3.

115. Spencer, *The Principles of Psychology*, 601.

116. Ibid., 602.

117. Ibid.

118. Ibid., 274.

119. Charles Darwin to Herbert Spencer, 2 February 1860, *CCD*, 8:66. On February 10, Spencer had written to a friend: "I am just reading Darwin's book (a copy of which has been searching for me since November and has only just come to hand) and want to send him the 'Population' to show how thoroughly his argument harmonizes with that I have used at the close of that essay" (ibid., 66n6).

120. Spencer, "The Social Organism," 121.

121. Spencer, *The Principles of Sociology*, 1:4.

122. Spencer, *The Principles of Biology 1*, 444.

123. Many thanks to Dr. Paul White at the Darwin Correspondence Project for his help on this point.

124. See p. 139.

125. Spencer, *The Principles of Sociology*, 2:571.

126. Ibid., 2:386–87.

127. Spencer, *The Principles of Psychology*, 620.

128. Spencer, *The Man versus The State.*

129. Spencer, *The Principles of Sociology*, 3:610.

130. Hawkins, *Social Darwinism in European and American Thought*, 88.

131. M. W. Taylor, *The Philosophy of Herbert Spencer*, 1.

Chapter Three

1. Schwartz, "Darwin, Wallace, and the *Descent of Man*," 271–72; Charles Darwin to Alfred Russel Wallace, 22 December 1857, *CCD*, 6:515.

2. Charles Darwin to Alfred Russel Wallace, 28 May 1864, *CCD*, 12:216–19.

3. Charles Darwin to Alfred Russel Wallace, 26 February [1867], *CCD*, 15:109.

4. Schwartz, "Darwin, Wallace and the *Descent of Man*," 288.

5. Charles Darwin to Alfred Russel Wallace, 28 [May 1864], *CCD*, 12:217.

6. Wallace, "The Origin of Human Races and the Antiquity of Man Deduced from the Theory of 'Natural Selection,'" clxii.

7. Bowler, *Evolution*, 207.

8. Comte, *System of Positive Polity*, vol. 1. Originally published in 1851 in French, Comte's work was influential among Ras Darwin's circle, especially with Martineau, Lewes, and John Stuart Mill. For Comte, the science of sociology required the spread of altruism, which was, he believed, the necessary key to human progress. Thomas Dixon (in *The Invention of Altruism*, 129–80) makes the point that Darwin did not frame *Descent of Man* in response to discussion of Comtean altruism per se; rather, he wrote in light of Enlightenment debates about the origin of the moral sentiments. It is notable that while this is the case, a number of Darwin's immediate contemporaries read his work as a contribution to debate about altruism.

9. Wallace, "Geological Climates and the Origin of Species," 391.

10. Ibid., 394.

11. Darwin, *The Descent of Man, and Selection in Relation to Sex*, 1:2–3.

12. Desmond and Moore, *Darwin's Sacred Cause.*

13. R. J. Richards, *Darwin and the Emergence of Evolutionary Theories of Mind and Behavior*, 272.

14. Darwin, *The Descent of Man, and Selection in Relation to Sex*, 1:3.

15. In *Descent of Man*, 1:93, Darwin cited "Mr. Bagehot's articles on the Importance of Obedience and Coherence to Primitive man, in the 'Fortnightly Review,' 1867, p. 529 and 1868, p. 457 & co." The referenced articles are the first two in the aforementioned series of essays entitled "Physics and Politics."

16. Greg, "On the Failure of 'Natural Selection' in the Case of Man."

17. Galton, "Herditary Talent and Character," 320.

18. R. J. Richards, *Darwin and the Emergence of Evolutionary Theories of Mind and Behavior*, 189; Darwin, *The Descent of Man, and Selection in Relation to Sex*, 1:3.

19. In the second edition of *The Descent of* Man, published in 1874, Darwin moved this treatment of morals to fit more comfortably into a later chapter.

20. Darwin, *The Descent of Man, and Selection in Relation to Sex*, 1:98.

21. R. J. Richards, *Darwin and the Emergence of Evolutionary Theories of Mind and Behavior*; Sober and Wilson, *Unto Others*; E. O. Wilson, *The Social Conquest of Earth.*

22. R. J. Richards, *Darwin and the Emergence of Evolutionary Theories of Mind and Behavior*, 14–19; Manier, *The Young Darwin and His Cultural Circle*, 138–46.

23. William Darwin to Charles Darwin, [April 1871?], *CCD*, 19:244.

24. Charles Darwin to John Morley, 14 April 1871, *More Letters of Charles Darwin*, 1:329.

25. See Darwin's reading notebooks, *CCD*, vol. 4, appendix 4, 434–573. Darwin read *On Liberty* on 21 March 1859 (ibid., 444); that he read *Utilitarianism* is evidenced by the footnotes for *The Descent of Man* (ibid., 1:71).

26. R. J. Richards notes that Darwin read Mackintosh starting in the summer of 1838, referring to it in his notebooks throughout much of 1839 and writing a short essay, dated 5 May 1839, on what he read there (see R. J. Richards, *Darwin and the Emergence of Evolutionary Theories of Mind and Behavior*, 114–18; the essay is reprinted as a part of the "Old and Useless Notes," in Gruber, *Darwin on Man*, 398–405).

27. Desmond and Moore, *Darwin*, 283.

28. Darwin, *Autobiography*, 55.

29. Darwin, "On the Moral Sense," 398.

30. Manier, *The Young Darwin and His Cultural Circle*, 142.

31. R. J. Richards, *Darwin and the Emergence of Evolutionary Theories of Mind and Behavior*, 116.

32. On Mackintosh, see Darwin's reading notebooks, *CCD*, vol. 4, appendix 4, 457. On Darwin reading Martineau, see ibid., 455. Darwin also read Martineau's *Hour and the Man* in December 1840 (ibid., 462); her *Eastern Travels* and *Society in America* in 1849, which he thought "curious and interesting" and "moderate," respectively (ibid., 478, 480); and her *Man's Nature and Development* in 1852 (ibid., 488).

33. Manier, *The Young Darwin and His Cultural Circle*, 140.

34. Ibid., 143.

35. Ibid., 112.

36. Ibid., 99.

37. Darwin, *The Descent of Man, and Selection in Relation to Sex*, 1:163.

38. Darwin, Notebook M, 155, in Gruber, *Darwin on Man*, 296.

39. Darwin, *The Descent of Man, and Selection in Relation to Sex*, 1:105.

40. Darwin, *Life and Letters*, 74. William Darwin is quoted in Gruber, *Darwin on Man*, 69. On Emma Darwin, see Loy and Loy, *Emma Darwin*, 58; and Litchfield, *Emma Darwin, A Century of Letters*, 1:45, 156–57.

41. Fawcett, "A Popular Exposition of Mr. Darwin on the Origin of Species."

42. Henry Fawcett to Charles Darwin, 16 July 1861, *CCD*, 9:204.

43. Charles Darwin to Asa Gray, 21 July 1861, *CCD*, 9:214.

44. Charles Darwin to Henry Fawcett, 20 July 1861, *CCD*, 9:212.

45. Darwin, *The Descent of Man, and Selection in Relation to Sex*, 1:71n5.

46. Ibid., 1:71.

47. Dixon, *The Invention of Altruism*, 137–38.

48. Darwin's reading notebook, *CCD*, vol. 4, appendix 2, 496.

49. Joseph Hamburger is keen to remind us that despite significant scholarly interest in Mill's theorization of "other-regarding" sentiments, as Richard Wollheim has pointed out, Mill never actually used the term (see Hamburger, *John Stuart Mill on Liberty and Control*, xii). Nevertheless, it is clear that this was what he intended by "virtue." Hamburger suggests that Mill was interested in the formulation of a social ethic as much as he was interested in the protection and extension of liberty per se. This fits well with my reading of Mill here.

50. Darwin, Notebook M, 108, in Gruber, *Darwin on Man*, 286.

51. John Stuart Mill, *On Liberty*, 65–174 (quotation, 132), in Mill and Comte, *"Utilitarianism," "Liberty," "Representative Government"* .

52. On Tocqueville's influence on Mill, see Halliday, "Some Recent Interpretations of John Stuart Mill."

53. John Stuart Mill, *Utilitarianism*, 1–64 (quotation, 46), in Mill and Comte, *"Utilitarianism," "Liberty," "Representative Government."* Recall, Darwin had charged that Mill had argued that the end of utilitarian ethics was the attainment of the general happiness of the society, not the general good. This passage is but one that suggests that there is plenty of reason to think that Mill was already alive to this point, emphasizing that ultimately the persistence of the community was the bottom line.

54. The emphasis upon a tendency to results is a reflection of an important development in utilitarian ethics that has come to be referred to as "rule utilitarianism," so named to distinguish it from "act utilitarianism." The former allows that actors might act in conformity with a rule—presumably based upon experience—that in "X" kind of circumstances, "Y" kind of actions tend to bring about the best results. This is contrasted with the actor having to attempt calculations for each and every action on its own merits. While for Mill both of these clearly

entailed a reasoned deliberation, action to rules learned by experience clearly did not require highly developed rational decision making. This would be appealing to someone approaching the issue from the perspective of natural history.

55. Darwin, *The Descent of Man, and Selection in Relation to Sex*, 1:71n5.

56. Jeremy Bentham had stated that "prejudice apart, the game of pushpin is of equal value with the arts and sciences of music and poetry" (see Bentham, *The Rationale of Reward*, 206). John Stuart Mill paraphrases this as: "Pushpin is as good as poetry" (see Mill, "Bentham," in *Dissertations and Discussions*, 1:389).

57. Semmel, *John Stuart Mill and the Pursuit of Virtue*, 18.

58. Mill, *The Subjection of Women*, 6.

59. John Stuart Mill, *Utilitarianism*, 24, in Mill and Comte, *"Utilitarianism," "Liberty," "Representative Government."*

60. Ibid., 46.

61. Reeves, *John Stuart Mill*, 333–34.

62. Mill, *The Subjection of Women*, 6.

63. Browne, *Charles Darwin: The Power of Place*, 297.

64. Charles and Emma stayed at Plas Caerdeon from June 12 to July 30 (see *CCD*, 17:277n4).

65. Browne, *Charles Darwin: The Power of Place*, 297.

66. *The Life of Frances Power Cobbe, by Herself*, 2:445. Evelleen Richards suggests that the whole exchange between Darwin and Cobbe took place as "a shouted discussion" of Mill "across a bramble patch," but Emma's comments suggest that the Darwins also hosted her at Plas Caerdeon (E. Richards, "Redrawing the Boundaries," 129).

67. Although both were animal lovers, Cobbe and Darwin were later to find themselves on opposite sides in the debate in the mid-to-late 1870s over the use of animals as research subjects in science.

68. *The Life of Frances Power Cobbe, by Herself*, 2:445. This point was echoed in correspondence between Darwin and Charles Kingsley on the same subject.

69. Cobbe, *An Essay on Intuitive Morals, Being an Attempt to Popularise Ethical Science*, vi.

70. *The Life of Frances Power Cobbe, by Herself*, 2:446. The vestiges of Darwin's having read Kant run through his chapter on the development of the moral sense.

71. Ibid.

72. Ibid., 2:445; Darwin, *The Descent of Man, and Selection in Relation to Sex*, 1:71.

73. Charles Darwin to R. F. Cooke, 30 January 1871, *CCD*, 19:49–50. In response, Murray wrote to Darwin imploring him not to allow Cobbe's review to appear before those of the book's other reviewers, who had not yet even received their review copies of the work (John Murray to Charles Darwin, 18 February 1871, *CCD*, 19:72–73).

74. *The Life of Frances Power Cobbe, by Herself*, 2:447.

75. Darwin, *The Descent of Man, and Selection in Relation to Sex*, 1:98.

76. Ibid., 1:71n5.

77. Smith's *Wealth of Nations* emphasized the benefits of self-interest throughout, and as noted in chapter 1, historians are agreed that it was the emphasis upon the individual that was the main thing Darwin took from reading Malthus.

78. Darwin, Notebook N, 109, in Gruber, *Darwin on Man*, 349.

79. A. Smith, *The Theory of Moral Sentiments*, 28, 33, 35.

80. Ibid., 28, 33.

81. A. Smith, *An Inquiry into the Nature and Causes of the Wealth of Nations*, 119.

82. Darwin, *The Descent of Man, and Selection in Relation to Sex*, 1:81. Darwin's allusion to "the sight of another person enduring hunger, cold, fatigue, revives in us some recollection of these states, which are painful even in idea" is a direct quote from Bain.

83. Ibid., 1:98.

84. Ibid., 1:163.

85. Presumably, one might add, without his first ascertaining whether the child was a close enough relative to make the risk to himself worth taking. J. B. S. Haldane later hypothesized this example in his 1932 book *The Causes of Evolution.*

86. Darwin, *The Descent of Man, and Selection in Relation to Sex*, 1:167.

87. G. Jones, *Social Darwinism in English Thought*, 42.

88. Bagehot, *Physics and Politics, or Thoughts on the Application of the Principles of 'Natural Selection,' and 'Inheritance' to Political Society*, 22.

89. Ibid., 109.

90. G. Jones, *Social Darwinism in English Thought*, 43.

91. Darwin, *The Descent of Man, and Selection in Relation to Sex*, 1:93–94. It is significant that these insights correspond not only with Hamilton's conception of inclusive fitness, but also with George Price's concerns about the evolution of a desire to inflict pain (see Harman, *The Price of Altruism*). The intolerance of outsiders was substantiated in both mice and rats by Konrad Lorenz in his famous study *On Aggression.*

92. Darwin, *The Descent of Man, and Selection in Relation to Sex*, 1:81.

93. A. Smith, *The Theory of Moral Sentiments*, 9.

94. Darwin, *The Descent of Man, and Selection in Relation to Sex*, 1:97–98.

95. A. Smith, *The Theory of Moral Sentiments*, 9.

96. Greta Jones has pointed out that Darwin and Wallace had always read Malthus differently. True to his Owenite upbringing, his deep appreciation for Herbert Spencer, and in line with his enthusiasm for the radical agenda proposed by the California land-reformer, Henry George, Wallace read the Malthusian struggle for existence as resulting from a less-than-perfect adaptation to environmental circumstance rather than overpopulation (Wallace, "The Origin of Human Races and the Antiquity of Man Deduced from the Theory of 'Natural Selection,'" clxii). The former could be resolved through better adaptation in nature or, in the case of humanity, with a better application of technology to nature coupled with a more fair distribution of wealth and land. Indeed—and quite contrary to what we have come to think of as a Darwinian reading of Malthus—in some instances Wallace believed that it was actually a want of population and a proper division of labor that resulted in an insufficiency of food (G. Jones, "Alfred Russel Wallace, Robert Owen, and the Theory of Natural Selection," 87–88). Wallace wrote to Darwin about George's insight into the limitations of Malthus's argument, explaining his belief that the "number of offspring is not *so important* an element in keeping up population of a species, as supply of food and other favourable conditions." He states too that he fears that this fact "will become a formidable weapon in the hands of the enemies of Nat. Selection" (Alfred Russel Wallace to Charles Darwin, 9 July 1881 and 8 [April] 1868, *CCD*, vol. 16, pt. 1, 389).

97. Wallace, "The Origin of Human Races and the Antiquity of Man Deduced from the Theory of 'Natural Selection,'" clxii.

98. Ibid.

99. Darwin, Notebook N, 109, in Gruber, *Darwin on Man*, 349. The notebook's transcriber, Paul Barrett, dates this entry to July 1839 (ibid., 379).

100. Darwin, *The Descent of Man, and Selection in Relation to Sex*, 1:84.

101. Darwin, *On the Origin of Species by Means of Natural Selection*, 90–91; Darwin, *The Descent of Man, and Selection in Relation to Sex*, 1:75.

102. Darwin, *The Descent of Man, and Selection in Relation to Sex*, 1:71.

103. Ibid., 1:71–72.

104. Ibid., 1:72.

105. Charles Kingsley to Charles Darwin, 31 January 1862, *CCD*, 10:62–64.

106. Argyll, *The Reign of Law*, 160.

107. Ibid., 162. Alfred Russel Wallace's review of the first edition of *The Reign of Law* is in Wallace, "Creation by Law".

108. Argyll, *The Reign of Law*, 160–61.

109. Charles Darwin to Charles Kingsley, 10 June 1867, *CCD*, 15:297.

110. While Darwin did find novelty in Wallace's suggestion that selection would come to operate on man's intellectual and moral faculties, this move away from Malthusian individualism was actually clearly present in Malthus's own 1826 essay, as he had incorporated some of the criticisms that Godwin had made of his earlier editions.

111. Darwin and Wallace wrote prior to any serious study of the social life of gorillas.

112. Darwin, *The Descent of Man, and Selection in Relation to Sex*, 1:155–57.

113. Charles Darwin to John Morley, 14 April 1871, *More Letters of Charles Darwin*, 1:329.

114. Darwin, *The Descent of Man, and Selection in Relation to Sex*, 1:163–64.

115. Ibid., 1:105.

116. Ibid., 1:98.

117. Ibid., 93–94.

118. Ibid., 1:166.

119. Darwin, *The Descent of Man, and Selection in Relation to Sex*, 1:167–68n10.

120. Browne, *Charles Darwin: Voyaging*, 78; Desmond and Moore, *Darwin*, 32.

121. Greg, "On the Failure of 'Natural Selection' in the Case of Man," 354.

122. Darwin, *The Descent of Man, and Selection in Relation to Sex*, 1:35.

123. Greg, "On the Failure of 'Natural Selection' in the Case of Man," 356.

124. Wallace, "The Origin of Human Races and the Antiquity of Man Deduced from the Theory of 'Natural Selection,'" clxii.

125. Greg, "On the Failure of 'Natural Selection' in the Case of Man."

126. Ibid., 358, 360.

127. Galton, "Hereditary Talent and Character," 319.

128. Greg, "On the Failure of 'Natural Selection' in the Case of Man," 362.

129. Galton, "Hereditary Talent and Character," 319.

130. Ibid.

131. Ibid., 325.

132. Darwin, *The Descent of Man, and Selection in Relation to Sex*, 1:159.

133. Ibid.

134. Ibid., 1:160.

135. Ibid., 1:164.

136. Ibid., 1:169.

137. Ibid., 1:172.

138. Ibid., 1:172–73.

139. Ibid., 1:177.

140. Paul, "Darwin, Social Darwinism, and Eugenics," 220.

141. Darwin, *The Descent of Man, and Selection in Relation to Sex,* 1:184.

142. Ibid., 1:170.

143. This is a partisan description of reciprocal altruism that presumes intelligent, reasoned, self-interested calculation on the part of the actors. I adopt this reading of it here because this is akin to the model of self-interested moral sentiment that Darwin sought to counter. We should recognize, however, that in the case of unconscious and unreasoning organisms what is referred to as "reciprocal altruism" is a post-hoc description—that is, that the behavior has evolved *because* organisms reciprocate. The organisms neither act out of a consciously self-seeking motive, nor simply because the act of reciprocity rewards the altruist does this make the action intrinsically "selfish." If this were the case, then any other-regarding sentiment that elicited a reciprocal response would have to be relabeled as self-interested.

144. Darwin, *The Descent of Man, and Selection in Relation to Sex,* 1:166.

145. Darwin, *The Descent of Man, and Selection in Relation to Sex,* 1:163.

146. Darwin, *On the Origin of Species by Means of Natural Selection,* 202; Darwin, *The Descent of Man, and Selection in Relation to Sex,* 1:163.

147. Darwin, *On the Origin of Species by Means of Natural Selection,* 237.

148. Darwin, *The Descent of Man, and Selection in Relation to Sex,* 1:161.

149. Ibid., 1:86; A. Smith, *The Theory of Moral Sentiments,* 13–14.

150. Darwin, *The Descent of Man, and Selection in Relation to Sex,* 1:164–66.

151. Darwin, *On the Origin of Species by Means of Natural Selection,* 87–90.

152. Darwin, *The Descent of Man, and Selection in Relation to Sex,* 2nd ed., 1:607.

153. Ibid., 2:338–54. See also Adrian Desmond and James Moore's introduction to *The Descent of Man* in the 2004 Penguin Classics edition of this work.

154. A wonderful treatment of the history of female choice as a component of sexual selection can be found in Milam, *Looking for a Few Good Males.*

155. Ibid., 46–47.

156. Darwin, *The Descent of Man, and Selection in Relation to Sex,* 2:358–60.

157. Ibid., 2:323.

158. Ibid., 2:358–59, 374–75. Evelleen Richards notes that this aspect of female agency was not taken up by many contemporary Darwinians or feminists (E. Richards, "Redrawing the Boundaries," 140n1).

159. Darwin, *The Descent of Man, and Selection in Relation to Sex,* 2:342.

160. Ibid., 2:373.

161. Ibid., 1:86.

162. Ibid., 1:100–101.

163. Ibid., 1:101.

164. Darwin, Notebook B: Transmutation (1837–8), CUL-DAR121, transcribed by Kees Rookmaker, p. 232, in van Wyhe, *The Complete Work of Charles Darwin Online,* http://darwin-online.org.uk/.

165. Cobbe, "Darwinism in Morals," 170–71. The political implications of Darwin's stance were to be made more explicit in the coming years as public debate erupted over vivisection beginning in the mid-1870s. Cobbe had appealed to Darwin to defend animals against such treatment on exactly the basis that animals and men share common ancestry. Darwin, although a passionate advocate of kindliness to animals, declined to endorse Cobbe's memorandum to this effect and worked hard to defend the rights of scientists to vivisect animals unregulated

when necessary. The best source on this remains French, *Antivivisection and Medical Science in Victorian Britain*, esp. 61–110—and for Darwin's refusal to endorse Cobbe's position, 70–71. The accusation that Darwinism was bereft of morality again became central to this debate as critics, including Cobbe ("Darwinism in Morals," 174), attempted to link evolutionary accounts of morals with the lack of humanity that they associated with the practice of vivisection.

166. Cobbe, "Darwinism in Morals," 174.

167. Robert J. Richards has argued that Darwin was ambitious of moving beyond utilitarianism. However, it seems more compelling that Darwin was anxious to move utilitarianism beyond individualism and self-interest. It is surely significant on this point that when Darwin mentions Herbert Spencer as "our greatest philosopher," it is in relation to Spencer's grounding of utility in the social instinct. It might not be inaccurate to suggest here that Darwin is directly comparing Spencer's philosophical account of the moral sense to that proposed by Mill.

168. *The Life of Frances Power Cobbe, by Herself*, 2:446–47. In a lighter vein, Cobbe also criticized the non-teleological bent of Darwin's argument. She somewhat wryly recorded an anecdote that Lyell had told her of the fact that even the most Darwinian of Darwinists could not help but use teleological language despite themselves: "I remember . . . Sir Charles telling me with much glee of two eminent agnostic friends of ours who had been discussing some question for a long time, when one said to the other, 'You are getting very *teleological*!' To which the friend responded, 'I can't help it!'" (ibid., 2:247).

169. No letter from Wedgwood on this subject has been found.

170. Emma Darwin to Frances Power Cobbe, [7 April 1871], *CCD*, 19:263.

171. Emma Darwin to Frances Power Cobbe, [25 February 1871], *CCD*, 19:106.

172. [Morley], "The Descent of Man." See also [Morley], "Mr. Darwin on Conscience."

173. Charles Darwin to Frederick Greenwood, 24 March [1871], *CCD*, 19:208. R. J. Richards misidentifies the editor as Morley himself, rather than Greenwood (see Richards, *Darwin and the Emergence of Evolutionary Theories of Mind and Behavior*, 223). Morley did not take over the editorship until 1880.

174. John Morley to Charles Darwin, 17 April 1871, *CCD*, 19:301–2.

175. Darwin, *On the Origin of Species by Means of Natural Selection*, 79, 490.

176. Darwin had wrestled with this very point though. As John Beatty has shown, while Darwin did believe that, on balance, selection had progressive outcomes—outcomes like the production of humans—of course he could not thereby say that the system itself was amenable to moral evaluation. The details—like ichneumon wasps, cats that play with mice, and children who die before their time—prevented him from it, and he felt that any explanation of these were "best left to chance" (Beatty, "'The Details Left to Chance'").

177. [Morley] "The Descent of Man"; [Morley], "Mr. Darwin on Conscience."

178. R. J. Richards, *Darwin and the Emergence of Evolutionary Theories of Mind and Behavior*, 222–23.

179. William E. Darwin to Charles Darwin [April 1871?], *CCD*, 19:244.

180. Darwin, *The Descent of Man, and Selection in Relation to Sex*, 1:2–3.

181. Ibid., 1:98.

Chapter Four

1. Darwin, *On the Origin of Species by Means of Natural Selection*, 482; Barton, "'Huxley, Lubbock, and Half a Dozen Others'"; Barton, "'An Influential Set of Chaps.'"

2. Hale, "The Search for Purpose in a Post-Darwinian Universe."

3. Freeden, *The New Liberalism.*

4. G. S. Jones, *Outcast London*, 10.

5. Disraeli, *Sybil; or, The Two Nations.*

6. Manier, *The Young Darwin and His Cultural Circle*, 96–101.

7. Matthew, "Smiles, Samuel (1812–1904)."

8. Joyce, *Visions of the People*, 57–58.

9. *The Leeds Times*, Spartacus Educational Website, s.v. Primary Sources, "Physical Force Chartism," http://www.spartacus.schoolnet.co.uk/PRLeedsTimes.htm.

10. Matthew, "Smiles, Samuel (1812–1904)."

11. Ibid.

12. Smiles, *Self Help; with Illustrations of Conduct and Perseverance*, 1.

13. Rooff, *A Hundred Years of Family Welfare.*

14. Matthew, "Smiles, Samuel (1812–1904)."

15. George Holyoake is quoted in Biagini, "Popular Liberals, Gladstonian Finance, and the Debate on Taxation," 142.

16. R. K. Webb, *Modern England*, 324–30.

17. Rees, *Poverty and Public Health.*

18. Joyce, *Visions of the People*, 57–58.

19. G. S. Jones, *Outcast London*, 13.

20. Fawcett, *Pauperism*, 111.

21. G. Pearson, *Hooligan*, 119.

22. J. E. Cairnes, *Some Leading Principles of Political Economy Newly Expounded* (1874), 348, quoted in G. S. Jones, *Outcast London*, 4.

23. Mowatt, *The Charity Organisation Society*, 1–18; G. S. Jones, *Outcast London*, 14–16.

24. Desmond and Moore, *Darwin*, 581.

25. Desmond, "Huxley, Thomas Henry (1825–1895)," 185. Desmond notes that a sheet would be sixteen pages.

26. Ibid., 192–93, 196–97.

27. Mills, "Forbes, Edward (1815–1854)." Forbes, whose health had long been precarious, died only six months later from overwork, much to the shock of the natural history community.

28. Thomas Huxley to Henrietta Huxley, 3 June 1854, quoted in Desmond, "Huxley, Thomas Henry (1825–1895)," 199.

29. Desmond, "Huxley, Thomas Henry (1825–1895)."

30. Ibid.

31. L. Huxley, *Life and Letters of Thomas Huxley*, 3:177n1.

32. Ibid., 3:177.

33. Hale, "Darwin's Other Bulldog."

34. Desmond, *Huxley*, 203.

35. Ibid., 386.

36. On Spencer's early work, see chap. 2; on Darwin's use of Spencer's phrase, see chap 1; on Huxley's response to the Education Act see Desmond, *Huxley*, 443; the Spencer quotation is from Spencer, *An Autobiogrpahy*, 2:232.

37. On the 1870 Education Act, see Murphy, *The Education Act 1870.*

38. See chap. 2 for the development of Spencer's views on education.

39. T. H. Huxley, "Administrative Nihilism," 251.

40. Ibid., 256.
41. Ibid., 257.
42. Ibid., 269.
43. Ibid., 258.
44. Ibid., 263.
45. Ibid., 268.
46. Ibid., 261.
47. Ibid., 270–71.
48. Ibid., 271.
49. Ibid.
50. Ibid., 272.
51. Immanuel Kant is quoted in ibid., 276.
52. Lightman, "Darwin and the Popularization of Evolution."
53. See Clifford, *Lectures and Essays.*
54. Clifford, "On the Scientific Basis of Morals," 123.
55. G. Jones, *Social Darwinism in English Thought,* 41.
56. Geddes and Thomson, *The Evolution of Sex,* 279.
57. MacCarthy, *William Morris,* 463.
58. J. Rae, "State Socialism," 380.
59. Wolfe, *From Radicalism to Socialism,* 16.
60. S. Yeo, "A New Life"; Pierson, *British Socialists.* This is also a clear thread through biographical studies of individuals in the movement, notably in E. P. Thompson's *William Morris: Romantic to Revolutionary;* in Laurence V. Thompson's, *Robert Blatchford: Portrait of an Englishman* and *The Enthusiasts: A Biography of John and Katherine Bruce Glasier;* and throughout the series of essays *Threads Through Time: Writings on History and Autobiography,* compiled by the socialist-feminist historian Sheila Rowbotham.
61. Wolfe, *From Radicalism to Socialism,* 16.
62. S. Yeo, "A New Life," 5; Waters, *British Socialists and the Politics of Popular Culture,* 29–30.
63. S. Yeo, "A New Life," 35–36.
64. J. R. Moore, "Deconstructing Darwinism"; Lightman, "Darwin and the Popularization of Evolution."
65. Besant, *Why I Am a Socialist,* 2.
66. Hale, "Of Mice and Men," 30–37.
67. Desmond's *The Politics of Evolution* demonstrates the prevalence of Lamarckian ideas in English Radicalism.
68. MacKenzie and MacKenzie, *The Fabians,* 191–206.
69. Hyndman, *The Historical Basis of Socialism in England.* Hyndman restated and developed these views over the next forty years, culminating in his book *The Evolution of Revolution.*
70. S. Yeo, "A New Life"; the William Morris quotation, quote in Yeo (p. 19), is from "Fabian Essays in Socialism" (review), *Commonweal,* 25 January 1890.
71. Waters, *British Socialists and the Politics of Popular Culture,* 44–50.
72. Morris, *News from Nowhere; or, An Epoch of Rest,* 134.
73. H. Taylor, *A Claim on the Countryside,* 85.
74. Harry Lowerism in *The Scout* (1895), quoted in Pye, *Fellowship Is Life,* 50.
75. Blatchford, *Merrie England,* 42; E. Carpenter, *Woman, and Her Free Place in Society,* 15.

76. I have made this point elsewhere (see Hale, "Of Mice and Men").

77. Hale, "William Morris, Human Nature, and the Biology of Utopia", 116.

78. Hale, "Of Mice and Men," 31–37.

79. Waters, *British Socialists and the Politics of Popular Culture*, 1–64, 97–130.

80. L. V. Thompson, *Robert Blatchford*, 56–58.

81. S. Yeo, "A New Life," 35–36.

82. Hyndman, *The Record of an Adventurous Life*, 273.

83. All subsequent references to *England for All*, both in the text and here in the Notes, refer to this cheap edition, published by E. W. Allen.

84. Pierson, *British Socialists*, 27.

85. Other significantly influential works include, of course, *England for All*, but also, in the next decade, Robert Blatchford's *Merrie England* (1891), William Morris's *News from Nowhere; or, An Epoch of Rest* (1891), and Edward Carpenter's *Towards Democracy*, which, initially published in four installments between 1882 and 1892, was published as a complete work in 1905. Among those who claim to have been inspired by George were Henry Hyndman, George Bernard Shaw, Robert Blatchford, and Edward Carpenter. Alfred Russel Wallace was also deeply impressed by George's book.

86. George, *Progress and Poverty*, 118.

87. Hyndman, *The Record of an Adventurous Life*; Tsuzuki, *H. M. Hyndman and British Socialism*.

88. Hyndman, *England for All*, 7–8.

89. Ibid., 76.

90. In *People's History and Socialist Theory*, Raphael Samuel has discussed the significance of cultural memory for the development of socialism in England.

91. I consider Morris and *News from Nowhere* in detail in chapter 6.

92. Marx and Engels, *The Communist Manifesto*, 84.

93. Hyndman, *England for All*, "Preface."

94. Periodicals surveyed throughout the 1880s and 1890s included *Justice*, *To-Day*, *Clarion*, and *Commonweal*.

95. Hyndman, *England for All*, 39.

96. Ibid., 34. Even well into the late nineteenth century, many areas of industry contracted out by piece-rate rather than moving production into the factories. Tailoring was one particularly egregious example of this. This is also the subject of Charles Kingsley's "condition of England" novel, *Alton Locke*, and is described tellingly in a chapter entitled "Why Men Turn Chartist." Such work practices continued for decades after Kingsley drew attention to them.

97. Hyndman, *England for All*, 52.

98. Ibid., 70.

99. Ibid., 74.

100. Marx, *Capital*, 263–98; Hyndman, *England for All*, 80–87.

101. Karl Marx to Ferdinand Lassalle, 16 January 1861, in *Karl Marx, Frederick Engels: Collected Works*, 41:245.

102. Marx, *Capital*, 263–98; Hyndman, *England for All*, 80–87.

103. Hyndman, *England for All*, 84–85.

104. Ibid., 93.

105. Ibid., "Preface."

106. Ibid., 107.

107. Pierson, "Bax, Ernest Belfort (1854–1926)."

108. William Morris, quoted in MacCarthy, *William Morris*, 463.

109. MacCarthy, *William Morris*, 462–65 (on the founding of the journal *Justice*, see p. 485).

110. Ibid., 463.

111. Hyndman, *The Historical Basis of Socialism in England*, viii.

112. Ibid., 4n.

113. Ibid., vii.

114. Ibid., 69n.

115. MacKenzie and MacKenzie, *The* Fabians, 16.

116. *The Correspondence of H. G. Wells*, 2:214.

117. MacKenzie and MacKenzie, *The Fabians*, 84.

118. Morris, "Fabian Essays in Socialism."

119. Shaw, *The Fabian Society*.

120. On the "Tory Gold" scandal, see Tsuzuki, *H. M. Hyndman and British Socialism*, 70–72; on agitation among the unemployed, ibid., 78–81. For Morris's perspective on the scandal, see MacCarthy, *William Morris*, 533.

121. E. P. Thompson, *William Morris*, 489–503.

122. McBriar, *Fabian Socialism and English Politics*, 20.

123. Hyndman, *The Economics of Socialism*, 4.

124. Manvell, *The Trial of Annie Besant and Charles Bradlaugh*, 1–10.

125. Bradlaugh and Besant, *Fruits of Philosophy*, 8.

126. Hyndman and Morris, *A Summary of the Principles of Socialism*, 46.

127. Ibid., 46–47.

128. Shaw, "Basis for Socialism: Economic," 2.

129. Ibid., 21–22.

130. B. Webb, *My Apprenticeship*, 20–22.

131. Ibid., 341–42.

132. Beilharz and Nyland, *The Webbs, Fabianism, and Feminism*, 10.

133. Ibid., 18.

134. MacDonald, *The Socialist Movement*, 148–49.

135. Beatrice Potter is quoted in Beilharz and Nyland, *The Webbs, Fabianism, and Feminism*, 30.

136. Peter Beilharz notes that both Sidney and Beatrice Webb were influenced by Auguste Comte, Herbert Spencer, Goethe, and Carlyle. It was from Spencer, however, that they took the notion of organic function; this was particularly so in Beatrice's case. Beatrice also read the utopian-socialist Robert Owen as insisting on the biological principle of functional adaptation in his own writings (see Beilharz and Nyland, *The Webbs, Fabianism, and Feminism*, 27–30; and on their felt obligation for service, ibid., 30).

137. Entry for 18 April 1896, *The Diary of Beatrice Webb*, 2:94.

138. S. Webb, *English Progress towards Social Democracy*, 13.

139. Beilharz and Nyland, *The Webbs, Fabianism, and Feminism*, 10; Den Otter, "Ritchie, David George (1853–1903)."

140. Ritchie, *Darwinism and Politics*, 1–3.

141. Ibid., 9–10.

142. Ibid., 12.

143. Ibid., 38–41.

144. Ibid., 22–23.

145. Ibid., 16.

146. Ibid., 68–69.

147. Ibid., 31.

148. See chap. 5.

149. Ritchie, *Darwinism and Politics*, 92–94.

150. Ibid., 74.

151. Ibid., 67.

152. Ibid., 78, 85.

153. Ibid., 80.

154. Ibid., 82.

155. Ibid., 100.

156. Ibid., 101.

157. Freeden, *The New Liberalism.*

Chapter Five

1. Barton, "'Huxley, Lubbock, and Half a Dozen Others'"; Barton, "'An Influential Set of Chaps'"; Jensen, "The X Club."

2. Barton, "'Huxley, Lubbock, and Half a Dozen Others,'" 430.

3. Ibid., 413. See also White, *Thomas Huxley.*

4. Desmond, *Huxley*, 385.

5. In her introduction to the Barnes and Noble edition of *"Evolution and Ethics" and Other Essays*, Sherrie Lyons has restated the view that Huxley was ardently opposed to reading politics from biology. Here, I argue that while he certainly denied that one should read the ways of nature as endorsing any form of human behavior, this did not mean that he did not conceive of humanity as having a human nature that was shaped by these forces and that needed restraining by education and law.

6. Todes, *Darwin without Malthus*, 123–42; also, see below.

7. Desmond, *Huxley*, 548.

8. Tsuzuki, *H. M. Hyndman and British Socialism.*

9. T. H. Huxley, "A Liberal Education and Where to Find It"; T. H. Huxley, "Administrative Nihilism," 253–57.

10. Beatty and Hale, "*Water Babies.*"

11. Hale, "Darwin's Other Bulldog."

12. Thomas Huxley to Henrietta Huxley, 22 March 1861, in L. Huxley, *Life and Letters of Thomas Henry Huxley*, 1:276.

13. H. G. Wells, a student of Huxley's in the mid-1880s, testifies to Huxley's refusal to address the subject in his lectures (see Wells, *Experiment in Autobiography*). See also Ruse, "Thomas Henry Huxley and the Status of Evolution as Science." On Huxley's ambivalence about publication, see Thomas Huxley to Charles Darwin, 2 December 1862, *CCD*, 10:579.

14. Charles Darwin to Thomas Huxley, 28 December 1862, *CCD*, 10:633–35; Charles Darwin to Thomas Huxley, 18 December 1862, ibid., 10:611–13; Charles Darwin to Thomas Huxley, 7 December 1862, ibid., 10:589–90.

15. Charles Darwin to Thomas Huxley, 18 December 1862, *CCD*, 10:611–13.

16. Colp, "The Contacts between Karl Marx and Charles Darwin," 329.

17. Thomas Huxley to Leonard Huxley, 19 October 1887, in L. Huxley, *Life and Letters of Thomas Henry Huxley*, 3:34.

18. Ibid., 3:35.

19. Ibid., 3:42–43.

20. Thomas Huxley to Charles Kingsley, 23 September 1860, ibid., 1:313–20. See also Beatty and Hale, "*Water Babies*."

21. Desmond, *Huxley*, 558.

22. Thomas Huxley to Herbert Spencer, 21 November 1887, in L. Huxley, *Life and Letters of Thomas Henry Huxley*, 3:344.

23. Ibid.

24. Thomas Huxley to Frederick Dyster, November 1887, ibid., 3:45.

25. Thomas Huxley to Herbert Spencer, 21 November 1887, ibid., 3:44.

26. Later editions of this essay appear under the title "The Struggle of Existence in Human Society."

27. T. H. Huxley, "The Struggle for Existence in Human Society," 205–6.

28. Ibid., 195–97.

29. For an analysis of Huxley's particular objections to George, see Douglas, "Huxley's Critique from Social Darwinism."

30. T. H. Huxley, "The Struggle for Existence in Human Society," 199–200.

31. Darwin, *On the Origin of Species by Means of Natural Selection*, 79.

32. Huxley "The Struggle for Existence in Human Society", 199.

33. Darwin, *On the Origin of Species by Means of Natural Selection*, 490.

34. T. H. Huxley, "The Struggle for Existence in Human Society," 199.

35. Thomas Huxley to Thomas Common, 23 March 1894, quoted in Paradis, "'*Evolution and Ethics*' in Its Victorian Context," 46–47.

36. T. H. Huxley, "The Struggle for Existence in Human Society," 202.

37. Ibid., 204.

38. Ibid., 199–200.

39. Ibid., 204.

40. Thomas Huxley to Charles Kingsley, 23 September 1860, in L. Huxley, *Life and Letters of Thomas Henry Huxley*, 1:318.

41. T. H. Huxley, "The Struggle for Existence in Human Society," 205.

42. Ibid.

43. Ibid., 205–6.

44. Ibid., 212.

45. Carlyle, *Past and Present*, 8–9.

46. Wells, *The Time Machine, an Invention*, 21.

47. T. H. Huxley, "The Struggle for Existence in Human Society," 213.

48. Ibid., 215–16.

49. Ibid., 228–31.

50. Ibid., 221.

51. Desmond, *Huxley*, 562.

52. Thomas Huxley to James Knowles, 14 December 1889, in L. Huxley, *Life and Letters of Thomas Henry Huxley*, 3:138–39.

53. Jenkins, "Henry George and the Dragon."

54. Kropotkin, *Memoirs of a Revolutionist*, 499.

55. The notable exceptions include Borrello, *Evolutionary Restraints*, 30–39; Harman, *The Price of Altruism*, 9–37; Todes, *Darwin without Malthus*, esp. 123–42; and Girón, "Kropotkin between Lamarck and Darwin." One might also include Stephen Jay Gould's short popular-science essay, "Kropotkin Was No Crackpot." Kropotkin gets a fair hearing in anarchist scholarship, of course, but there, an adequate treatment of the scientific context is often wanting.

56. There is a good survey of the ways in which historians of science have treated Kropotkin in Fulmer, "Political Biology."

57. Todes, *Darwin without Malthus*, 123–25; Borrello, *Evolutionary Restraints*, 30. Borrello identifies Joel Schwartz's article "Robert Chambers and Thomas Henry Huxley, Science Correspondents" as exemplary, but Schwartz is far from alone in this.

58. Todes, *Darwin without Malthus*, 126.

59. Ibid., 127.

60. Kropotkin, *Memoirs of a Revolutionist*, 97–98.

61. Borrello, *Evolutionary Restraints*, 30.

62. Kropotkin, *Memoirs of a Revolutionist*, 235.

63. Ibid., 49–59.

64. Ibid., 216.

65. Ibid., 217.

66. Ibid., 116–17.

67. Ibid., 218–328.

68. Ibid., 208–27.

69. Ibid., 330–34.

70. Ibid., 343, 360.

71. Ibid., 351.

72. *Freedom* is still published today.

73. Walter, "Wilson, Charlotte Mary (1854–1944)."

74. Kropotkin, "Charles Darwin."

75. Henry Walter Bates to Peter Kropotkin, quoted in Kropotkin, *Mutual Aid*, 17.

76. These included "The Theory of Evolution and Mutual Aid"; "The Direct Action of Environment on Plants"; "The Response of Animals to Their Environment"; "Inheritance of Acquired Characters"; "Inherited Variation in Plants"; and "Inherited Variation in Animals." They have been reprinted in Kropotkin, *Evolution and Environment*, and also in Kropotkin, "Direct Action of Environment and Evolution."

77. See Spencer, "The Inadequacy of 'Natural Selection'"; Weismann, "The All-Sufficiency of Natural Selection. A Reply to Herbert Spencer"; and Romanes, "Mr. Herbert Spencer on 'Natural Selection.'" See also Churchill, "The Weismann-Spencer Controversy over the Inheritance of Acquired Characters."

78. Girón, "Kropotkin between Lamarck and Darwin."

79. Although, as Girón notes, E. Ray Lankester, an ardent neo-Darwinian, questioned Kropotkin's competence as a biologist over the matter (Girón, "Kropotkin between Lamarck and Darwin," 205). Still, Knowles, the editor of *Nineteenth Century*, was not above allowing Kropotkin to refute neo-Darwinian conclusions in the pages of that journal (E. Ray Lankester, "Heredity and the Direct Action of the Environment," 484).

80. Lebedev, introduction to *Ethics, Origin and* Development, by Peter Kropotkin.

81. Brackman, *A Delicate Arrangement*, 300.

82. Kropotkin, *Mutual Aid*, 24.

83. Ibid., 25. See also Todes, *Darwin without Malthus*, 104.

84. Todes, *Darwin without Malthus*, 110.

85. Daniel Todes has exposed the extent to which Kropotkin's views were not unique in this. The vast majority of Russian naturalists thought that the English school was much too taken up with the political economy of Malthus and that it colored their objectivity (see Todes, "Darwin's Malthusian Metaphor and Russian Evolutionary Thought"; and Todes, *Darwin without Malthus*).

86. Kropotkin, "Mutual Aid among Animals," 26.

87. Darwin, *On the Origin of Species by Means of Natural Selection*, 121.

88. Todes, *Darwin without Malthus*, 129; S. J. Gould, "Kropotkin Was No Crackpot," 333; Borrello, *Evolutionary Restraints*, 32; Kropotkin, *Mutual Aid*, 27 (quotation).

89. Kropotkin, *Mutual Aid*, 13. As Todes has pointed out, Kropotkin acknowledged that he had been led to his conclusions in part by the sparse environment (Todes, *Darwin without Malthus*, 129). However, as these passages show, it is evident that Kropotkin saw mutual aid even in densely populated landscapes.

90. Whether selection acts at the level of the species, the group, the individual, or the gene subsequently became a central question in evolutionary biology, especially following the publication of William Hamilton's work in the mid-1960s. It has been in the context of writing the history of this debate that recent historians have found cause to return to Kropotkin. Most notable here are Borrello, *Evolutionary Restraints*, and Harman, *The Price of Altruism*. I shall return to this below.

91. Bernard Lightman, in "Darwin and the Popularization of Evolution," has recently observed that Darwin struggled to control the interpretation and application of his ideas by a whole range of popularizers, arguing that they are therefore not accurately described as "Darwinian." It is notable, too, that David Stack in his *The First Darwinian Left* argues that Malthusianism is an essential part of Darwinian theory, and thus that those who denied Malthus were not real Darwinians. Significantly, this definition serves his agenda to legitimate the Fabians while marginalizing both Marxists and those socialists who, denying Malthus, based their strategies on Lamarckian and neo-Lamarckian biology.

92. Kropotkin, *Mutual Aid*, 21.

93. Darwin, *On the Origin of Species by Means of Natural Selection*, 62.

94. Kropotkin, *Mutual Aid*, 64.

95. Ibid., 63.

96. Darwin, *On the Origin of Species by Means of Natural Selection*, 75.

97. Ibid., 65.

98. Huxley, "The Origin of Species," 77.

99. Kropotkin, *Mutual Aid*, 65–66.

100. Ibid.

101. Ibid., 63.

102. Ibid., 73.

103. Ibid., 27.

104. Darwin, *The Descent of Man, and Selection in Relation to Sex*, 1:71–72.

105. Kropotkin, *Mutual Aid*, 76–77.

106. Charles Darwin to John Morley, 14 April 1871, *More Letters of Charles Darwin*, 1:329.

107. Kropotkin, "The Morality of Nature," 410–11.

108. Kropotkin, *Mutual Aid,* 15. Kropotkin also acknowledged that he had found the work of other naturalists useful on this point too—notably, Alfred Espinas, *Des Sociétés animales: Etude de psychologie comparée* (1887), and a lecture by J. L. Lanessan, *La Lutte pour l'existence et l'association pour la lute* (1881).

109. Kropotkin, *Mutual Aid,* 15–16.

110. Ibid., 16.

111. Kropotkin, "The Morality of Nature," 411.

112. Ibid.

113. Kropotkin, *Mutual Aid,* 16, 58.

114. Kropotkin, "The Morality in Nature," 411.

115. While this is not a study on the effects of individual psychology on science, it is notable that in writing about heredity Darwin had considered the views of his grandfather, consulted his father, and studied his children, whereas Kropotkin had little in common with his own father but became increasingly close to his brother, Alexander.

116. Darwin, *On the Origin of Species by Means of Natural Selection,* 237.

117. See chap. 3.

118. It is important to note that although in the context of current genetic understandings of kin relationships Kropotkin's criticisms appear misplaced, they were certainly not so in the context in which he made them. Historians have usually credited William Hamilton or in some cases J. B. S. Haldane with fully realizing the importance of broader kin relations in the preservation of traits. It is clear from this discussion that both Darwin and Kropotkin were aware of their importance long before the question was reframed in the context of the genetics of altruism.

119. Wallace, "The Origin of Human Races and the Antiquity of Man Deduced from the Theory of 'Natural Selection,'"; T. H. Huxley, "Further Remarks upon the Human Remains from the Neanderthal"; Lubbock, *Pre-historic Times.*

120. Kropotkin, *Mutual Aid,* 83.

121. Ibid., 45.

122. Ibid., 84–85.

123. Ibid., 76; Lubbock, *Pre-historic Times.*

124. Kropotkin, *Mutual Aid,* 76.

125. Ibid., 99.

126. Ibid., 100. This is an important element of Kropotkin's theory of mutual aid which has been overlooked by those historians who have considered him as an early advocate of "group selection" in the context of the "levels of selection" debate. While Kropotkin certainly did advocate group selection, this and similar comments suggest that he did not simply conceive of selection occurring indiscriminately at the species level. The appearance of an occasional lack of clarity on this point is indicative more of the fact that this is a theoretical question that was framed almost a century after Kropotkin wrote, than of any woolly thinking on his part. The same might be said of Darwin.

127. Ibid.

128. Ibid., 102.

129. Kropotkin, "An Appeal to the Young," 277.

130. Kropotkin, *Mutual Aid,* 105.

131. Ibid., 108.

132. Ibid., 113.

133. Ibid., 103.

134. Ibid., 114.

135. Henry Maine, *International Law* (1888), quoted in Kropotkin, *Mutual Aid*, 117–18.

136. Kropotkin, *Mutual Aid*, 181–82.

137. Hobbes, *Leviathan*, 97.

138. Kropotkin, *Mutual Aid*, 183.

139. Borrello, *Evolutionary Restraints*, 34. See also Todes, *Darwin without Malthus*, 138–39.

140. Darwin to Hooker, 5 June 1860, quoted in Kropotkin, "Evolution and Mutual Aid," 125. For the full text, see *CCD*, 8:237–40, esp. 238.

141. Kropotkin, "The Theory of Evolution and Mutual Aid", esp. 127–30.

142. Darwin, *On the Origin of Species by Means of Natural Selection*, 6.

143. Kropotkin, "Evolution and Mutual Aid," 124.

144. Charles Darwin to Joseph Dalton Hooker, 23 November 1856, quoted in Kropotkin, "Evolution and Mutual Aid," 124. For the full text, see *CCD*, 6:281–82.

145. Charles Darwin to Joseph Dalton Hooker, 12 June 1860, *CCD*, 8:251–52.

146. Charles Darwin to Joseph Dalton Hooker, 5 June 1860, in Kropotkin, "The Theory of Evolution and Mutual Aid," 125. For the full text, see *CCD*, 8:237–40.

147. Kropotkin, "The Theory of Evolution and Mutual Aid," 125.

148. Ibid., 127.

149. Charles Darwin to Horace Dobell, February 1863, quoted in ibid., 128.

150. Kropotkin, "The Theory of Evolution and Mutual Aid," 129–30.

151. Ibid., 128.

152. Ibid., 117.

153. Ibid., 117–18.

154. Kropotkin, *Mutual Aid*, 98.

155. Churchill, "The Weismann-Spencer Controversy over the Inheritance of Acquired Characters."

156. Weismann, "On Heredity," 1:70.

157. Kropotkin, *The Conquest of Bread*, 210, 222.

158. Ibid., 171–72.

Chapter Six

1. Churchill, "The Spencer-Weismann Controversy," 452.

2. Ibid., 463.

3. Alfred Russel Wallace was consistently opposed to the inheritance of acquired characters; however, he did not join the socialist movement until 1889, when he read Edward Bellamy's book *Looking Backward*.

4. *Unto This Last* was first published serially in the *Cornhill Magazine* for 1860.

5. Weiner, *English Culture and the Decline of the Industrial Spirit*, 32.

6. Ruskin, *The Nature of Gothic*, 23.

7. William Morris, quoted in MacCarthy, *William Morris*, 463.

8. S. Yeo, "A New Life," 7.

9. Wiener, *English Culture and the Decline of the Industrial Spirit*, 58–59, 119.

10. Morris, *News from Nowhere; or, An Epoch of Rest*, 75.

11. Morris, "How I Became a Socialist," 382.

12. Erin McLaughlin Jenkins has argued that many radicals first began to lose their faith when Huxley attacked Henry George in 1890 (see Jenkins, "Henry George and the Dragon," 31–51). However, Huxley's position in "The Struggle for Existence in Human Society" began the disaffection that later became widespread.

13. Willis-Harris, "The Survival of the Fittest."

14. Frederick Rockell, "The Last of the Great Victorians: Special Interview with Dr. Alfred Russel Wallace," *The Millgate Monthly* 7, no. 83 (1912), 657–63 (quotation, 632), which is quoted in Lester, "Uneasy Bedfellows."

15. Raby, *Alfred Russel Wallace*, 256–57.

16. Hyndman and Morris, *A Summary of the Principles of Socialism*, 46.

17. Ibid., 46–47.

18. Morris, *News from Nowhere and Other Writings*, 158.

19. When Kropotkin and Charlotte Wilson began the publication of *Freedom* in October, they were given use of the *Commonweal*'s press facilities (see Hulse, *Revolutionists in London*, 90–99). For more on Morris and Kropotkin, see Kinna, "Morris, Anti-Statism, and Anarchy."

20. Glasier, *William Morris and the Early Days of the Socialist Movement*, 3; Hale, "William Morris, Human Nature, and the Biology of Utopia," 109.

21. Hulse, *Revolutionists in London*, 91; Kropotkin, Review of *News from Nowhere; or, An Epoch of Rest*.

22. Morris, *News from Nowhere; or, An Epoch of Rest*, 69, 107–11; Kropotkin, *Mutual Aid*, 106–13.

23. Morris, *News from Nowhere; or, An Epoch of Rest*, 135.

24. Ibid., 159.

25. William Morris, "Fabian Essays in Socialism" (review), *Commonweal*, 25 January 1890, quoted in S. Yeo, "A New Life," 19.

26. Hale, "William Morris, Human Nature, and the Biology of Utopia"; Hale, "Of Mice and Men."

27. Morris, *News from Nowhere; or, An Epoch of Rest*, 123.

28. Ibid., 228.

29. Ibid., 133–34.

30. Ibid., 95. Patrick Parrinder, in "Eugenics and Utopia," suggests that this passage is indicative that Morris accepted prevailing eugenic ideas.

31. Robert Owen had emphasized the importance of the environment upon the development of character. A utopian-socialist, Owen was also an industrialist and founded the town of New Lanarck in South Lanarckshire, Scotland, in which he hoped to demonstrate his ideas. His most famous work was *A New View of Society, or, Essays on the Principle of the Formation of the Human Character, and the Application of the Principle to Practice* (1813). In contrast to Morris, Kropotkin, Spencer, and others who emphasized the importance of human agency in the formation of character, Owen had argued that a man's character was made for him.

32. Morris, *News from Nowhere; or, An Epoch of Rest*, 63.

33. Ibid., 118.

34. Ibid., 291.

35. Carlyle, *Past and Present*, 226.

36. Morris, *News from Nowhere; or, An Epoch of Rest*, 75.

37. Whereas Parrinder speculates that the "mully-grubs" might be a mild form of leprosy (as indeed the character Hammond speculates in *News from Nowhere*), Wilmer traces the term

to a contemporary colloquialism for depression or hypochondria. Given the context, Wilmer's explanation is the more convincing (see ibid., 411n18).

38. Ibid., 76.

39. Ibid., 75.

40. Morris, "A Factory As It Might Be."

41. Ibid. See also Morris, *News from Nowhere; or, An Epoch of Rest.*

42. Morris, "The Art of the People," 22:38.

43. Unless otherwise noted, all biographical information in this chapter is from Wells, *An Experiment in Autobiography.*

44. On the 1862 revised code, see Stephens. *Education in Britain*, 7, 18.

45. Wells, *Experiment in Autobiography*, 159.

46. Ibid., 161.

47. Mackenzie and Mackenzie, *The Time Traveler*, 51.

48. Wells, *Experiment in Autobiography*, 161.

49. John Collier was married to Mady Huxley at the time of the portrait. Two years after Mady's death in 1887, he married her younger sister, Ethel Huxley.

50. Wells, *Experiment in Autobiography*, 163.

51. Ibid., 160.

52. Bowler, "Lankester, Sir (Edwin) Ray (1847–1929)"; Hale, "Of Mice and Men," 40.

53. Wells, *Experiment in Autobiography*, 163.

54. Ibid., 165.

55. Obituary announcement, *Nature*, 28 October 1886, 625–27.

56. Wells, *Experiment in Autobiography*, 194.

57. Ibid., 193.

58. MacCarthy, *William Morris*, 522. A flyer advertising forthcoming lectures by prominent socialists, which included notice of a lecture by Besant, is reproduced on p. 523 of this work.

59. Wells, *Experiment in Autobiography*, 193.

60. Ibid.; D. C. Smith, *H. G. Wells*, 13.

61. Wells, "A Slip under the Microscope."

62. Wells, *Experiment in Autobiography*, 192.

63. Morris, "The Lesser Arts," 24–25.

63. Kropotkin, *An Appeal to the Young*, 264–67. On Kropotkin in the periodical *Justice*, see Jenkins, "Henry George and the Dragon," 457.

64. Wells, *Anticipations of the Reaction of Mechanical and Scientific Progress upon Human Life and Thought*, 288.

65. Not only did Wells reject Morris's vision of the future explicitly in *A Modern Utopia* (1905), he also did so in *Time Machine* (1895) and in *The Island of Doctor Moreau* (1896), among others.

66. Wells, *A Modern Utopia*, 1.

67. P. C. Gould, *Early Green Politics*; Marsh, *Back to the Land.*

68. Morris, *News from Nowhere; or, An Epoch at Rest*, 200.

69. T. H. Huxley, "The Struggle for Existence in Human Society," 196.

70. Wells, *A Modern Utopia*, 145, 180. See also T. H. Huxley, "The Struggle for Existence in Human Society," 329–30.

71. Wells, *A Modern Utopia*, 1–6, 78–80.

72. Wells, *A Modern Utopia*, 18.

73. Hale, "Labor and the Human Relationship with Nature, 262–69.

74. Wells, "The Biological Problem of Today"; Wells, "Incidental Thoughts on a Bald Head."

75. Weismann, "The Supposed Transmision of Mutilations," 421–48. Wells wrote about Weismann's work in his science journalism, as well as in his more explicitly political and sociological works, as explained below, and Shaw commented on what he perceived to be Weismann's cruelty and ignorance in the preface to his play *Back to Methuselah* (see below).

76. Mayr, "What Is Darwinism Today?"

77. Wells, "Human Evolution, An Artificial Process," 211.

78. Wells, *A Modern Utopia*, 18–19.

79. Weismann, "On Heredity," 1:90.

80. Barnett, "Education or Degeneration," 211n18. Barnett suggests that Lankester's interest in degeneration was initiated by Weismann's theory of panmixia, but the chronology here is surely wrong. Darwin and Huxley had already paid significant attention to the fact of degeneration among marine invertebrates, and this was Lankester's specialty.

81. Morris and Shaw have each been identified as among the various guests that assembled to hear the Time Traveller's tale.

82. Wells, *The Time Machine, an Invention*, 49.

83. Herbert George Wells to Thomas Henry Huxley, May 1895, in *The Correspondence of H. G. Wells*, 1:238.

84. Wells, "Human Evolution, An Artificial Process" (quotations, 211, 217).

85. H. G. Wells, J. Huxley, and G. P. Wells, *The Science of Life*, 787–90.

86. Barnett, "Education or Degeneration," 205.

87. Ibid., 205, 213–14 (Lankester quotation, 211).

88. Freeden, "Eugenics and Progressive Thought," 656. See also Semmel, "Karl Pearson," 112; Ray, "Eugenics, Mental Deficiency, and Fabian Socialism between the Wars," 215; and Partington, *Building Cosmopolis*, 54–56.

89. Searle, *Eugenics and Politics in Britain*, 3, 32.

90. Paul, "Eugenics and the Left," 586; and on the latter point, Freeden, "Eugenics and Progressive Thought," 656.

91. Wells, *Mankind in the Making*, 145, ix.

92. T. H. Huxley, "*Evolution and Ethics*," 23, in Huxley, Paradis, and Williams, *Evolution and Ethics*, 81.

93. Wells, *Mankind in the Making*, 40.

94. T. H. Huxley, "Evolution and Ethics," 22–23, in Huxley, Paradis, and Williams, *Evolution and Ethics*, 81.

95. T. H. Huxley, "Evolution and Ethics," 39, ibid., 97.

96. G. S. Jones, *Outcast London*, passim; Ray, "Eugenics, Mental Deficiency, and Fabian Socialism between the Wars," 215.

97. T. H. Huxley, "*Evolution and Ethics*," 39, in Huxley, Paradis, and Williams, *Evolution and Ethics*, 97.

98. Mackenzie and Mackenzie, *The Fabians*, 328–38.

99. Partington, *Building Cosmopolis*, 57.

100. Hale, "Labor and the Human Relationship with Nature," 262–63.

101. Kevles, *In the Name of Eugenics*, 164.

102. Wells, *Mankind in the Making*, 40.

103. Wells, *A Modern Utopia*, 107–8.

104. Ibid., 114. For the context of "voluntary" sterilization, see Searle, *Eugenics and Politics in Britain.*

105. Wells, *Mankind in the Making*, 305.

106. Julian Huxley is quoted in Dronamraju, *If I Am to be Remembered*, 43.

107. Hunt, *Equivocal Feminists.*

108. Wells, *A Modern Utopia*, 156.

109. Ibid., 237.

110. Ibid., 139–70.

111. Ibid., 151–56. It is interesting to note that exactly this schema was portrayed in the 2002 science fiction film *Gattaca.*

112. Ibid., 156, 152.

113. Wells, *Ann Veronica, A Modern Love Story*, 168–70.

114. Wells, *Men Like Gods*, 79.

115. Wells is quoted in Hyde, "The Socialism of H. G. Wells in the Early Twentieth Century," 234.

116. H. G. Wells, J. Huxley, and G. P. Wells, *The Science of Life*, 875.

117. Ibid., 877.

118. Coupland, "H. G. Wells's 'Liberal Fascism,'" 541.

119. Wells, *Experiment in Autobiography*, 554.

120. Wells, *Mind at the End of Its Tether*, 18.

121. Shaw, preface to *Back to Methuselah*, xi. Given the diversity of opinion among contemporary men of science, Shaw is by no means accurate in his perception that neo-Darwinism was the dominant view among contemporary scientists (see Bowler, *The Eclipse of Darwinism*).

122. Bowler, "The Spectre of Darwinism," 50.

123. Shaw, preface to *Back to Methuselah*, xlvii; Shaw and Weintraub, *Shaw: An Autobiography Selected from His Writings*, 1:145–46; Shaw and Wells, *Bernard Shaw and H. G.* Wells, 104.

124. Shaw, preface to *Back to Methuselah*, lii.

125. Ibid., xlii.

126. Ibid., xxiv.

127. H. G. Wells, J. Huxley, and G. P. Wells, *The Science of Life*, 263.

128. Shaw, preface to *Back to Methuselah*, xxii.

129. Bowler, *Evolution*, 318.

130. Hale, "The Search for Purpose in a Post-Darwinian Universe."

131. Bowler, *Evolution*, 321. For a more detailed history of the history of vitalism in biology, see Allen, *Life Science in the Twentieth Century.*

132. Shaw and Wells, *Bernard Shaw and H. G.* Wells, 104–5.

133. Shaw, preface to *Back to Methuselah*, lii, 1, xxiii.

134. Darwin, *On the Origin of Species by Means of Natural Selection*, 6th ed., 129. A wonderful resource created by Ben Fry that charts the development of each edition of *Origin* is available at http://benfry.com/traces/.

135. Rignano, *Upon the Inheritance of Acquired Characters*, 166–68.

136. Kropotkin, "Inheritance of Acquired Characters," 513–14.

137. It later transpired that epileptic guinea pigs will not only gnaw off their own toes, but will gnaw off those of their litter. It seems that this aspect of Brown-Séquard's experiments was not as careful as had been presumed. On Weismann's mice, see Weismann, "The Supposed Transmission of Mutilations," 1:444–45.

138. Shaw, preface to *Back to Methuselah*, xxiii.

139. Shaw, *Collected Letters*, 728–29.

140. Shaw, preface to *Back to Methuselah*, xxiii.

141. Ibid.

142. Ibid., xxix.

143. Shaw, *Man and Superman, A Comedy and a Philosophy*, xix.

144. Shaw, *Morris As I Knew Him*, xxxix.

145. Shaw, preface to *Back to Methuselah*, xix.

146. Hawkins, *Social Darwinism in European and American Thought*, 167; Shaw, preface to *Back to Methuselah*, xi, xiii–xvii; Irvine, *The Universe of G.B.S.*, 236.

147. Shaw, preface to *Back to Methuselah*, xxv.

148. Desmond, *The Politics of Evolution*, chaps. 7, 8.

149. S. J. Gould, *Ontogeny and Phylogeny*; R. J. Richards, *The Meaning of Evolution*, 17–55.

150. Robert J. Richards discusses this reading of embryological recapitulation in the context of transcendental morphology in *The Meaning of Evolution*, 42–55. Dov Ospovat, in *The Development of Darwin's Theory*, has shown that Von Baer's divergent reading of embryological development was popular among scientists much earlier than had often been argued in the literature (he cites William Coleman's *Biology in the Nineteenth Century* as exemplary of the view that Von Baer had little influence in his own lifetime). Indeed, Müller, Owen, Milne Edwards, W. B. Carpenter, and of course Herbert Spencer, each incorporated Von Baer's views into their work (see Ospovat, *The Development of Darwin's Theory*, 11; and also Ospovat, "The Influence of Karl Ernst von Baer's Embryology").

151. It was only after the publication of translations of Von Baer's work by Thomas Huxley that embryology began to be read in England as evidence of common descent rather than in this literal and linear recapitulation of the natural history of life (see T. H. Huxley, "Fragments Relating to Philosophical Zoology").

152. Shaw, preface to *Back to Methuselah*, xxviii.

153. Ibid., xxvi.

154. Ibid., xxvii

155. Ibid., xxix.

156. Ibid., xviii.

157. Weismann, "The Duration of Life," 8–9.

158. Ibid., 20.

159. Ibid., 24–25.

160. Shaw and Wells, *Bernard Shaw and H. G.* Wells, 105. Wells had summarized Weismann's article in a *Saturday Review* article on 23 February 1895 (see Wells, "The Duration of Life"). Despite the fact that Wells does not draw attention to this aspect of Weismann's essay in his review, it is clear from the 1894 version of *Time Machine*, serialized in the *National Observer*, that he had not missed this point, incorporating a similar explanation of longevity into the novel and, echoing Weismann, having the Time Traveller observe of the Eloi: "The average duration of life was about nineteen or twenty years. Well—what need of longer? People live nowadays to threescore and ten because of their excessive vitality, and because of the need there has been of guarding, rearing, and advising a numerous family" (74). The 1894 edition of *Time Machine* appears in *Early Writings in Science and Science Fiction by H. G. Wells*, ed. Robert Philmus and David Y. Hughes, 47–104. However, this passage is notable by its absence from the 1895 text.

161. Shaw, preface to *Back to Methuselah*, xix. It should be noted here that two of Shaw's biographers, Archibald Henderson (*Bernard Shaw: Playboy and Prophet*) and William Irvine (*The*

Universe of G.B.S.), do not find Shaw's conclusions in this regard overly outlandish, Henderson going so far as to endorse Shaw's views.

162. Shaw, preface to *Back to Methuselah*, ci.

163. Ibid., xix. Despite the ridicule that Shaw received from the likes of Julian Huxley and J. B. S. Haldane as to the merit of his evolutionary speculations about the future (see Hale, "The Search for Purpose in a Post-Darwinian Universe"), it is notable that Haldane later speculated not only about the possibility of an increase in human longevity—tellingly, to reach three thousand years—but also of the existence of an evolutionary "life force" (Haldane, "The Last Judgement," 295). For the intellectual context of Haldane's "visionary biology," see Adams, "The Last Judgment"; and Adams, "The Quest for Immortality."

164. Ray, "Eugenics, Mental Deficiency, and Fabian Socialism between the Wars," 216.

165. Freeden, "Eugenics and Progressive Thought," 648.

166. Hale, "The Search for Purpose in a Post-Darwinian Universe," 206–7.

167. Shaw, "The Case for Equality," 15–16.

168. Ibid., 13–19.

169. Hawkins, *Social Darwinism in European and American Thought*, 167.

170. George Bernard Shaw to Lady Mary Murray, quoted in Holroyd, *Bernard Shaw*, 2: 393–94.

171. Mackenzie and Mackenzie, *The Fabians*, 162.

172. Arnot, *Bernard Shaw and William Morris*, 23–24.

Chapter Seven

1. Ruse, *Monad to Man*, 224.

2. Lankester, *Degeneration*, 26 (emphsis in the original).

3. Ibid., 30, 29.

4. Churchill, "The Weismann-Spencer Controversy over the Inheritance of Acquired Characters," 453.

5. Ibid., 456–57.

6. Historians who have made note of panmixia, although only as a side issue in the broader debate over neo-Lamarckism and neo-Darwinism, are Churchill, "The Weismann-Spencer Controversy over the Inheritance of Acquired Characters," and more recently, Cock and Forsdyke, *Treasure Your Exceptions*, 111–13.

7. Churchill, "The Weismann-Spencer Controversy over the Inheritance of Acquired Characters," 457.

8. See chap. 6.

9. Weismann, "The All-Sufficiency of Natural Selection," 335.

10. Peter Bowler, for instance, does not mention the term "panmixia" in any of the three editions of his comprehensive text, *Evolution: The History of an Idea.*

11. Churchill, "The Weismann-Spencer Controversy over the Inheritance of Acquired Characters," 451.

12. Bowler, *The Non-Darwinian Revolution*, 92; Bowler, *The Eclipse of Darwinism*; Bowler, *Evolution*, 224. Still the best monograph on the synthesis is Smocovitis, *Unifying Biology*, although see also Joe Cain's important revisionist work in which he argues that we should abandon the concept of the "modern synthesis" because it blinds us to the broader complexities at stake in biology during the 1920s and 1930s (Cain, "Rethinking the Synthesis Period in Evolutionary Studies").

13. Thomas Huxley to Charles Darwin, 23 November 1859, *CCD*, 7:390–91; Charles Darwin to Thomas Huxley, 22 November 1860, *CCD*, 8:487.

14. Provine, *The Origins of Theoretical Population Genetics*, 14–15.

15. Galton, *Natural Inheritance*, 28.

16. Bowler, *Evolution*, 267; Olby, "Bateson, William (1861–1926)." Bowler notes that the claims of a third rediscoverer, Erich von Tschermak, are no longer accepted because it is evident that he did not understand the laws that were central to Mendel's claims.

17. Provine, *Population Genetics*, chaps. 3, 4; Bowler, *Evolution*, 256–73.

18. Cock and Forsdyke, *Treasure Your Exceptions*, 11.

19. Romanes, "The Spencer-Weismann Controversy."

20. George John Romanes, Letter to the Editor ("Panmixia"), *Nature*, 13 March 1890, 437–39.

21. George John Romanes, Letter to the Editor ("Mr. Herbert Spencer on 'Natural Selection'"), *Contemporary Review* 63 (April 1893): 499–517. See also Romanes, Letter to the Editor ("The Spencer-Weismann Controversy"), *Contemporary Review* 64 (July 1893): 50–60; and Romanes, Letter to the Editor ("A Note on Panmixia"), *Contemporary Review* 64 (October 1893): 611–13.

22. Poulton had asked Wallace to proofread the manuscript of his translation of Weismann's *Essays on Heredity*; Poulton footnoted the fact that Wallace had sent him a note he had written some years before in which he made exactly the same point as Weismann regarding a species' longevity being an outcome of natural selection (see the editor's note in Weismann, "The Duration of Life," 1:23n1).

23. Wallace, "The Future of Civilisation," 549.

24. Alfred Russel Wallace to Edward B. Poulton, 16 June 1892, in *Alfred Russel Wallace: Letters and Reminiscences*, 305.

25. Wallace, "The Future of Civilisation," 549.

26. Alfred Russel Wallace to Edward B. Poulton, 29 August 1892, in *Alfred Russel Wallace: Letters and Reminiscences*, 306.

27. Wallace, "The Future of Civilisation," 549.

28. Alfred Russel Wallace, Letter to the Editor ("Panmixia and Natural Selection"), *Nature*, 28 June 1894, 196–97.

29. Lightman, "Darwin and the Popularization of Evolution," 15–16.

30. Crook, *Benjamin Kidd*, 85.

31. Lightman, "Darwin and the Popularization of Evolution," 15; Crook, *Benjamin Kidd*, 98–103. On Kidd's meeting with Kropotkin, see Crook, *Benjamin Kidd*, 86.

32. Crook, *Benjamin Kidd*, 38.

33. Kidd, *Social Evolution*, 211.

34. Ibid., 243.

35. Ibid., 245.

36. Ibid., 245–46.

37. Wallace, "The Future of Civilisation," 549.

38. Crook, *Benjamin Kidd*, 210–11.

39. Kidd, *Social Evolution*, 210.

40. Ibid., 292.

41. Wallace, "The Future of Civilisation," 549.

42. Spencer, "The Inadequacy of 'Natural Selection,'" 153.

43. Peter Chalmers Mitchell, Letter to the Editor ("The Spencer-Weismann Controversy"), *Nature*, 15 February 1894, 373–74.

44. George John Romanes, Letter to the Editor ("Panmixia"), *Nature*, 26 April 1894, 599–600.

45. Ibid. Romanes had already published a lengthy critical analysis of Weismann's views in his 1893 work *An Examination of Weismannism.*

46. Ibid.

47. Magnello, "Weldon, Walter Frank Raphael (1860–1906)."

48. Cock and Forsdyke, *Treasure Your Exceptions*, 12.

49. Magnello, "Weldon, Walter Frank Raphael (1860–1906); Cock and Forsdyke, *Treasure Your Exceptions*, 101–2. Weldon was appointed to the Jodrell Chair in Zoology in 1890, taking up the post in 1891.

50. Bowler, *Evolution*, 258–60.

51. W. F. R. Weldon, Letter to the Editor ("Panmixia"), *Nature* 50, 3 May 1894, 5.

52. Bowler, Evolution, 259.

53. George John Romanes, Letter to the Editor ("Panmixia"), *Nature*, 26 April 1894, 599–600; W. F. R. Weldon, Letter to the Editor ("Panmixia"), *Nature* 50, 3 May 1894, 5.

54. W. F. R. Weldon, Letter to the Editor ("Panmixia"), *Nature* 50, 3 May 1894, 5.

55. Ibid.

56. George John Romanes, Letter to the Editor ("Panmixia"), *Nature*, 10 May 1894, 28–29 (quotation, 28).

57. K. Pearson, "The Philosophy of Natural Science," 2.

58. Porter, *Karl Pearson.*

59. Porter, "Statistical Utopianism in an Age of Aristocratic Efficiency," 210. The fact that Pearson does not appear at all in Stanley Pierson's *British Socialists* is not misrepresentative of histories of British socialism.

60. Semmel, "Karl Pearson."

61. K. Pearson, "Socialism in Theory and Practice," 362.

62. Porter, "Statistical Utopianism in an Age of Aristocratic Efficiency."

63. On the spilt of the Social Democratic Federation and the founding of the Socialist League, see MacCarthy, *William Morris*, 504–15; on Morris's Icelandic sagas, 278–310.

64. Theordore Porter argues that while Pearson was involved in the British socialist movement throughout the 1880s, he was indifferent to Morris's views, being much closer to both George Bernard Shaw and to the anarchist Charlotte Wilson (Porter, *Karl Pearson*, 108–9). However, while Pearason was certainly on more familiar terms with both Shaw and Wilson, this overlooks the ideological similarities between Shaw, Wilson, and Morris, and the closer affinities between Morris's socialism and Pearson's in the 1880s when compared to either Shaw's or Wilson's views, not to mention Pearson's acknowledgment of Morris in his own work (see K. Pearson, "Socialism in Theory and Practice," 346).

65. K. Pearson, "Socialism and Natural Selection," 107.

66. Ibid., 103–7 (quotation, 105).

67. Crook, *Benjamin Kidd*, 37.

68. K. Pearson, "Socialism and Natural Selection," 104.

69. Ibid., 105.

70. Ibid., 107.

71. See Magnello, "Karl Pearson's Gresham Lectures"; Bowler, *Evolution*, 259; Salsburg, *The Lady Tasting Tea*, 22–23; and Porter, *Karl Pearson*, 239. Magnello points out that prior to her own study, scholarship focused on Pearson as the developer of Galton's ideas on regression. While in this study, I too focus on Pearson's development of Galton's work, I emphasize that he reached very different conclusions to Galton and that in this as well as in his later work he can by no means be adequately characterized as a "Galtonian."

72. Magnello, "Weldon, Walter Frank Raphael (1860–1906)."

73. [Pearson and Shipley], "Walter Frank Raphael Weldon, 1866–1906."

74. Magnello, "Weldon, Walter Frank Raphael (1860–1906)."

75. Weldon, "Presidential Address," 887–902 (quotations, 897).

76. Ibid., 898–99.

77. Ibid., 899–900.

78. K. Pearson, "Socialism and Natural Selection," 110.

79. Ibid., 118.

80. Ibid.

81. Ibid., 114.

82. Darwin is quoted in ibid., 127.

83. See chap. 3.

84. K. Pearson, "Socialism and Natural Selection," 111.

85. Ibid., 121.

86. Ibid., 125.

87. Ibid., 121.

88. Ibid., 129–39, esp. 131.

89. Pearson, *The Grammar of Science*, 26.

90. Ibid., 26–27.

91. T. H. Huxley, *"Evolution and Ethics"*, 50, Huxley, Paradis, and Williams, *Evolution and Ethics*, 108.

92. K. Pearson, *The Grammar of Science*, 26, 20–21.

93. Ibid., 26.

94. K. Pearson, *The Problem of Practical Eugenics.*

95. K. Pearson, *The Grammar of Science*, 26–27.

96. Wells, *Anticipations of the Reaction of Mechanical and Scientific Progress upon Human Life and Thought*, 288.

97. K. Pearson, *National Life from the Standpoint of Science*, 106.

98. Ibid., 21.

99. Kevles, *In the Name of Eugenics*, 39.

100. Kevles, *In the Name of Eugenics*, 74; S. Webb, *The Decline in the Birth Rate*; Heron, *On the Relations of Fertility in Man to Social Status and on the Changes in This Relation That Have Taken Place during the Last Fifty Years.*

101. Kevles, *In the Name of Eugenics*, 38.

102. K. Pearson, *The Grammar of Science*, 26.

103. Ibid., 27.

104. Ibid.

105. Ibid., 26n2.

106. K. Pearson, "The Moral Basis of Socialism," 307–8.

107. K. Pearson, *The Grammar of Science*, 21. This echoes almost word for word the sentiments expressed by E. O. Wilson in *Sociobiology*.

Conclusion

1. Nelson, "Harriet Martineau's Political Economy." This is also the subject of Nelson's ongoing work at the University of Oklahoma.

2. Ibid.

3. Desmond, *The Politics of Evolution*, 10, 209.

4. See Desmond, *The Politics of Evolution*; and Elliott, *The Derby Philosophers.*

5. Thompson, *The Making of the English Working Class.*

6. R. J. Richards, *Darwin and the Emergence of Evolutionary Theories of Mind and Behavior.*

7. Darwin, *On the Origin of Species by Means of Natural Selection*, 241.

8. Schabas, *The Natural Origins of Economics*; Winch, *Riches and Poverty.*

9. Grant Allen, "The New Theory of Heredity: Our Scientific Causerie," *Review of Reviews* 1 (June 1890): 537–38, cited in Crook, *Benjamin Kidd*, 37.

10. Crook, *Benjamin Kidd*, 38.

11. Ibid., 59.

12. G. R. Searle, *The Quest for National Efficiency.*

13. Innes, "Utopian Apocalypses, Shaw, War, and H. G. Wells," 39–43.

14. Shaw, preface to *Back to Methuselah*, xi.

15. Clark, *J.B.S.: The Life and Work of J. B. S. Haldane*, 35.

16. This now-apocryphal quotation has variously been described as having been a response to a Christian questioner at a meeting, as I have it here, or as the punch line to a joke in a pub. Either is fully plausible, but my preference for the former comes from the testimony of Krishna Dronamraju, one of Haldane's former students.

17. Smith and Price, "The Logic of Animal Conflict."

18. On Wynne-Edwards, see Borrello, *Evolutionary Restraints.*

19. Swenson, "The Nature of Selfishness and the Genetics of Altruism," 10.

20. W. D. Hamilton, "The Evolution of Altruistic Behaviour." See also W. D. Hamilton, "The Genetical Evolution of Social Behaviour," pts. 1 and 2.

21. Axelrod and Hamilton, "The Evolution of Cooperation"; Swenson, "The Nature of Selfishness and the Genetics of Altruism," 52.

22. See Richard Dawkins, "The Descent of Edward Wilson," *Prospect Magazine*, 24 May 2012; Steven Pinker, "The False Allure of Group Selection," *Edge*, 18 June 2012; and David Sloan Wilson, "Clash of Paradigms," *Huffington Post*, 15 July 2012, all of which are cited in Swenson, "The Nature of Selfishness and the Genetics of Altruism," 52.

Afterword

1. Darwin, *On the Origin of Species By Means of Natural Selection*, 488.

Bibliography

Adams, Mark B. "The Last Judgment: The Visionary Biology of J. B. S. Haldane." *Journal of the History of Biology* 33, no. 3 (2000): 457–91.

———. "The Quest for Immortality: Visions and Presentiments in Science and Literature." In *The Fountain of Youth: Cultural, Scientific, and Ethical Perspectives on a Biomedical Goal*, edited by Stephen G. Post and Robert H. Brinstock, 38–71. Oxford: Oxford University Press, 2004.

Albrecht, William P. "Hazlitt and Malthus." *Modern Language Notes* 60, no. 4 (1945): 215–26.

Allen, Garland. *Life Science in the Twentieth Century*. New York: Wiley, 1975.

Amundson, Ron. "Richard Owen and Animal Form." In *On the Nature of Limbs: A Discourse*, by Richard Owen; with a preface by Brian K. Hall; with introductory essays by Ron Amundson, Kevin Padian, Mary P. Winsor, and Jennifer Coggon, xv–li. Chicago: University of Chicago Press, 2007.

Andelson, Robert V. *Critics of Henry George: A Centenary Appraisal of Their Strictures on "Progress and Poverty."* Rutherford, N.J.: Fairleigh Dickinson University Press, 1979.

Anderson, Perry. "The Origin of the Present Crisis." *New Left Review* 1, no. 23 (1964): 26–53.

Argyll, George Douglas Campbell. *The Reign of Law*. New York: A. L. Burt, 1868.

Arnhart, Larry. *Darwinian Conservatism*. Exeter, U.K.: Imprint Academic, 2005.

Arnot, Robert Page. *Bernard Shaw and William Morris*. [Folcroft, Pa.]: Folcroft Library Editions, 1973.

Axelrod, Robert, and William D. Hamilton. "The Evolution of Cooperation." *Science* 211, no. 4489 (1981): 1390–96.

Bagehot, Walter. "Physics and Politics, No. I: The Pre-Economic Age." *Fortnightly Review* 8 (November 1867): 518–38.

———. "Physics and Politics, No. II: The Age of Conflict." *Fortnightly Review* 9 (April 1868): 452–71.

———. "Physics and Politics, No. III: Nation Making." *Fortnightly Review* 6 (July 1869): 58–72.

———. "Physics and Politics, No. IV: Nation Making." *Fortnightly Review* 16 (December 1871): 696–717.

———. "Physics and Politics, No. V: The Age of Discussion." *Fortnightly Review* 17 (January 1872): 46–70.

———. *Physics and Politics, or Thoughts on the Application of the Principles of 'Natural Selection,' and 'Inheritance' to Political Society.* London: H. S. King and Co., 1872.

Banton, Michael. *Darwinism and the Study of Society: A Centenary Symposium.* London: Tavistock Publications, 1961.

Barnett, Richard. "Education or Degeneration: E. Ray Lankester, H. G. Wells, and the Outline of History." *Studies in the History and Philosophy of Biological and Biomedical Sciences* 37, no. 2 (2006): 203–29.

Barr, Alan P. *Thomas Henry Huxley's Place in Science and Letters: Centenary Essays.* Athens: University of Georgia Press, 1997.

Barton, Ruth. "'An Influential Set of Chaps': The X Club and Royal Society Politics." *British Journal for the History of Science* 23, no. 1 (1990): 53–81.

———. "'Huxley, Lubbock, and Half a Dozen Others': Professionals and Gentlemen in the Formation of the X Club, 1851–1864." *Isis* 89, no. 3 (1998): 410–44.

———. "'Men of Science': Language, Identity, and Professionalization in the Mid-Victorian Scientific Community." *History of Science* 41, no. 1 (2003): 73–119.

Beatty, John. "'The Details Left to Chance': Evolutionary Contingency and Its Broader Implications in the Work of Charles Darwin and Stephen Jay Gould." Lecture given at The Darwinian Revolution: Presidential Dream Course Lecture Series, University of Oklahoma, 12 March 2009.

———. "Speaking of Species: Darwin's Strategy." In *Units of Evolution: Essays on the Nature of Species,* edited by M. Ereshevsky, 227–46. Cambridge, Mass.: MIT Press, 1992.

Beatty, John, and Piers J. Hale. "*Water Babies*: An Evolutionary Fairy Tale." *Endeavour* 32, no. 4 (2008): 141–46.

Beilharz, Peter, and Chris Nyland. *The Webbs, Fabianism, and Feminism.* London: Ashgate, 1998.

Bennett, Phillippa, and Rosie Miles. *William Morris in the Twenty-First Century.* Oxford: Peter Lang, 2010.

Bentham, Jeremy. *The Rationale of Reward.* London: R. Heward, 1830.

———. *An Introduction to the Principles of Morals and Legislation.* 1823. Reprint, New York: Hafner, 1948.

Besant, Annie. *Why I Am a Socialist.* London: Printed by Annie Besant and Charles Bradlaugh, 1886.

Biagini, Eugenio F. "Popular Liberals, Gladstonian Finance and the Debate on Taxation, 1860–1874." In *Currents of Radicalism: Popular Radicalism, Organised Labour, and Party Politics in Britain, 1850–1914,* edited by Eugenio Biagini and Alastair Reid, 134–62. Cambridge: Cambridge University Press, 1991.

Biber, George Eduard. *Henry Pestalozzi and His Plan of Education: Being an Account of His Life and Writings; with Copious Extracts from His Works, and Extensive Details Illustrative of the Practical Parts of His Method.* London: J. Souter, 1831.

Blatchford, Robert. *Merrie England.* New York: Monthly Review Press, 1966.

Borrello, Mark E. *Evolutionary Restraints: The Contentious History of Group Selection.* Chicago: University of Chicago Press, 2010.

Bourdieu, Pierre. *Distinction: A Social Critique of the Judgement of Taste.* Cambridge, Mass: Harvard University Press, 1984.

Bowler, Peter J. *The Eclipse of Darwinism: Anti-Darwinian Theories in the Decades around 1900.* Baltimore: Johns Hopkins University Press, 1983.

———. *Evolution: The History of an Idea.* 3rd ed. Berkeley: University of California Press, 2003.

———. *The Non-Darwinian Revolution: Reinterpreting a Historical Myth.* Baltimore: Johns Hopkins University Press, 1988.

———. "The Spectre of Darwinism: The Popular Image of Darwinism in Early Twentieth-Century Britain." In *Darwinian Heresies*, edited by Abigail Lustig, Robert J. Richards, and Michael Ruse, 48–68. Cambridge: Cambridge University Press, 2004.

———. "Lankester, Sir (Edwin) Ray (1847–1929)." *Oxford Dictionary of National Biography.* Oxford: Oxford University Press, 2004.

Brackman, Arnold C. *A Delicate Arrangement: The Strange Case of Charles Darwin and Alfred Russel Wallace.* New York: Times Books, 1980.

Bradlaugh, Charles, and Annie Besant. *Fruits of Philosophy: A Treatise on the Population Question.* San Francisco: Reader's Library, 1891.

Brewster, David. Review of *Vestiges of the Natural History of Creation*, by Robert Chambers. *North British Review*, no. 3 (1845): 471.

Briggs, Asa. *The Age of Improvement, 1783–1867.* Harlow, Eng.: Longman, 2000.

———. *Victorian Cities.* London: Odhams Press, 1963.

Browne, E. J. *Charles Darwin: The Power of Place.* New York: Alfred A. Knopf, 2002.

———. *Charles Darwin: Voyaging: A Biography.* Princeton, N.J.: Princeton University Press, 1996.

Cain, Joe. "Rethinking the Synthesis Period in Evolutionary Studies." *Journal of the History of Biology* 42, no. 4 (2009): 621–48.

Calhoun, Craig J. *The Question of Class Struggle: Social Foundations of Popular Radicalism during the Industrial Revolution.* Chicago: University of Chicago Press, 1982.

Cantor, Geoffrey, Gowan Dawson, Graeme Gooday, Richard Noakes, Sally Shuttleworth, and Jonathan R. Topham. *Science in the Nineteenth-Century Periodical: Reading the Magazine of Nature.* Cambridge: Cambridge University Press, 2004.

Carlyle, Thomas. *Past and Present.* New York: C. Scribner's Sons, 1918.

Carpenter, Edward. *Towards Democracy.* New York: M. Kennerley, 1922.

———. *Woman, and Her Free Place in Society.* Manchester: Labour Press Society, 1894.

Carpenter, William B. "Darwin on the Origin of Species." *National Review* 10 (January 1860): 188–214.

———. *Principles of General and Comparative Physiology Intended as an Introduction to the Study of Human Physiology and as a Guide to the Philosophical Pursuit of Natural History.* London: John Churchill, 1839.

———. "The Theory of Development in Nature." *British and Foreign Medico-Chirurgucal Review*, no. 25 (1860): 367–404.

Chambers, Robert. *Vestiges of the Natural History of Creation and Other Evolutionary Writings.* Edited with a new introduction by James A. Secord. Chicago: University of Chicago Press, 1994.

Chase, Malcolm. *"The People's Farm": English Radical Agrarianism, 1775–1840.* Oxford: Clarendon Press, 1988.

Chauvin, Rémy. *Animal Societies: From the Bee to the Gorilla.* New York: Hill and Wang, 1968.

Churchill, Frederick B. "The Weismann-Spencer Controversy over the Inheritance of Acquired Characters." In *Human Implications of Scientific Advance: Proceedings of the 15th International Congress of the History of Science, Edinburgh, 10–15 August 1977*, edited by Eric G. Forbes, 451–68. Edinburgh: Edinburgh University Press, 1978.

Clark, Ronald William. *J.B.S.: The Life and Work of J. B. S. Haldane.* Oxford: Oxford University Press, 1984.

Clifford, William Kingdon. *Lectures and Essays by William Kingdon Clifford.* Edited by Leslie Stephen and Frederick Pollock. 2 vols. London: Macmillan, 1879.

———. "On the Scientific Basis of Morals." In *Lectures and Essays by William Kingdon Clifford*, edited by Leslie Stephen and Frederick Pollock, 1:106–23. London: Macmillan, 1879.

Cobbe, Frances Power. "Darwinism in Morals." *Theological Review: A Quarterly Journal of Religious Thought and Life* 8, no. 33 (1871): 167–92.

———. *An Essay on Intuitive Morals, Being an Attempt to Popularise Ethical Science.* London: Longman, 1855.

———. *The Life of Frances Power Cobbe, by Herself.* 2 vols. Boston: Houghton, Mifflin, 1895.

Cobbett, William. *Advice to Young Men and (Incidentally) to Young Women, in the Middle and Higher Ranks of Life in a Series of Letters Addressed to a Youth, a Bachelor, a Lover, a Husband, a Father, a Citizen, or a Subject.* London: Published by the author, 1829.

———. *Rural Rides in the Counties of Surrey, Kent, Sussex . . . with Economical and Political Observations Relative to Matters Applicable to, and Illustrated by, the State of Those Counties Respectively.* London: W. Cobbett, 1830.

Cock, Alan G., and Donald R. Forsdyke. *Treasure Your Exceptions: The Science and Life of William Bateson.* New York: Springer, 2008.

Collini, Stefan. *Liberalism and Sociology: L. T. Hobhouse and Political Argument in England, 1880–1914.* Cambridge: Cambridge University Press, 1979.

Colp, Ralph, Jr. "The Contacts Between Karl Marx and Charles Darwin." *Journal of the History of Ideas* 35, no. 2 (1974): 329–38.

Comte, Auguste. *System of Positive Polity.* 4 vols. London: Longmans, Green and Co., 1875–77.

Coupland, Philip. "H. G. Wells's 'Liberal Fascism,'" *Journal of Contemporary History* 35, no. 4 (2000): 541–58.

Creath, Richard, and Jane Maienschein. *Biology and Epistemology.* Cambridge: Cambridge University Press, 2000.

Crook, D. P. *Benjamin Kidd: Portrait of a Social Darwinist.* Cambridge: Cambridge University Press, 1984.

Cuvier, Georges. *An Essay on the Theory of the Earth.* Edited by Robert Jameson. New York: Kirk and Mercein, 1818.

Darwin, Charles. *The Autobiography of Charles Darwin, 1809–1882: With Original Omissions Restored.* Edited by Nora Barlow. New York: Norton, 1969.

———. *Charles Darwin's "Beagle" Diary.* Edited by R. D. Keynes. Cambridge: Cambridge University Press, 1988.

———. *Charles Darwin's Notebooks, 1836–1844.* Transcribed and edited by Paul H. Barrett, Peter J. Gautrey, Sandra Herbert, David Kohn, and Sydney Smith. Cambridge: Cambridge University Press, 2008.

———. *Charles Darwin's Notebooks from the Voyage of the "Beagle."* Edited by G. R. Chancellor and John Van Wyhe. Cambridge: Cambridge University Press, 2009.

———. *The Correspondence of Charles Darwin.* Edited by Frederick Burkhardt, Sydney Smith et al. 18 vols. Cambridge: Cambridge University Press, 1985–2010.

———. *The Descent of Man, and Selection in Relation to Sex.* 1st ed. 2 vols. London: John Murray, 1871.

———. *The Descent of Man.* 2nd ed. Amherst, N.Y.: Prometheus Books, 1998. [The 2nd edition was originally published in 1874.]

———. *Journal of Researches into the Natural History and Geology of the Countries Visited dur-*

ing the Voyage of H.M.S. "Beagle" Round the World, Under the Command of Capt. Fitz Roy, R.N. 4 vols. New York: Harper and Bros., 1846.

———. *The Life and Letters of Charles Darwin, Including an Autobiographical Chapter.* Edited by Francis Darwin. 2 vols. New York: D. Appleton and Co., 1897.

———. *Narrative of the Surveying Voyages of His Majesty's Ships "Adventure" and "Beagle," between the Years 1826 and 1836: Describing Their Examination of the Southern Shores of South America, and the "Beagle's" Circumnavigation of the Globe . . . .* Vol. 3, *Journal and Remarks, 1832–1836.* London: Henry Colburn, Great Marlborough Street, 1839.

———. *More Letters of Charles Darwin, a Record of His Work in a Series of Hitherto Unpublished Letters.* Edited by Francis Darwin. 2 vols. New York: D. Appleton and Co., 1903.

———. "On the Moral Sense." In *Darwin on Man: A Psychological Study of Scientific Creativity,* by Howard E. Gruber, 398–405. New York: E. P. Dutton, 1974.

———. *On the Origin of Species by Means of Natural Selection, or Preservation of Favoured Races in the Struggle for Life.* 1st ed. London: John Murray, Albemarle Street, 1859.

———. *On the Origin of Species by Means of Natural Selection, or, The Preservation of Favoured Races in the Struggle for Life.* 2nd ed. London: J. Murray, 1860.

———. *On the Origin of Species by Means of Natural Selection, or, The Preservation of Favoured Races in the Struggle for Life.* 5th ed. London: J. Murray, 1869.

———. *The Origin of Species: By Means of Natural Selection, or the Preservation of Favoured Races in the Struggle for Life.* 6th ed. London: John Murray, 1872.

———. *The Red Notebook of Charles Darwin.* Edited by Sandra Herbert. [London]: British Museum (Natural History), 1979.

———. *The Variation of Animals and Plants under Domestication.* London: John Murray, 1868.

———. *The Voyage of the "Beagle."* Washington, D.C.: National Geographic Society, 2004.

———. *The Zoology of the Voyage of HMS "Beagle."* In *The Complete Work of Charles Darwin Online,* edited by John van Wyhe, 2002–. http://darwin-online.org.uk/EditorialIntroductions/Freeman_ZoologyOfBeagle.html.

Darwin, Erasmus. *The Temple of Nature; or, The Origin of Society: A Poem, with Philosophical Notes.* London: Jones and Co., 1825.

Darwin, Francis, ed. *The Life of Charles Darwin.* Reading: Senate, 1995.

Dawkins, Richard. *The Selfish Gene.* Oxford: Oxford University Press, 1976.

De Beer, G. R. "The Origin of Darwin's Ideas on Evolution and Natural Selection." *Proceedings of the Royal Society of London* 155 (1961): 321–78.

Den Otter, S. M. "Ritchie, David George (1853–1903)." *Oxford Dictionary of National Biography.* Oxford: Oxford University Press, 2004.

Depew, David J., and Bruce H. Weber. *Darwinism Evolving: Systems Dynamics and the Genealogy of Natural Selection.* Cambridge, Mass.: MIT Press, 1995.

Desmond, Adrian J. *Huxley: From Devil's Disciple to Evolution's High Priest.* Reading, Mass.: Perseus, 1997.

———. "Huxley, Thomas Henry (1825–1895)." In *Oxford Dictionary of National Biography.* Oxford University Press, 2004.

———. *The Politics of Evolution: Morphology, Medicine, and Reform in Radical London.* Chicago: University of Chicago Press, 1989.

Desmond, Adrian J., and James R. Moore. *Darwin: The Life of a Tormented Evolutionist.* New York: Warner, 1991.

———. *Darwin's Sacred Cause: How a Hatred of Slavery Shaped Darwin's Views on Human Evolution.* Boston: Houghton Mifflin Harcourt, 2009.

———. Introduction to *The Descent of Man, and Selection in Relation to Sex,* by Charles Darwin, xi–lxvi. Penguin Classics. London: Penguin, 2004.

Disraeli, Benjamin. *Sybil; or, The Two Nations.* London: Henry Colburn, 1845.

Dixon, Thomas. *The Invention of Altruism: Making Moral Meanings in Victorian Britain.* Oxford: Published for the British Academy by Oxford University Press, 2008.

Douglas, Roy. "Huxley's Critique from Social Darwinism." In *Critics of Henry George: A Centenary Appraisal of Their Strictures on "Progress and Poverty,"* edited by Robert V. Andelson, 137–52. London: Blackwell, 1979.

Dronamraju, Krishna R., ed. *Haldane and Modern Biology.* Baltimore: Johns Hopkins University Press, 1968.

———. *If I Am to be Remembered: The Life and Work of Julian Huxley with Selected Correspondence.* Singapore: World Scientific Pub. Co., 1993.

Drummond, Henry. *The Ascent of Man.* New York: Pott, 1894.

Dugatkin, Lee Alan. *The Altruism Equation: Seven Scientists Search for the Origins of Goodness.* Princeton, N.J.: Princeton University Press, 2006.

———. *The Prince of Evolution: Peter Kropotkin's Adventures in Science and Politics.* Lexington, Ky.: Printed by CreateSpace, an Amazon.com Company, 2011.

Duke of Argyll (George Douglas Campbell). *The Reign of Law.* New York: Burt, 1868.

Duster, Troy. The *Backdoor to Eugenics.* New York: Routledge, 1990.

Elliott, Paul A. *The Derby Philosophers: Science and Culture in British Urban Society, 1700–1850.* Manchester: Manchester University Press, 2009.

———. "'The Derbyshire Darwinians': The Persistence of Erasmus Darwin's Influence on a British Provincial Library and Scientific Community, c. 1780–1850." In *The Genius of Erasmus Darwin,* edited by C. U. M. Smith and Robert Arnott, 179–92. Aldershot, Eng.: Ashgate, 2005.

———. "Erasmus Darwin, Herbert Spencer, and the Origins of the Evolutionary Worldview in British Provincial Scientific Culture, 1770–1850." *Isis* 94, no. 1 (2003): 1–29.

Engels, Friedrich. *The Part Played by Labor in the Transition from Ape to Man.* New York: International, 1950.

Ereshevsky, M., ed. *Units of Evolution: Essays on the Nature of Species.* Cambridge, Mass.: MIT Press, 1992.

Evans, Eric J. *Britain before the Reform Act: Politics and Society, 1815–1832.* London: Longman, 1989.

———. *The Forging of the Modern State: Early Industrial Britain, 1783–1870.* London: Longman, 1983.

Faulkner, Peter, and Peter Preston, eds. *William Morris: Centenary Essays: Papers from the Morris Centenary Conference Organized by the William Morris Society at Exeter College Oxford, 30 June–3 July 1996.* Exeter, U.K.: University of Exeter Press, 1999.

Fawcett, Henry. *Pauperism: Its Causes and Remedies.* London: Macmillan and Co., 1871.

———. "A Popular Exposition of Mr. Darwin on the Origin of Species." *Macmillan's Magazine* 3 (1860): 81–92.

Fichman, Martin. *An Elusive Victorian: The Evolution of Alfred Russel Wallace.* Chicago: University of Chicago Press, 2004.

Fitzroy, Robert. *Narrative of the Surveying Voyages of His Majesty's Ships "Adventure" and "Beagle," between the Years 1826 and 1836: Describing Their Examination of the Southern Shores of South America, and the "Beagle's" Circumnavigation of the Globe*. . . . Vol. 2, *Proceedings of the Second Expedition, 1831–1836, under the Command of Captain Robert Fitz-Roy, R.N.* London: Henry Colburn, Great Marlborough Street, 1839.

Forbes, Eric G., ed. *Human Implications of Scientific Advance: Proceedings of the 15th International Congress of the History of Science, Edinburgh, 10–15 August 1977*. Edinburgh: Edinburgh University Press, 1978.

Francis, Mark. *Herbert Spencer and the Invention of Modern Life*. Ithaca, N.Y.: Cornell University Press, 2007.

Freeden, Michael. "Eugenics and Progressive Thought: A Study in Ideological Affinity." *The Historical Journal* 22, no. 3 (1979): 645–71.

———. *The New Liberalism: An Ideology of Social Reform*. Oxford: Clarendon, 1978.

Freeman, Derek. "The Evolutionary Theories of Charles Darwin and Herbert Spencer" [with comments and replies]. *Current Anthropology* 15, no. 3 (1974): 211–37.

Freeman, R. B. Bibliographical Introduction to *Journal of Researches*. In *The Complete Works of Charles Darwin Online*, edited by John Van Wyhe. 2002–. http://darwin-online.org.uk/EditorialIntroductions/Freeman_JournalofResearches.html.

French, Richard D. *Antivivisection and Medical Science in Victorian Society*. Princeton, N.J.: Princeton University Press, 1975.

Fukuyama, Francis. *Our Posthuman Future: Consequences of the Biotechnology Revolution*. New York: Farrar, Straus and Giroux, 2002.

Fulmer, Brian. "Political Biology: Peter Kropotkin, Evolution and Socialism." BA thesis, Colby College, 2007.

Galton, Francis. *Hereditary Genius: An Inquiry into Its Laws and Consequences*. London: Macmillan, 1869.

———. "Hereditary Genius: The Judges of England between 1660 and 1865." *Macmillan's Magazine* 16 (1865): 424–31.

———. "Hereditary Talent and Character." *Macmillan's Magazine* 12 (1865): 157–66, 318–27.

———. *Natural Inheritance*. London: Macmillan, 1889.

Gayon, Jean. *Darwin's Struggle for Survival: Heredity and the Hypothesis of Natural Selection*. Cambridge: Cambridge University Press, 1992.

Geddes, Patrick, and John Arthur Thomson. *The Evolution of Sex*. London: W. Scott, 1889.

George, Henry. *Progress and Poverty, an Inquiry into the Cause of Industrial Depressions and of Increase of Want with Increase of Wealth; the Remedy*. New York: Modern Library, 1938.

Gillespie, Neal C. "The Duke of Argyll, Evolutionary Anthropology, and the Art of Scientific Controversy." *Isis* 68, no. 1 (1977): 40–54.

Girón, Álvaro. "Kropotkin between Lamarck and Darwin: The Impossible Synthesis." *Asclepio* 55, no. 1 (2003): 189–213.

Glasier, J. Bruce. *William Morris and the Early Days of the Socialist Movement; Being Reminiscences of Morris' Work as a Propagandist, and Observations on His Character and Genius, with Some Account of the Persons and Circumstances of the Early Socialist Agitation, Together with a Series of Letters Addressed by Morris to the Author*. London: Longmans, Green, and Co., 1921.

Godwin, William. *The Anarchist Writings of William Godwin*. Edited by Peter H. Marshall. London: Freedom Press, 1986.

———. *Of Population: An Enquiry Concerning the Power of Increase in the Numbers of Mankind, Being an Answer to Mr. Malthus's Essay on That Subject*. New York: A. M. Kelly, 1964.

Gould, Peter C. *Early Green Politics: Back to Nature, Back to the Land, and Socialism in Britain, 1880–1900.* Brighton, Sussex: Harvester Press, 1988.

Gould, Stephen J. "Darwin's Sea Change, or Five Years at the Captain's Table." In *Ever since Darwin: Reflections in Natural History*, 28–38. New York: Norton, 1977.

———. "Kropotkin Was No Crackpot." In *Bully for Brontosaurus: Reflections in Natural History*, 325–39. New York: Norton, 1991.

———. *The Mismeasure of Man.* New York: Norton, 1996.

———. *Ontogeny and Phylogeny.* Cambridge, Mass: Belknap Press of Harvard University Press, 1977.

Green, Thomas Hill. *T. H. Green: Lectures on the Principles of Political Obligation, and Other Writings.* Edited by Paul Harris and John Morrow. Cambridge: Cambridge University Press, 1986.

Greene, John C. "Darwin as a Social Evolutionist." *Journal of the History of Biology* 10, no. 1 (1977): 1–27.

Greg, W. R. "On the Failure of 'Natural Selection' in the Case of Man." *Fraser's Magazine* 78 (September 1868): 353–62.

Gruber, Howard E. *Darwin on Man: A Psychological Study of Scientific Creativity.* Together with Darwin's early and unpublished notebooks, transcribed and annotated by Paul H. Barrett; foreword by Jean Piaget. New York: E. P. Dutton, 1974.

Gruber, Jacob W. "Who Was the *Beagle*'s Naturalist?" *British Journal for the History of Science* 4, no. 3 (1969): 266–82.

Habermas, Jürgen. *The Structural Transformation of the Public Sphere: An Inquiry into a Category of Bourgeois Society.* Cambridge, Mass: MIT Press, 1991.

Haldane, J. B. S. *The Causes of Evolution.* Princeton, N.J.: Princeton University Press, 1990.

———. *The Inequality of Man, and Other Essays.* London: Chatto and Windus, 1932.

———. "The Last Judgement." In *Possible Worlds*, by J. B. S. Haldane, 287–312. London: Library Press, 1927.

Hale, Piers J. "Darwin's Other Bulldog: Charles Kingsley and the Popularization of Evolution in England." *Science and Education* 21, no. 7 (2012): 977–1014.

———. "Labor and the Human Relationship with Nature: The Naturalization of Politics in the Work of Thomas Henry Huxley, Herbert George Wells, and William Morris." *Journal of the History of Biology* 36, no. 2 (2003): 249–84.

———. "Of Mice and Men: Evolution and the Socialist Utopia. William Morris, H. G. Wells, and George Bernard Shaw." *Journal of the History of Biology* 43, no. 1 (2010): 17–66.

———. "The Search for Purpose in a Post-Darwinian Universe: George Bernard Shaw, 'Creative Evolution,' and Shavian Eugenics: 'The Dark Side of the Force.'" *History and Philosophy of the Life Sciences* 28, no. 2 (2006): 191–214.

———. "William Morris, Human Nature, and the Biology of Utopia." In *William Morris in the Twenty-First Century*, edited by Phillippa Bennett and Rosie Miles, 107–27. Oxford: Peter Lang, 2010.

Hale, Piers J., and Jonathan Smith, eds. *Negotiating Boundaries.* Pt. 1, vol. 1 of *Victorian Science and Literature*, edited by Bernard Lightman and Gowan Dawson. London: Pickering and Chatto, 2011.

———. "William Whewell: The Philosophy of the Inductive Sciences." In *Negotiating Boundaries.* Pt. 1, vol. 1 of *Victorian Science and Literature*, edited by Bernard Lightman and Gowan Dawson. London: Pickering and Chatto, 2011.

Halliday, R. J. "Some Recent Interpretations of John Stuart Mill." *Philosophy* 43, no. 163 (1968): 1–17.

Hamburger, Joseph. *John Stuart Mill on Liberty and Control.* Princeton, N.J.: Princeton University Press, 1999.

Hamilton, Gail, ed. *The Insuppressible Book: A Controversy between Herbert Spencer and Frederic Harrison.* Boston: S. E. Cassino and Co., 1885.

Hamilton, William D. "The Evolution of Altruistic Behavior." *American Naturalist* 97, no. 896 (1963): 354–56.

———. "The Genetical Evolution of Social Behaviour." Pts. 1 and 2. *Journal of Theoretical Biology* 7, no. 1 (1964): 1–16; 7, no. 1 (1964): 17–52.

Harman, Oren S. *The Price of Altruism: George Price and the Search for the Origins of Kindness.* New York: Norton, 2010.

Harvey, Paul H., and Linda Partridge, eds. *Oxford Surveys in Evolutionary Biology* 6 (1989).

Hawkins, Mike. *Social Darwinism in European and American Thought, 1860–1945: Nature as Model and Nature as Threat.* Cambridge: Cambridge University Press, 1997.

Hazlitt, William. *A Reply to the Essay on Population in a Series of Letters, to Which are Added, Extracts from the Essay, with Notes by the Rev. T. R. Malthus.* London: Longman, Hurst, Rees, and Orme, 1807.

———. *The Spirit of the Age, or, Contemporary Portraits.* New York: Derby and Jackson, 1859.

Henderson, Archibald. *Bernard Shaw: Playboy and Prophet.* New York: Appleton and Co., 1932.

Herbert, Sandra. "The Place of Man in the Development of Darwin's Theory of Transmutation. Part II." *Journal for the History of Biology* 10, no. 2 (1977): 155–227.

Heron, David. *On the Relations of Fertility in Man to Social Status and On the Changes in This Relation That Have Taken Place during the Last Fifty Years.* Draper's Company Research Memoirs, Studies in Natural Deterioration, 1. London: Dulau Co., 1906.

Hesketh, Ian. *Of Apes and Ancestors: Evolution, Christianity, and the Oxford Debate.* Toronto: University of Toronto Press, 2009.

Hey, Jody. "Regarding the Confusion between the Population Concept and Mayr's 'Population thinking.'" *Quarterly Review of Biology* 86, no. 4 (2011): 253–64.

Hilton, Boyd. *The Age of Atonement: The Influence of Evangelicalism on Social and Economic Thought, 1795–1865.* Oxford: Clarendon Press, 1988.

Himmelfarb, Gertrude. *Darwin and the Darwinian Revolution.* New York: Norton, 1986.

Hobbes, Thomas. *Leviathan.* Oxford: Clarendon, 1909.

Hodge, Jonathan. "The Notebook Programme and Projects of Darwin's London Years." In *The Cambridge Companion to Darwin,* edited by M. J. S. Hodge and Gregory Radick, 40–68. Cambridge: Cambridge University Press, 2003.

Hofstadter, Richard. *Social Darwinism in American Thought.* Boston: Beacon Press, 1955.

Holroyd, Michael. *Bernard Shaw.* 4 vols. New York: Random House, 1888–92.

Houghton, Walter E. *The Victorian Frame of Mind, 1830–1870.* New Haven, Conn.: Yale University Press, 1957.

Hulse, James W. *Revolutionists in London: A Study of Five Unorthodox Socialists.* Oxford: Clarendon Press, 1970.

Hume, David. *An Enquiry Concerning Human Understanding: and Selections from "A Treatise of Human Nature"; with Hume's Autobiography and a Letter from Adam Smith.* New York: Barnes and Noble, 2004.

———. *A Treatise on Human Nature.* New York: Barnes and Noble, 2005.

Hunt, Karen. *Equivocal Feminists: The Social Democratic Federation and the Woman Question, 1884–1911.* New York: Cambridge University Press, 1996.

Hunter, T. Russell. "Re-Thinking Asa Gray's 'Natural Selection Not Inconsistent With Natural Theology.'" Master's thesis, University of Oklahoma, 2009.

Huxley, Julian. *Evolution: The Modern Synthesis.* New York: Harper and Bros., 1942.

Huxley, Leonard. *Life and Letters of Thomas Henry Huxley*, by his son. 3 vols. London: Macmillan and Co., 1913.

Huxley, Thomas Henry. "Administrative Nihilism." In *Collected Essays*, vol. 1 *Method and Results*, 251–89. London: Macmillan, 1894.

———. *Collected Essays.* 9 vols. London: Macmillan, 1894.

———. "Fragments Relating to Philosophical Zoology, Selected from the Works of K. E. von Baer." In *Scientific Memoirs Selected from the Transactions of Foreign Academies of Science, and from Foreign Journals. Natural History*, edited by Arthur Henfry and Thomas Henry Huxley, 176–238. London: Taylor and Francis, 1853.

———. "Further Remarks upon the Human Remains from the Neanderthal." *Natural History Review*, n.s., 4, no. 1 (1864): 429–46.

———. "A Liberal Education and Where to Find It". In *Collected Essays*, vol. 3, *Science and Education*, 76–110. London, Macmillan: 1894.

———. "Mr. Darwin's Critics." In *Collected Essays*, Vol. 2, *Darwiniana*, 120–86. London: Macmillan, 1894.

———. "The Origin of Species" (book review).. In *Collected Essays*, vol. 2, *Darwiniana*, 22–79. London: Macmillan, 1894.

———. "The Struggle for Existence: A Programme." *The Nineteenth Century* 23, no. 132 (1888): 161–80.

———. "The Struggle for Existence in Human Society." In *Collected Essays*, vol. 9 *Essays in Science*, 193–236. London: Macmillan, 1894.

Huxley, Thomas Henry, James Paradis, and G. C. Williams. *Evolution and Ethics: T. H. Huxley's "Evolution and Ethics" with New Essays on Its Victorian and Sociobiological Context.* Princeton, N.J.: Princeton University Press, 1989. [Huxley's "Evolution and Ethics" was originally published in London by Macmillan in 1894. The version reproduced in this volume is a facsimile of that original edition.]

Hyde, William J. "The Socialism of H. G. Wells in the Early Twentieth Century." *Journal of the History of Ideas* 17, no. 2 (1956): 217–34.

Hyndman, H. M. *The Economics of Socialism.* Boston: Small, Maynard and Co., 1921.

———. *The Evolution of Revolution.* New York: M. Y. Boni, 1921.

———. *The Historical Basis of Socialism in England.* London: K. Paul, Trench and Co., 1883.

———. *The Record of an Adventurous Life.* London: Macmillan, 1911.

———. *England for All.* London: Gilbert and Rivington, 1881.

———. *The Text-Book of Democracy: England for All.* London: E. W. Allen, 1881. [This is the so-called cheap edition.]

Hyndman, H. M., and William Morris. *A Summary of the Principles of Socialism, Written for the Democratic Federation.* London: The Modern Press, 1884.

Innes, C. D. "Utopian Apocalypses, Shaw, War, and H. G. Wells." *Shaw: The Annual of Bernard Shaw Studies*, no. 23 (2003): 37–56.

Irvine, William. *The Universe of G.B.S.* New York: McGraw-Hill, 1949.

Jameson, Robert. Introduction to *An Essay on the Theory of the Earth*, by Georges Cuvier, v–xv. New York: Kirk and Mercein, 1818.

Jenkins, Erin M. "Henry George and the Dragon: T. H. Huxley's Response to *Progress and Poverty*." In *Henry George's Legacy in Economic Thought*, edited by John Laurant, 31–50. Cheltenham, U.K.: Edward Elgar, 2005.Jenson, Vernon. "The X Club: Fraternity of Victorian Scientists." *British Journal for the History of Science* 5, no. 1 (1970): 63–72.

Jones, Gareth Stedman. *Outcast London: A Study in the Relationship between Classes in Victorian Society*. New York: Pantheon, 1984.

Jones, Greta. "Alfred Russel Wallace, Robert Owen, and the Theory of Natural Selection." *British Journal for the History of Science* 35, no. 1 (2002): 73–96.

———. *Social Darwinism in English Thought*. Sussex: Harvester, 1980.

———. "Spencer and His Circle." In *Herbert Spencer: The Intellectual Legacy*, edited by Greta Jones and Robert A. Peel, 1–16. London: Galton Institute, 2004.

Jones, Greta, and Robert A. Peel, eds. *Herbert Spencer: The Intellectual Legacy*. London: Galton Institute, 2004.

Jones, Peter. *Industrial Enlightenment: Science, Technology, and Culture in Birmingham and the West Midlands, 1760–1820*. Manchester, U.K.: Manchester University Press, 2008.

Joyce, Patrick. *Visions of the People: Industrial England and the Question of Class, 1848–1914*. Cambridge: Cambridge University Press, 1991.

Kass, Leon. *Life, Liberty, and the Defense of Dignity: The Challenge for Bioethics*. San Francisco: Encounter Books, 2002.

Kellogg, Vernon L. *Headquarters Nights; a Record of Conversations and Experiences at the Headquarters of the German Army in France and Belgium*. Boston: Atlantic Monthly Press, 1917.

Kevles, Daniel J. *In the Name of Eugenics: Genetics and the Uses of Human Heredity*. New York: Knopf, 1985.

Kidd, Benjamin. *Social Evolution*. New York: Macmillan and Co., 1894.

Kinna, Ruth. "Morris, Anti-Statism, and Anarchy." In *William Morris: Centenary Essays: Papers from the Morris Centenary Conference Organized by the William Morris Society at Exeter College Oxford, 30 June–3 July 1996*, edited by Peter Faulkner and Peter Preston, 215–18. Exeter, U.K.: University of Exeter Press, 1999.

Kohn, David. "Theories To Work By: Rejected Theories, Reproduction, and Darwin's Path to Natural Selection." *Studies in the History of Biology*, no. 4 (1980): 67–170.

Kottler, Malcolm Jay. "Alfred Russel Wallace, *The Origin of Man*, and Spiritualism." *Isis* 65, no. 2 (1974): 145–92.

Kropotkin, Peter Alekseevich. "An Appeal to the Young." In *Kropotkin's Revolutionary Pamphlets*, edited with introduction, biographical sketch, and notes by Roger N. Baldwin, 26–82. New York: Dover, 1970.

———. "Charles Darwin." *Le Revolte*, 29 April 1882, 1.

———. *The Conquest of Bread*. Edited by Paul Avrich. New York: New York University Press, 1972.

———. "Direct Action of Environment and Evolution." *The Nineteenth Century*, January 1919, 70–89.

———. "The Direct Action of Environment on Plants." *The Nineteenth Century*, July 1910, 58–77.

———. *Ethics, Origin and Development*. Authorized translation from the Russian by Louis S. Friedland and Joseph R. Piroshnikoff. New York: Dial Press, 1936 (© 1924).

———. *Evolution and Environment*. Edited by George Woodcock. Vol. 11 of *The Collected Works of Petr Kropotkin*. Montreal: Black Rose Books, 1995.

———. "Inheritance of Acquired Characters: Theoretical Difficulties." *The Nineteenth Century*, March 1912, 511–31.

———. "Inherited Variation in Animals." *The Nineteenth Century*, November 1915, 1124–44.

———. "Inherited Variation in Plants." *The Nineteenth Century*, October 1914, 816–36.

———. *Memoirs of a Revolutionist.* New York: Grove Press, 1968.

———. "The Morality of Nature." *Nineteenth Century* 57, no. 337 (1905): 407–26.

———. *Mutual Aid: A Factor of Evolution.* London: Freedom Press, 1986.

———. "Mutual Aid among Animals" (1890). In *Mutual Aid: A Factor of Evolution*, by Peter Kropotkin, 21–73. London: Freedom Press, 1986.

———. "The Response of Animals to Their Environment." *The Nineteenth Century*, November 1910, 856–1059.

———. Review of *News from Nowhere; or, An Epoch of Rest*, by William Morris. *Freedom*, November 1896, 109.

———. "The Theory of Evolution and Mutual Aid" (1910). In *Evolution and Environment*, 117–38. Montreal: Black Rose Books, 1995.

Lankester, E. Ray. *Degeneration: A Chapter in Darwinism.* London: Macmillan and Co., 1880.

———. "Heredity and the Direct Action of the Environment." *Nineteenth Century and After* 68, no. 403 (1910): 483–91.

———. "The Transmission of Acquired Characters and Panmixia." *Nature*, 27 March 1890, 487–88.

Latham, R. G. *Man and His Migrations.* New York: C. B. Norton, 1852.

Laurent, John. *Henry George's Legacy in Economic Thought.* Cheltenham, U.K.: Edward Elgar, 2005.

Lebedev, N. Introduction to *Ethics, Origin and Development*, by Peter Kropotkin. Authorized translation from the Russian by Louis S. Friedland and Joseph R. Piroshnikoff. New York: Dial Press, 1936 (© 1924).

The Leeds Times. Spartacus Educational Website, http://www.spartacus.schoolnet.co.uk/PRLeedsTimes.htm.

Lester, Ahren. "Uneasy Bedfellows: Alfred Russel Wallace and Nineteenth-Century 'Socialist Darwinism.' *Reinvention: An International Journal of Undergraduate Research* 3, no. 1 (2010). www.warwick.ac.uk/go/reinventionjournal/issues/volume3issue1/lester.

Lightman, Bernard V. "Darwin and the Popularization of Evolution." *Notes and Records of the Royal Society* 64, no. 1 (2010): 5–24.

———. "Robert Elsmere and the Agnostic Crisis of Faith." In *Victorian Faith in Crisis: Essays on Continuity and Change in Nineteenth-Century Religious Belief*, edited by R. J. Helmstadter and Bernard Lightman, 283–311. Basingstoke, U.K.: Macmillan, 1990.

———. *Victorian Science in Context.* Chicago: University of Chicago Press, 1997.

Limoges, Camille. *La selection naturelle: Etude sur la premiere constitution d'un concept, 1837–1859.* Paris: Presses Universitaires de France, 1970.

Litchfield, Henrietta Emma Darwin. *Emma Darwin, A Century of Family Letters, 1792–1896.* 2 vols. New York: D. Appleton and Co., 1915.

Lorenz, Konrad. *On Aggression.* London: Routledge, 2002.

Loy, James D., and Kent M. Loy. *Emma Darwin: A Victorian Life.* Gainesville: University Press of Florida, 2010.

Lubbock, John. *Pre-historic Times, as Illustrated by Ancient Remains, and the Manners and Customs of Modern Savages.* London: Williams and Norgate, 1865.

Lustig, Abigail, Robert J. Richards, and Michael Ruse, eds. *Darwinian Heresies.* Cambridge: Cambridge University Press, 2004.

Lyell, Charles. *The Geological Evidences of the Antiquity of Man, with Remarks on Theories of the Origin of Species by Variation.* London: J. Murray, 1863.

———. *Principles of Geology.* 3 vols. Chicago: University of Chicago Press, 1991.

Lynch, John M. *Darwin's Theory of Natural Selection: British Responses, 1859–1871.* Bristol: Thoemmes, 2001.

Lyons, Sherrie. Introduction to *"Evolution and Ethics": And Other Essays,* by Thomas Henry Huxley, vii–xv. New York: Barnes and Noble, 2006.

MacCarthy, Fiona. *William Morris: A Life for Our Time.* New York: Knopf, 1995.

MacDonald, James Ramsay. *The Socialist Movement.* New York: H. Holt and Co., 1911.

MacIntyre, Alasdair C. *After Virtue: A Study in Moral Theory.* Notre Dame, Ind.: University of Notre Dame Press, 1984.

MacKenzie, Norman Ian, and Jeanne MacKenzie. *The Fabians.* New York: Simon and Schuster, 1977.

———. *The Time Traveller: The Life of H. G. Wells.* New York: Simon and Schuster, 1973.

Magnello, Eileen M. "Karl Pearson's Gresham Lectures: W. F. R. Weldon, Speciation, and the Origins of Pearsonian Statistics." *British Journal of the History of Science* 29, no. 1 (1996): 43–63.

———. "Weldon, Walter Frank Raphael (1860–1906)." *Oxford Dictionary of National Biography.* Oxford: Oxford University Press, 2004.

Malthus, Thomas Robert. *An Essay on the Principle of Population, As It Affects the Future Improvement of Society, with Remarks on the Speculations of Mr. Godwin, M. Condorcet, and Other Writers.* London: J. Johnson, 1798.

———. *An Essay on the Principle of Population, or, A View of Its Past and Present Effects on Human Happiness: With an Inquiry into Our Prospects Respecting the Future Removal or Mitigation of the Evils Which it Occasions.* 6th ed. 2 vols. London: John Murray, 1826.

Manier, Edward. *The Young Darwin and His Cultural Circle: A Study of Influences which Helped Shape the Language and Logic of the First Drafts of the Theory of Natural Selection.* Dordrecht, Holland: D. Reidel, 1978.

Manvell, Roger. *The Trial of Annie Besant and Charles Bradlaugh.* New York: Horizon Press, 1976.

Marsh, Jan. *Back to the Land: The Pastoral Impulse in England, from 1880 to 1914.* London: Quartet Books, 1982.

Marshall, Alfred. *Principles of Economics: An Introductory Volume.* London: Macmillan, 1920.

Martineau, Harriet. *Autobiography.* Edited by Linda H. Peterson. Peterborough, Ont.: Broadview, 2007.

———. *How to Observe Morals and Manners.* London: C. Knight and Co., 1838.

Marx, Karl. *Capital: An Abridged Edition.* Edited with an introduction by David McLellan. Oxford: Oxford University Press, 1995.

Marx, Karl, and Friedrich Engels. *The Communist Manifesto.* London: Penguin Books, 1985.

———. *Karl Marx and Frederick Engels, Selected Works.* 3 vols. Moscow: Progress Publishers, 1969–70.

———. *Karl Marx, Frederick Engels: Collected Works.* 50 vols. New York: International Publishers, 1975–2004.

Matthew, H. C. G. "Smiles, Samuel (1812–1904)." *Oxford Dictionary of National Biography.* Oxford: Oxford University Press, 2004.

Mayr, Ernst. *One Long Argument: Charles Darwin and the Genesis of Modern Evolutionary Thought.* Cambridge, Mass: Harvard University Press, 1991.

———. "What Is Darwinism Today?" *Proceedings of the Biennial Meeting of the Philosophy of Science Association* 2 (1984): 145–56.

McBriar, A. M. *Fabian Socialism and English Politics, 1884–1918.* Cambridge: Cambridge University Press, 1962.

McCalman, Iain. 2009. *Darwin's Armada: Four Voyages and the Battle for the Theory of Evolution.* New York: Norton, 2009.

McKibben, Bill. *Enough: Staying Human in an Engineered Age.* New York: Times Books, 2003.

McNeil, Maureen. *Under the Banner of Science: Erasmus Darwin and His Age.* Manchester, U.K.: Manchester University Press, 1987.

Michael Holdroyd. *Bernard Shaw.* 2 vols. London: Chatto and Windus, 1989.

Milam, Erika Lorraine. *Looking for a Few Good Males: Female Choice in Evolutionary Biology.* Baltimore: Johns Hopkins University Press, 2010.

Mill, John Stuart. *Autobiography of John Stuart Mill.* New York: New American Library, 1964.

———. *Dissertations and Discussions: Political, Philosophical, and Historical.* 2 vols. London: John W. Parker and Son, 1859.

———. *The Subjection of Women.* New York: D. Appleton and Co., 1869.

Mill, John Stuart, and Auguste Comte. "*Utilitarianism,*" "*Liberty,*" "*Representative Government*": *Selections from Auguste Comte and Positivism.* Edited by H. B. Acton. London: Dent, 1983.

Mills, Eric L. "Forbes, Edward (1815–1854)." *Oxford Dictionary of National Biography.* Oxford University Press, 2004.

[Mivart, St. George Jackson.] Review of "The Descent of Man, and Selection in Relation to Sex, by Charles Darwin FRS etc." *Quarterly Review* 131, no. 261 (1871): 47–90.

———. *On the Genesis of Species.* New York: D. Appleton and Co., 1871.

Moore, G. E. *Principia Ethica.* Cambridge: Cambridge University Press, 1929.

Moore, James R. "Deconstructing Darwinism: The Politics of Evolution in the 1860s." *Journal for the History of Biology* 24, no. 3 (1991): 353–408.

———. "Herbert Spencer's Henchmen." In *Darwinism and Divinity,* edited by John R. Durant, 76–100. Oxford: Basil Blackwell, 1985.

———. *The Post-Darwinian Controversies: A Study of the Protestant Struggle to Come to Terms with Darwin in Great Britain and America, 1870–1900.* Cambridge: Cambridge University Press, 1979.

[Morley, John.] "The Descent of Man" (book review). *Pall Mall Gazette,* 21 March 1871, 11–12.

———. "Mr. Darwin on Conscience." *Pall Mall Gazette,* 12 April 1871, 10–11.

———. *Recollections.* 2 vols. New York: Macmillan, 1917.

Morris, William. "The Art of the People." In vol. 22 of *The Collected Works of William Morris,* 28–50. London: Elibron Classics, 2006.

———. "Fabian Essays in Socialism" (review). *Commonweal* 6, no. 211 (25 January 1890): 28–29.

———. *News from Nowhere and Other Writings.* London: Penguin, 1993.

———. "A Factory As It Might Be." *Justice,* 17 May 1884, 2.

———. "How I Became a Socialist." In *News from Nowhere and Other Writings,* 379–83. London: Penguin, 1993. Originally published in *Justice,* 16 June 1894.

———. "The Lesser Arts." In vol. 22 of *The Collected Works of William Morris,* 3–27. London: Elibron Classics, 2006.

———. *News from Nowhere; or, An Epoch of Rest: Being Some Chapters from a Utopian Romance.* London: Routledge and K. Paul, 1970.

Mowatt, Charles Loch. *The Charity Organisation Society, 1869–1913; Its Ideas and Work*. London: Methuen, 1961.

Murphy, James. *The Education Act 1870: Text and Commentary*. New York: Barnes and Noble, 1972.

Nairn, Tom. "The English Working Class." *New Left Review* 1, no. 24 (1964): 43–57.

Nelson, Ashley Nicole. "Harriet Martineau's Political Economy." Master's thesis, University of Central Oklahoma, 2012.

Olby, Robert. "Bateson, William (1861–1926)." *Dictionary of National Biography*. Oxford: Oxford University Press, 2004.

Oldroyd, D. R. *Darwinian Impacts: An Introduction to the Darwinian Revolution*. Atlantic Highlands, N.J.: Humanities Press, 1980.

Ospovat, Dov. *The Development of Darwin's Theory: Natural History, Natural Theology, and Natural Selection, 1838–1859*. Cambridge: Cambridge University Press, 1981.

———. "The Influence of Karl Ernst von Baer's Embryology, 1828–1859: A Reappraisal in Light of Richard Owen's and William B. Carpenter's "Palaeontological Application of 'Von Baer's Law.'" *Journal of the History of Biology* 9, no. 1 (1976): 1–28.

Owen, Robert. *A New View of Society and Other Writings*. Edited by Gregory Claeys. London: Penguin Books, 1991.

Paine, Thomas. *The Age of Reason: Being an Investigation of True and Fabulous Theology*. Philadelphia, 1794.

Paradis, James G. "'*Evolution and Ethics*' in Its Victorian Context." In *Evolution and Ethics: T. H. Huxley's "Evolution and Ethics" with New Essays on Its Victorian and Sociobiological Context*, by Thomas Henry Huxley, James Paradis, and G. C. Williams. Princeton, N.J.: Princeton University Press, 1989.

———. *Samuel Butler, Victorian against the Grain: A Critical Overview*. Toronto: University of Toronto Press, 2007.

Parrinder, Patrick. "Eugenics and Utopia: Sexual Selection from Galton to Morris." *Utopian Studies* 8, no. 2 (1997): 1–12.

Partington, John S. *Building Cosmopolis: The Political Thought of H. G. Wells*. Aldershot: Ashgate, 2003.

Paul, Diane B. "Darwin, Social Darwinism, and Eugenics." In Hodge, M. J. S., and Gregory Radick. *The Cambridge Companion to Darwin*, edited by M. J. S. Hodge and Gregory Radick, 214–39. Cambridge: Cambridge University Press, 2003.

———. "Eugenics and the Left." *Journal of the History of Ideas* 45, no. 4 (1984): 567–90.

Paul, Diane B., and John Beatty. "Discarding Dichotomies, Creating Community: Sam Schweber and Darwin Studies." In *Positioning the History of Science*, edited by J. Renn and K. Gavrugla, 113–18. Dordrecht: Springer, 2007.

Paul, Diane B., and Benjamin Day. "John Stuart Mill, Innate Differences, and the Regulation of Reproduction," *Studies in History and Philosophy of Biological and Biomedical Sciences* 39, no. 2 (2008): 222–31.

Pearson, Geoffrey. *Hooligan: A History of Respectable Fears*. London: Macmillan, 1983.

Pearson, Karl. *The Chances of Death, and Other Studies in Evolution*. 2 vols. London: E. Arnold, 1897.

———. *The Grammar of Science*. London: J. M. Dent and Sons, 1937.

———. "The Moral Basis of Socialism." In *The Ethic of Freethought and Other Addresses and Essays*, 301–29. 2nd ed. London: Adam and Charles Black, 1901.

———. *National Life from the Standpoint of Science*. London: A. and C. Black, 1905.

———. "The Philosophy of Natural Science: Studien zu Methodenlehre und Erkenntnisskritik Erkenntnistheoretische Grundzüge der Naturwissenschaften und ihre Beziehungen zuw Geistesleben der Gegenwart Allgemein wissenschafteliche Vorträge." *Nature*, 5 November 1896, 1–4.

———. *The Problem of Practical Eugenics.* London: Dulau, 1912.

———. "Socialism and Natural Selection." *Fortnightly Review* 56 (July 1894): 1–21.

———. "Socialism in Theory and Practice." In *The Ethic of Free Thought: A Selection of Essays and Lectures*, by Karl Pearson, 346–69. London: Unwin, 1888.

[Pearson Karl, and A. E. Shipley]. "Walter Frank Raphael Weldon, 1866–1906." *Proceedings of the Royal Society of London, Series B*, no. 30 (1908): xxv–xli. (This obituary was abstracted by A. E. Shipley from a longer essay by Karl Pearson and A. E. Shipley that appeared in *Biometrika.*)

Peel, J. D. Y. *Herbert Spencer: The Evolution of a Sociologist.* New York: Basic Books, 1971.

Perkin, Harold James. *The Origins of Modern English Society, 1780–1880.* London: Routledge and K. Paul, 1969.

Peterson, William. "Malthus and the Intellectuals." *Population and Development Review* 5, no. 3 (1979): 469–78.

———. "The Malthus-Godwin Debate, Then and Now." *Demography* 8, no. 1 (1971): 13–26.

Pierson, Stanley. "Bax, Ernest Belfort (1854–1926)." *Oxford Dictionary of National Biography.* Oxford: Oxford University Press, 2004.

———. *British Socialists: The Journey from Fantasy to Politics.* Cambridge, Mass: Harvard University Press, 1979.

Porter, Theodore M. *Karl Pearson: The Scientific Life in a Statistical Age.* Princeton, N.J.: Princeton University Press, 2004.

———. "Statistical Utopianism in an Age of Aristocratic Efficiency." *Osiris*, 2nd ser., 17 (2002): 210–17.

Post, Stephen Garrard, and Robert H. Binstock. *The Fountain of Youth: Cultural, Scientific, and Ethical Perspectives on a Biomedical Goal.* Oxford: Oxford University Press, 2004.

Provine, William B. *The Origins of Theoretical Population Genetics.* Chicago: University of Chicago Press, 2001.

Pye, Denis. *Fellowship Is Life: The National Clarion Cycling Club, 1885–1995.* Bolton: Clarion, 1995.

Raby, Peter. *Alfred Russel Wallace: A Life.* Princeton, N.J.: Princeton University Press, 2001.

Radick, Gregory. "Is the Theory of Natural Selection Independent of Its History?" In *The Cambridge Companion to Darwin*, edited by Jonathan Hodge and Gregory Radick, 147–72. Cambridge: Cambridge University Press, 2009.

Rae, John. "State Socialism." *Contemporary Review* 54 (September 1888): 224–45, 378–92.

Ray, L. J. "Eugenics, Mental Deficiency, and Fabian Socialism between the Wars." *Oxford Review of Education* 9, no. 3 (Mental Handicap and Education issue) (1983): 213–22.

Rees, Rosemary. *Poverty and Public Health, 1815–1948.* Oxford: Heinemann, 2001.

Reeves, Richard. *John Stuart Mill: Victorian Firebrand.* London: Atlantic Books, 2007.

Reid, Loren. *Charles James Fox: A Man for the People.* Columbia: University of Missouri Press, 1969.

Renn, Jürgen, and Kōstas Gavroglou. *Positioning the History of Science.* Dordrecht: Springer, 2007.

Richards, Evelleen. "Redrawing the Boundaries: Darwinian Science and Victorian Women Intellectuals." In *Victorian Science in Context*, edited by Bernard Lightman, 119–42. Chicago: University of Chicago Press, 1997.

Richards, Robert J. *Darwin and the Emergence of Evolutionary Theories of Mind and Behavior.* Chicago: University of Chicago Press, 1987.

———. "Darwin on Mind, Morals, and Emotions." In *The Cambridge Companion to Darwin*, edited by Jonathan Hodge and Gregory Radick, 96–119. Cambridge: Cambridge University Press, 2003.

———. *The Meaning of Evolution: The Morphological Construction and Ideological Reconstruction of Darwin's Theory.* Chicago: University of Chicago Press, 1992.

Ridley, Matt. *The Origins of Virtue: Human Instincts and the Evolution of Cooperation.* London: Penguin, 1996.

Rignano, Eugenio. *Upon the Inheritance of Acquired Characters: A Hypothesis of Heredity, Development, and Assimilation.* Translated by Basil Coleman Hyatt Harvey. Chicago: Open Court Pub. Co., 1911.

Ritchie, David George. *Darwinism and Politics, with Two Additional Essays on Human Evolution.* London: S. Sonnenschein and Co., 1891.

Romanes, George John. *An Examination of Weismannism.* 2nd ed. Chicago: Open Court Publishing Co., 1899. First edition published in 1893.

Rooff, Madeline. *A Hundred Years of Family Welfare: A Study of the Family Welfare Association (formerly Charity Organisation Society), 1869–1969.* London: Joseph, 1972.

Rowbotham, Sheila. *Threads Through Time: Writings on History and Autobiography.* London: Penguin Books, 1999.

Rudwick, M. J. S. *Bursting the Limits of Time: The Reconstruction of Geohistory in the Age of Revolution.* Chicago: University of Chicago Press, 2005.

———. *Georges Cuvier, Fossil Bones, and Geological Catastrophes: New Translations and Interpretations of the Primary Texts.* Chicago: University of Chicago Press, 1997.

Ruse, Michael. "Alfred Russel Wallace, the Discovery of Natural Selection, and the Origins of Humankind." In *Rebels, Mavericks, and Heretics in Biology*, edited by Oren S. Harman, Michael R. Dietrich, 20–36. New Haven, Conn.: Yale University Press, 2008.

———. "Charles Darwin and Group Selection" *Annals of Science* 37 (1980): 615–30.

———. *Darwin and Design: Does Evolution Have a Purpose?* Cambridge, Mass: Harvard University Press, 2003.

———. *The Darwinian Revolution: Science Red in Tooth and Claw.* Chicago: University of Chicago Press, 1979.

———. "Darwin's Debt to Philosophy: An Examination of the Influence of the Philosophical Ideas of John F. W. Herschel and William Whewell on the Development of Charles Darwin's Theory of Evolution." *Studies in the History and Philosophy of Science* 6, no. 1 (1975): 159–81.

———. *Monad to Man: The Concept of Progress in Evolutionary Biology.* Cambridge, Mass: Harvard University Press, 1996.

———. "Thomas Henry Huxley and the Status of Evolution as Science" (1997). In *Thomas Henry Huxley's Place in Science and Letters: Centenary Essays*, edited by Alan P. Barr, 140–58. Athens: University of Georgia Press, 1997.

Ruskin, John. *Modern Painters: Their Superiority in the Art of Landscape Painting to All the Ancient Masters, Proved by Examples of the True, the Beautiful, and the Intellectual from the Works of Modern Artists, Especially from those of J. M. W. Turner.* London: Smith, Elder, 1843.

———. *The Nature of Gothic: A Chapter of "The Stones of Venice."* Hammersmith: Printed by William Morris at the Kelmscott Press and published by George Allen, London, 1892.

Salsburg, David. *The Lady Tasting Tea. How Statistics Revolutionized Science in the Twentieth Century.* New York: Henry Holt, 2001.

Samuel, Raphael, ed. *People's History and Socialist Theory.* London: Routledge and Kegan Paul, 1981.

Schabas, Margaret. "The Greyhound and the Mastiff: Darwinian Themes in Mill and Marshall." In *Natural Images in Economic Thought,* edited by P. Mirowski, 322–35. Cambridge: Cambridge University Press, 1994.

———. "Ricardo Naturalized: Lyell and Darwin on the Economy of Nature." In *Classicals, Marxians, and Neo-Classicals: Selected Papers from the History of Economics Society Conference, 1988,* 40–49. Aldershot, Hants, England: Published for the History of Economics Society by E. Elgar, 1990.

———. *The Natural Origins of Economics.* Chicago: University of Chicago Press, 2005.

Schiebinger, Londa. *The Mind Has No Sex? Women in the Origins of Modern Science.* Cambridge, Mass.: Harvard University Press, 1991.

Schwartz, Joel S. "Darwin, Wallace, and *The Descent of Man.*" *Journal for the History of Biology* 17, no. 2 (1984): 271–89.

———. "Robert Chambers and Thomas Henry Huxley, Science Correspondents: The Popularization and Dissemination of Nineteenth Century Natural Science." *Journal for the History of Biology* 32, no. 2 (1999): 342–83.

Schweber, Silvan S. "Darwin and the Political Economists: Divergence of Character." *Journal of the History of Biology* 13, no.2 (1980): 195–289.

———. "The Origin of 'Origin' Revisited." *Journal for this History of Biology* 10, no. 2 (1977): 231–32.

———. "Scientists as Intellectuals: The Early Victorians." In *Victorian Science and Victorian Values: Literary Perspectives,* edited by James Paradis and Thomas Postlewait, 1–37. New Brunswick, N.J.: Rutgers University Press, 1985.

Searle, G. R. *Eugenics and Politics in Britain, 1900–1914.* Leyden: Noordhoff International Pub., 1976.

———. *The Quest for National Efficiency: A Study in British Politics and Political Thought, 1899–1914.* Berkeley, Calif.: University of California Press, 1971.

Secord, James A. *Victorian Sensation: The Extraordinary Publication, Reception, and Secret Authorship of Vestiges of the Natural History of Creation.* Chicago: University of Chicago Press, 2000.

Secord, Anne. "Corresponding Interests: Artisans and Gentlemen in Nineteenth-Century Natural History." *British Journal for the History of Science* 27, no. 4 (1994): 383–408.

Semmel, Bernard. *John Stuart Mill and the Pursuit of Virtue.* New Haven, Conn.: Yale University Press, 1984.

———. "Karl Pearson: Socialist and Darwinist." *British Journal of Sociology* 9, no. 2 (1958): 111–25.

Shapin, Steven. "A Man with a Plan." *New Yorker,* 13 August 2007.

Shaw, George Bernard. *Back to Methuselah.* In vol. 2 of *Bernard Shaw, Complete Plays, with Prefaces.* New York: Dodd, Mead, 1963.

———. "Basis for Socialism: Economic." In *Fabian Essays in Socialism,* edited by George Bernard Shaw et al., 3–29. London: The Fabian Society, 1889.

———. "The Case for Inequality." In *Shavian Tracts,* 15–16. 6 vols. in 1. Berkeley: Johnson Reprint Co., 1968.

———. *Collected Letters [of] Bernard Shaw, 1898–1910.* Edited by Dan H. Laurence. London: M. Reinhardt, 1972.

———. *The Fabian Society: Its Early History.* London: Fabian Society, 1892.

———. *Man and Superman, A Comedy and a Philosophy.* Westminster: Archibald Constable and Co., 1903.

———. *Morris As I Knew Him.* London: William Morris Society, 1966.

———. Preface to *Back to Methuselah.* In vol. 2 of *Bernard Shaw, Complete Plays, with Prefaces.* New York: Dodd, Mead, 1963.

Shaw, George Barnard, and Stanley Weintraub. *Shaw: An Autobiography 1856–1898; Selected from His Writings.* 2 vols. New York: Weybright Talley, 1929.

Shaw, George Bernard, and H. G. Wells. *Bernard Shaw and H. G. Wells.* Edited by Percy J. Smith. Selected Correspondence of Bernard Shaw. Toronto: University of Toronto Press, Scholarly Publishing Division, 1995.

Shaw, George Bernard, et al. *Fabian Essays in Socialism.* Edited by G. Bernard Shaw. London: The Fabian Society, 1889.

Shermer, Michael. *In Darwin's Shadow: The Life and Science of Alfred Russel Wallace: A Biographical Study on the Psychology of History.* Oxford: Oxford University Press, 2002.

Smiles, Samuel. *Self Help; with Illustrations of Conduct and Perseverance.* London: John Murray, 1859.

Smith, Adam. *An Inquiry into the Nature and Causes of the Wealth of Nations.* London: Penguin, 1999.

———. *The Theory of Moral Sentiments.* Oxford: Clarendon Press, 1976.

Smith, C. U. M. "Evolution and the Problem of the Mind: Part 1. Herbert Spencer." *Journal of the History of Biology* 15, no. 1 (1982): 55–88.

Smith, C. U. M., and Robert Arnott, eds. *The Genius of Erasmus Darwin.* Aldershot, Hampshire, England: Ashgate, 2005.

Smith, David C. *H. G. Wells: Desperately Mortal: A Biography.* New Haven, Conn.: Yale University Press, 1986.

Smith, John Maynard, and George Price. "The Logic of Animal Conflict." *Nature,* 246, 2 November 1973, 15–18.

Smith, Roger. "William Carpenter (1813–1885)." *Oxford Dictionary of National Biography.* Oxford: Oxford University Press, 2004.

Smith, Sidney. "The Origin of 'The Origin.'" *Advancement of Science* 16, no. 64 (1960): 391–401.

Smocovitis, Betty. *Unifying Biology: The Evolutionary Synthesis and Evolutionary Biology.* Princeton, N.J.: Princeton University Press, 1996.

Snyder, Laura J. *The Philosophical Breakfast Club: Four Remarkable Friends Who Transformed Science and Changed the World.* New York: Broadway Books, 2011.

———. *Reforming Philosophy: A Victorian Debate on Science and Society.* Chicago: University of Chicago Press, 2006.

Sober, Elliott, and David Sloan Wilson. *Unto Others: The Evolution and Psychology of Unselfish Behavior.* Cambridge, Mass: Harvard University Press, 1998.

Spencer, Herbert. *An Autobiography.* 2 vols. New York: D. Appleton, 1904.

———. *Education: Intellectual, Moral, and Physical.* New York: Appleton and Co., 1885.

———. "The Inadequacy of 'Natural Selection.'" *Contemporary Review* 63 (February 1893): 153–66; 63 (March 1893): 439–456.

———. *The Principles of Biology 1.* London: Williams and Norgate, 1864.

———. The *Principles of Sociology.* 3 vols. New York: D. Appleton and Co., 1881.

———. *The Principles of Psychology.* London: Longman, Brown, Green and Longmans, 1855.

———. "The Social Organism." *Westminster Review* 17, no. 1 (1860): 90–121.

———. *Social Statics: or, The Conditions Essential to Human Happiness Specified, and the First of Them Developed.* London: John Chapman, 1851.

———. *The Man versus The State.* 1884. Reprint, Caldwell, Idaho: Caxton Printers, 1940.

———. "A Theory of Population, Deduced from the General Law of Animal Fertility." *Westminster Review* 57, no. 112 (1852): 468–501.

Spencer, Herbert, and Frederic Harrison. *The Insuppressible Book. A Controversy between Herbert Spencer and Frederic Harrison,* from the Nineteenth Century and Pall Mall Gazette, with comments by Gail Hamilton [pseud.]. Boston: S. E. Cassino and Co., 1885.

Stack, David A. "The First Darwinian Left: Radical and Socialist Responses to Darwin, 1859–1914." *History of Political Theory* 21, no. 4 (2000): 682–710.

———. *The First Darwinian Left: Socialism and Darwinism, 1859–1914.* London: New Clarion Press, 2003.

Stedman Jones, Gareth. *Outcast London: A Study in the Relationship between Classes in Victorian Society.* Oxford: Clarendon Press, 1971.

Stephen, Leslie. *The Science of Ethics.* London: Smith, Elder, 1882.

Stephens, W. B. *Education in Britain, 1750–1914.* London: Macmillan, 1998.

Stewart, Dugald. "'An Account of the Life and Writings of Adam Smith LL.D.,' from the Transactions of the Royal Society of Edinburgh, read by Mr. Stewart, January 21, and March 18, 1973." In vol. 7 of *The Works of Dugald Stewart,* 1–75. Cambridge, Mass.: Hilliard and Brown, 1829.

Stocking, George W. *Victorian Anthropology.* New York: Free Press, 1987.

Sulloway, Frank. "Geographic Isolation in Darwin's Thinking: The Vicissitudes of a Crucial Idea." In *Studies in the History of Biology,* edited by William Coleman and Camille Limoges, 23–65. Baltimore: John Hopkins University Press, 1979.

Swenson, Sarah. "The Nature of Selfishness and the Genetics of Altruism: The Role of W. D. Hamilton in Discerning Myth from Reality." Msc. thesis, Oxford University, 2012.

Taylor, Harvey. *A Claim on the Countryside: A History of the British Outdoor Movement.* Edinburgh: Keele University Press, 1997.

Taylor, M. W. *The Philosophy of Herbert Spencer.* London: Continuum, 2007.

Thompson, E. P. *Customs in Common.* London: Merlin Press, 1991.

———. *The Making of the English Working Class.* New York: Pantheon Books, 1963.

———. "The Moral Economy of the English Crowd in the Eighteenth Century." *Past and Present,* no. 50 (February 1971): 76–136.

———. "The Peculiarities of the English." *Socialist Register* 2 (1965): 311–62.

———. 1968. *The Poverty of Theory and Other Essays.* New York: Monthly Review Press, 1968.

———. *William Morris: Romantic to Revolutionary.* New York: Pantheon Books, 1977.

Thompson, Laurence Victor. *The Enthusiasts: A Biography of John and Katharine Bruce Glasier.* London: Gollancz, 1971.

———. *Robert Blatchford: Portrait of an Englishman.* London: Gollancz, 1951.

Thomson, Keith Stewart. *The Young Charles Darwin.* New Haven, Conn.: Yale University Press, 2009.

Todes, Daniel Philip. "Darwin's Malthusian Metaphor and Russian Evolutionary Thought, 1859–1917." *Isis* 78, no. 4 (1987): 537–51.

———. *Darwin without Malthus: The Struggle for Existence in Russian Evolutionary Thought.* Oxford: Oxford University Press, 1989.

Toynbee, Arnold, and Benjamin Jowett. *Lectures on the Industrial Revolution of the 18th Century in England: Popular Addresses, Notes, and Other* Fragments. Together with a Short Memoir by B. Jowett. 2nd ed. London: Rivingtons, 1887.

Trivers, Robert L. "The Evolution of Reciprocal Altruism." *Quarterly Review of Biology* 46, no. 1 (1971): 35–57.

———. *The Folly of Fools: The Logic of Deceit and Self-Deception in Human Life.* New York: Basic Books, 2011.

Tsuzuki, Chūshichi. *H. M. Hyndman and British Socialism.* Edited by Henry Pelling. [London]: Oxford University Press, 1961.

Uglow, Jennifer S. *The Lunar Men: Five Friends Whose Curiosity Changed the World.* New York: Farrar, Straus, and Giroux, 2002.

Van Wyhe, John, ed. *The Complete Work of Charles Darwin Online.* 2002–. http://darwin-online.org.uk/.

———. "Mind the Gap: Did Darwin Avoid Publishing His Theory for Many Years?" *Notes and Records of the Royal Society of London* 61, no. 2 (2007): 177–205.

———. "'My Appointment Received the Sanction of the Admiralty': Why Charles Darwin Really Was the Naturalist on HMS *Beagle.*" *Studies in the History and Philosophy of the Biological and Biomedical Sciences* 44, no. 3 (2013): 316–26.

Wallace, Alfred Russel. *Alfred Russel Wallace: Letters and Reminiscences.* Edited by James Marchant. New York: Harper, 1916.

———. *Contributions to the Theory of Natural Selection: A Series of Essays.* London: Macmillan and Co., 1870.

———. "Creation by Law." *Quarterly Journal of Science* 4, no. 16 (1867): 471–88.

———. "Development of the Human Races under the Law of Natural Selection." In *Contributions to the Theory of Natural Selection: A Series of Essays,* 313. London: Macmillan and Co., 1870.

———. "The Future of Civilisation." Review of *Social Evolution,* by Benjamin Kidd. *Nature,* 12 April 1894, 549–51.

———. "Geological Climates and the Origin of Species." Review of *Principles of Geology* (10th ed.) and *Elements of Geology* (6th ed.), by Sir Charles Lyell. *Quarterly Review* 126, no. 252 (1869): 391–94.

———. "The Limits of Natural Selection as Applied to Man." In *Contributions to the Theory of Natural Selection: A Series of Essays,* 332–71. London: Macmillan and Co., 1870.

———. *My Life: A Record of Events and Opinions.* London: Chapman and Hall, 1905.

———. "The Origin of Human Races and the Antiquity of Man Deduced from the Theory of 'Natural Selection.'" *Journal of the Anthropological Society of London,* no. 2 (1864): clviii–clxxxvii. (Pp. clxx–clxxxvii present a general discussion of the paper by members of the Society.)

Walter, Nicolas. "Wilson, Charlotte Mary (1854–1944)." *Oxford Dictionary of National Biography.* Oxford: Oxford University Press, 2004.

Waters, Chris. *British Socialists and the Politics of Popular Culture, 1884–1914.* Stanford, Calif.: Stanford University Press, 1990.

Webb, Beatrice. *My Apprenticeship.* Cambridge: Cambridge University Press, 1979.

———. *The Diary of Beatrice Webb.* Edited by Norman Ian MacKenzie and Jeanne MacKenzie. 4 vols. Cambridge, Mass.: Belknap Press of Harvard University Press, 1982.

Webb, R. K. *Modern England: From the Eighteenth Century to the Present.* New York: Dodd, Mead, 1968.

Webb, Sidney. *The Decline in the Birth Rate.* Fabian Tract No. 131. London: Fabian Society, 1907.

———. *English Progress towards Social Democracy.* Manchester: John Heywood, 1890.

Weikhart, Richard. *From Darwin to Hitler: Evolutionary Ethics, Eugenics, and Racism in Germany.* New York: Palgrave Macmillan, 2004.

———. *Hitler's Ethic: The Nazi Pursuit of Evolutionary Progress.* New York: Palgrave Macmillan, 2009.

Weismann, Friedrich Leopold August. "The All-Sufficiency of Natural Selection. A Reply to Herbert Spencer." *Contemporary Review* 64 (September 1893): 309–39, 596–610.

———. "The Duration of Life" (1881). In vol. 1 of *Essays upon Heredity and Kindred Biological Problems,* edited by Edward Bagnell Poulton, 3–65. 2nd ed. Oxford: Clarendon Press, 1891–92.

———. "On Heredity" (1883). In vol. 1 of *Essays upon Heredity and Kindred Biological Problems,* edited by Edward Bagnell Poulton, 69–105. 2nd ed. Oxford: Clarendon Press, 1891–92.

———. "The Supposed Transmission of Mutilations." In vol. 1 of *Essays upon Heredity and Kindred Biological Problems,* edited by Edward Bagnell Poulton, 421–48. 2nd ed. Oxford: Clarendon, 1891–92.

Weldon, W. F. R. "Presidential Address." Section D, Zoology (Including Animal Physiology). In *Report of the Sixty-Eighth Meeting of the British Association for the Advancement of Science; Held at Bristol in September 1898,* 887–92. London: John Murray, 1899.

Wells, H. G. *Ann Veronica, a Modern Love Story.* New York: Harper and Bros., 1909.

———. *Anticipations of the Reaction of Mechanical and Scientific Progress upon Human Life and Thought.* London: Chapman and Hall, 1902.

———. "The Biological Problem of Today." *Saturday Review,* 29 December 1894, 703–4.

———. *The Complete Short Stories.* London: Phoenix Giant, 1999.

———. *The Correspondence of H. G. Wells.* Vol. 1, *1880–1903.* Edited by David C. Smith. London: Pickering and Chatto, 1998.

———. *The Correspondence of H. G. Wells.* Vol. 2, *1904–1918.* Edited by David C. Smith; consulting editor Patrick Parrinder. London: Pickering and Chatto, 1998.

———. "The Duration of Life." In *H. G. Wells: Early Writings in Science and Science Fiction,* edited by Robert M. Philmus and David Y. Hughes, 132–35. Berkeley: University of California Press, 1975. Previously published in the *Saturday Review,* 23 February 1895.

———. *Early Writings in Science and Science Fiction by H. G. Wells.* Edited by Robert Philmus and David Y. Hughes. Berkeley: University of California Press, 1975.

———. *Experiment in Autobiography; Discoveries and Conclusions of a Very Ordinary Brain (since 1866).* New York: The Macmillan Co., 1934.

———. "Human Evolution, An Artificial Process." In *H. G. Wells: Early Writings in Science and Science Fiction,* edited by Robert M. Philmus and David Y. Hughes, 211–19. Berkeley: University of California Press, 1975. Originally published in *Fortnightly Review,* n.s., 60 (October 1896): 590–95.

———. "Incidental Thoughts on a Bald Head." *Pall Mall Gazette,* 1 March 1895.

———. *Mankind in the Making.* London: Chapman and Hall, 1904.

———. *Men Like Gods: A Novel.* New York: The Macmillan Co., 1927.

———. *Mind at the End of Its Tether.* London: W. Heinemann, 1945.

———. *A Modern Utopia.* Thirsk, Scotland: House of Stratus, 2002.

———. "A Slip under the Microscope." In *The Complete Stories of H. G. Wells,* edited by John Hammond, 250–66. London, 1998.

———. *The Time Machine.* Edited by Patrick Parrinder. London: Penguin, 2005.

———. *Time Machine: An Invention: A Critical Text of the 1895 London First Edition, with an Introduction and Appendices*. Edited by Leon Stover. London: MacFarland and Co., 1996.

Wells, H. G., Julian Huxley, and G. P. Wells. *The Science of Life*. Garden City, N.Y.: Doubleday, Doran and Co., 1931.

Whewell, William. *Astronomy and General Physics Considered with Reference to Natural Theology*. London: W. Pickering, 1833.

———. *Indications of the Creator: Extracts, Bearing upon Theology, from the History and the Philosophy of the Inductive Sciences*. London: J. W. Parker, 1846.

———. *The Philosophy of the Inductive Sciences, Founded upon Their History*. London: J. W. Parker, 1840.

White, Paul. *Thomas Huxley: Making the "Man of Science."* Cambridge: Cambridge University Press, 2003.

Wiener, Martin J. *English Culture and the Decline of the Industrial Spirit, 1850–1980*. Cambridge: Cambridge University Press, 1981.

Wilkins, John S. *Species: A History of the Idea*. Berkeley: University of California Press, 2009.

Williams, George C. *Adaptation and Natural Selection: A Critique of Some Current Evolutionary Thought*. Princeton, N.J.: Princeton University Press, 1992.

Willis-Harris, H. "The Survival of the Fittest." *Justice*, 28 April 1888, 2.

Wilson, David Sloan. *The Neighborhood Project: Using Evolution to Improve My City One Block at a Time*. New York: Little, Brown, 2011.

Wilson, David Sloan, and Elliott Sober. "Reintroducing Group Selection to the Human Behavioral Sciences." *Behavioral and Brain Sciences* 17, no. 4 (1994): 585–654.

Wilson, David Sloan, and E. O. Wilson. "Rethinking the Theoretical Boundaries of Sociobiology." *Quarterly Review of Biology* 82, no. 4 (2007): 327–48.

Wilson, E. O. *The Social Conquest of Earth*. New York: Norton, 2012.

———. *Sociobiology*. Cambridge, Mass.: The Belknap, 1980.

Winch, Donald. *Riches and Poverty: An Intellectual History of Political Economy in Britain, 1750–1834*. Cambridge: Cambridge University Press, 1996.

Wolfe, Willard. *From Radicalism to Socialism: Men and Ideas in the Formation of Fabian Socialist Doctrines, 1881–1889*. New Haven, Conn.: Yale University Press, 1975.

Wollstonecraft, Mary. *Posthumous Works of the Author of A Vindication of the Rights of Women*. Edited by William Godwin. 4 vols. London: J. Johnson and G. G. and J. Robinson, 1798.

Woram, John. *Ship's Company, Passenger List, and Fate of HMS "Beagle."* Rockville Center, N.Y.: Rockville Press, 2011. http://www.rockvillepress.com/tierra/texts/CREWLIST.HTM.

Wynne-Edwards, Vero Copner. *Animal Dispersion in Relation to Social Behaviour*. New York: Hafner Pub. Co., 1962.

Yeo, Richard. *Defining Science: William Whewell, Natural Knowledge, and Public Debate in Early Victorian England*. Cambridge: Cambridge University Press, 1993.

———. "An Idol of the Marketplace: Baconism in Nineteenth-Century Britain." *History of Science* 23, no. 61 (1985): 251–98.

Yeo, Stephen. "A New Life: The Religion of Socialism in Britain, 1883–1896." *History Workshop Journal* 4, no. 1 (1977): 5–56.

Young, Robert M. *Darwin's Metaphor: Nature's Place in Victorian Culture*. Cambridge: Cambridge University Press, 1985.

———. "Malthus and the Evolutionists: The Common Context of Biological and Social Theory." In *Darwin's Metaphor: Nature's Place in Victorian Culture*, by Robert M. Young, 23–55. Cambridge: Cambridge University Press, 1985.

Index